Springer Texts in Business and Economics

More information about this series at http://www.springer.com/series/10099

Erik Mooi • Marko Sarstedt • Irma Mooi-Reci

Market Research

The Process, Data, and Methods Using Stata

Springer

Erik Mooi
Department of Management
and Marketing
University of Melbourne
Parkville, Victoria, Australia

Marko Sarstedt
Chair of Marketing
Otto-von-Guericke-University
Magdeburg, Sachsen-Anhalt, Germany

Irma Mooi-Reci
School of Social and Political Sciences
University of Melbourne
Parkville, Victoria, Australia

ISSN 2192-4333 ISSN 2192-4341 (electronic)
Springer Texts in Business and Economics
ISBN 978-981-10-5217-0 ISBN 978-981-10-5218-7 (eBook)
DOI 10.1007/978-981-10-5218-7

Library of Congress Control Number: 2017946016

Printed on acid-free paper

This Springer imprint is published by Springer Nature
The registered company is Springer Nature Singapore Pte Ltd.
The registered company address is: 152 Beach Road, #21-01/04 Gateway East, Singapore 189721, Singapore

To Irma
– Erik Mooi

To Johannes
– Marko Sarstedt

To Erik
– Irma Mooi-Reci

Preface

In the digital economy, data have become a valuable commodity, much in the way that oil is in the rest of the economy (Wedel and Kannan 2016). Data enable market researchers to obtain valuable and novel insights. There are many new sources of data, such as web traffic, social networks, online surveys, and sensors that track suppliers, customers, and shipments. A Forbes (2015a) survey of senior executives reveals that 96% of the respondents consider data-driven marketing crucial to success. Not surprisingly, data are valuable to companies who spend over $44 billion a year on obtaining insights (Statista.com 2017). So valuable are these insights that companies go to great lengths to conceal the findings. Apple, for example, is known to carefully hide that it conducts a great deal of research, as the insights from this enable the company to gain a competitive advantage (Heisler 2012).

This book is about being able to supply such insights. It is a valuable skill for which there are abundant jobs. Forbes (2015b) shows that IBM, Cisco, and Oracle alone have more than 25,000 unfilled data analysis positions. Davenport and Patil (2012) label data scientist as the sexiest job of the twenty-first century.

This book introduces market research, using commonly used quantitative techniques such as regression analysis, factor analysis, and cluster analysis. These statistical methods have generated findings that have significantly shaped the way we see the world today. Unlike most market research books, which use SPSS (we've been there!), this book uses Stata. Stata is a very popular statistical software package and has many advanced options that are otherwise difficult to access. It allows users to run statistical analyses by means of menus and directly typed commands called *syntax*. This syntax is very useful if you want to repeat analyses or find that you have made a mistake. Stata has matured into a user-friendly environment for statistical analysis, offering a wide range of features.

If you search for market(ing) research books on Google or Amazon, you will find that there is no shortage of such books. However, this book differs in many important ways:

- This book is a bridge between the theory of conducting quantitative research and its execution, using the market research process as a framework. We discuss market research, starting off by identifying the research question, designing the data collection process, collecting, and describing data. We also introduce essential data analysis techniques and the basics of communicating the results, including a discussion on ethics. Each chapter on quantitative methods describes key theoretical choices and how these are executed in Stata. Unlike most other books, we do not discuss theory *or* application but link the two.

- This is a book for nontechnical readers! All chapters are written in an accessible and comprehensive way so that readers without a profound background in statistics can also understand the introduced data analysis methods. Each chapter on research methods includes examples to help the reader gain a hands-on feeling for the technique. Each chapter concludes with an illustrated case that demonstrates the application of a quantitative method.
- To facilitate learning, we use a single case study throughout the book. This case deals with a customer survey of a fictitious company called Oddjob Airways (familiar to those who have seen the James Bond movie Goldfinger!). We also provide additional end-of-chapter cases, including different datasets, thus allowing the readers to practice what they have learned. Other pedagogical features, such as keywords, examples, and end-of-chapter questions, support the contents.
- Stata has become a very popular statistics package in the social sciences and beyond, yet there are almost no books that show how to use the program without diving into the depths of syntax language.
- This book is concise, focusing on the most important aspects that a market researcher, or manager interpreting market research, should know.
- Many chapters provide links to further readings and other websites. Mobile tags in the text allow readers to quickly browse related web content using a mobile device (see section "How to Use Mobile Tags"). This unique merger of offline and online content offers readers a broad spectrum of additional and readily accessible information. A comprehensive web appendix with information on further analysis techniques and datasets is included.
- Lastly, we have set up a Facebook page called *Market Research: The Process, Data, and Methods*. This page provides a platform for discussions and the exchange of market research ideas.

Find us on Facebook

How to Use Mobile Tags

In this book, there are several mobile tags that allow you to instantly access information by means of your mobile phone's camera if it has a mobile tag reader installed. For example, the following mobile tag is a link to this book's website at http://www.guide-market-research.com.

Several mobile phones come with a mobile tag reader already installed, but you can also download tag readers. In this book, we use QR (quick response) codes, which can be accessed by means of the readers below. Simply visit one of the following webpages or download the App from the iPhone App Store or from Google Play:

- Kaywa: http://reader.kaywa.com/
- i-Nigma: http://www.i-nigma.com/

Once you have a reader app installed, just start the app and point your camera at the mobile tag. This will open your mobile phone browser and direct you to the associated website.

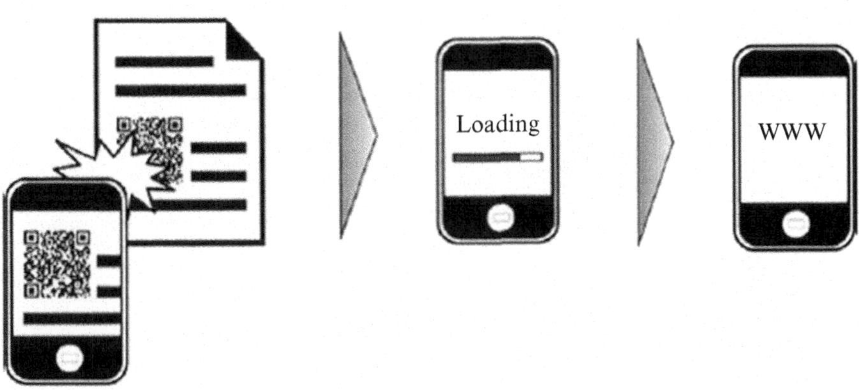

Step 1
Point at a mobile tag and take a picture

Step 2
Decoding and loading

Step 3
Website

How to Use This Book

The following will help you read this book:

- Stata commands that the user types or the program issues appear in a different `font`.
- Variable or file names in the main text appear in *italics* to distinguish them from the descriptions.
- Items from Stata's interface are shown in **bold**, with successive menu options separated while variable names are shown in *italics*. For example, the text could read: "Go to ► Graphics ► Scatterplot matrix and enter the variables *s1*, *s2*, and *s3* into the **Variables** box." This means that the word **Variables** appears in the Stata interface while *s1*, *s2*, and *s3* are variable names.
- Keywords also appear in **bold** when they first appear in the main text. We have used many keywords to help you find everything quickly. Additional index terms appear in *italics*.
- If you see ⭳ Web Appendix → Downloads in the book, please go to https://www.guide-market-research.com/stata/ and click on downloads.

In the chapters, you will also find boxes for the interested reader in which we discuss details. The text can be understood without reading these boxes, which are therefore optional. We have also included mobile tags to help you access material quickly.

For Instructors

Besides the benefits described above, this book is also designed to make teaching as easy as possible when using this book. Each chapter comes with a set of detailed and professionally designed PowerPoint slides for educators, tailored for this book, which can be easily adapted to fit a specific course's needs. These are available on the website's instructor resources page at http://www.guide-market-research.com. You can gain access to the instructor's page by requesting log-in information under Instructor Resources.

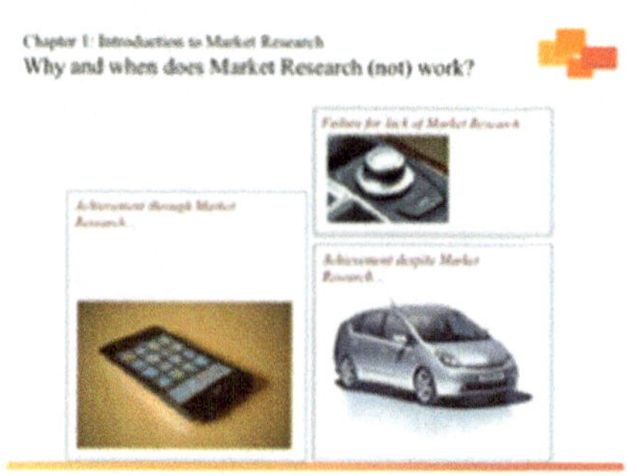

The book's web appendices are freely available on the accompanying website and provide supplementary information on analysis techniques not covered in the book and datasets. Moreover, at the end of each chapter, there is a set of questions that can be used for in-class discussions.

If you have any remarks, suggestions, or ideas about this book, please drop us a line at erik.mooi@unimelb.edu.au (Erik Mooi), marko.sarstedt@ovgu.de (Marko Sarstedt), or irma.mooi@unimelb.edu.au (Irma Mooi-Reci). We appreciate any feedback on the book's concept and contents!

Parkville, VIC, Australia — Erik Mooi
Magdeburg, Germany — Marko Sarstedt
Parkville, VIC, Australia — Irma Mooi-Reci

Acknowledgments

Thanks to all the students who have inspired us with their feedback and constantly reinforce our choice to stay in academia. We have many people to thank for making this book possible. First, we would like to thank Springer and particularly Stephen Jones for all their help and for their willingness to publish this book. We also want to thank Bill Rising of StataCorp for providing immensely useful feedback. Ilse Evertse has done a wonderful job (again!) proofreading the chapters. She is a great proofreader and we cannot recommend her enough! Drop her a line at stpubus@gmail.com if you need proofreading help. In addition, we would like to thank the team of current and former doctoral students and research fellows at Otto-von-Guericke-University Magdeburg, namely, Kati Barth, Janine Dankert, Frauke Kühn, Sebastian Lehmann, Doreen Neubert, and Victor Schliwa. Finally, we would like to acknowledge the many insights and 1 suggestions provided by many of our colleagues and students. We would like to thank the following:

Ralf Aigner of Wishbird, Mexico City, Mexico
Carolin Bock of the Technische Universität Darmstadt, Darmstadt, Germany
Cees J. P. M. de Bont of Hong Kong Polytechnic University, Hung Hom, Hong Kong
Bernd Erichson of Otto-von-Guericke-University Magdeburg, Magdeburg, Germany
Andrew M. Farrell of the University of Southampton, Southampton, UK
Sebastian Fuchs of BMW Group, München, Germany
David I. Gilliland of Colorado State University, Fort Collins, CO, USA
Joe F. Hair Jr. of the University of South Alabama, Mobile, AL, USA
Jörg Henseler of the University of Twente, Enschede, The Netherlands
Emile F. J. Lancée of Vrije Universiteit Amsterdam, Amsterdam, The Netherlands
Tim F. Liao of the University of Illinois Urbana-Champaign, USA
Peter S. H. Leeflang of the University of Groningen, Groningen, The Netherlands
Arjen van Lin of Vrije Universiteit Amsterdam, Amsterdam, The Netherlands
Leonard J. Paas of Massey University, Albany, New Zealand

Sascha Raithel of FU Berlin, Berlin, Germany
Edward E. Rigdon of Georgia State University, Atlanta, GA, USA
Christian M. Ringle of Technische Universität Hamburg-Harburg, Hamburg, Germany
John Rudd of the University of Warwick, Coventry, UK
Sebastian Scharf of Hochschule Mittweida, Mittweida, Germany
Tobias Schütz of the ESB Business School Reutlingen, Reutlingen, Germany
Philip Sugai of the International University of Japan, Minamiuonuma, Niigata, Japan
Charles R. Taylor of Villanova University, Philadelphia, PA, USA
Andrés Trujillo-Barrera of Wageningen University & Research
Stefan Wagner of the European School of Management and Technology, Berlin, Germany
Eelke Wiersma of Vrije Universiteit Amsterdam, Amsterdam, The Netherlands
Caroline Wiertz of Cass Business School, London, UK
Michael Zyphur of the University of Melbourne, Parkville, Australia

References

Davenport, T. H., & Patil, D. J. (2012). Data scientist. The sexiest job of the 21st century. *Harvard Business Review*, *90*(October), 70–76.

Forbes. (2015a). *Data driven and customer centric: Marketers turning insights into impact*. http://www.forbes.com/forbesinsights/data-driven_and_customer-centric/. Accessed 21 Aug 2017.

Forbes. (2015b). *Where big data jobs will be in 2016*. http://www.forbes.com/sites/louiscolumbus/2015/11/16/where-big-data-jobs-will-be-in-2016/#68fece3ff7f1/. Accessed 21 Aug 2017.

Heisler, Y. (2012). *How Apple conducts market research and keeps iOS source code locked down*. Network world, August 3, 2012, http://www.networkworld.com/article/2222892/wireless/how-apple-conducts-market-research-and-keeps-iossource-code-locked-down.html. Accessed 21 Aug 2017.

Statista.com. (2017). *Market research industry/market – Statistics & facts*. https://www.statista.com/topics/1293/market-research/. Accessed 21 Aug 2017.

Wedel, M., & Kannan, P. K. (2016). Marketing analytics for data-rich environments. *Journal of Marketing*, *80*(6), 97–121.

Contents

1 Introduction to Market Research 1
- 1.1 Introduction 1
- 1.2 What Is Market and Marketing Research? 2
- 1.3 Market Research by Practitioners and Academics 3
- 1.4 When Should Market Research (Not) Be Conducted? 4
- 1.5 Who Provides Market Research? 5
- 1.6 Review Questions 8
- 1.7 Further Readings 8
- References 9

2 The Market Research Process 11
- 2.1 Introduction 11
- 2.2 Identify and Formulate the Problem 12
- 2.3 Determine the Research Design 13
  - 2.3.1 Exploratory Research 14
  - 2.3.2 Uses of Exploratory Research 15
  - 2.3.3 Descriptive Research 17
  - 2.3.4 Uses of Descriptive Research 17
  - 2.3.5 Causal Research 18
  - 2.3.6 Uses of Causal Research 21
- 2.4 Design the Sample and Method of Data Collection 23
- 2.5 Collect the Data 23
- 2.6 Analyze the Data 23
- 2.7 Interpret, Discuss, and Present the Findings 23
- 2.8 Follow-Up 23
- 2.9 Review Questions 24
- 2.10 Further Readings 24
- References 25

3 Data 27
- 3.1 Introduction 28
- 3.2 Types of Data 28
  - 3.2.1 Primary and Secondary Data 31
  - 3.2.2 Quantitative and Qualitative Data 32

3.3 Unit of Analysis . . . 33
3.4 Dependence of Observations . . . 34
3.5 Dependent and Independent Variables . . . 35
3.6 Measurement Scaling . . . 35
3.7 Validity and Reliability . . . 37
3.7.1 Types of Validity . . . 39
3.7.2 Types of Reliability . . . 40
3.8 Population and Sampling . . . 41
3.8.1 Probability Sampling . . . 43
3.8.2 Non-probability Sampling . . . 45
3.8.3 Probability or Non-probability Sampling? . . . 46
3.9 Sample Sizes . . . 47
3.10 Review Questions . . . 47
3.11 Further Readings . . . 48
References . . . 49

4 Getting Data . . . 51
4.1 Introduction . . . 51
4.2 Secondary Data . . . 52
4.2.1 Internal Secondary Data . . . 53
4.2.2 External Secondary Data . . . 54
4.3 Conducting Secondary Data Research . . . 58
4.3.1 Assess Availability of Secondary Data . . . 58
4.3.2 Assess Inclusion of Key Variables . . . 60
4.3.3 Assess Construct Validity . . . 60
4.3.4 Assess Sampling . . . 61
4.4 Conducting Primary Data Research . . . 62
4.4.1 Collecting Primary Data Through Observations . . . 62
4.4.2 Collecting Quantitative Data: Designing Surveys . . . 64
4.5 Basic Qualitative Research . . . 82
4.5.1 In-Depth Interviews . . . 82
4.5.2 Projective Techniques . . . 84
4.5.3 Focus Groups . . . 84
4.6 Collecting Primary Data Through Experimental Research . . . 86
4.6.1 Principles of Experimental Research . . . 86
4.6.2 Experimental Designs . . . 87
4.7 Review Questions . . . 89
4.8 Further Readings . . . 90
References . . . 91

5 Descriptive Statistics . . . 95
5.1 The Workflow of Data . . . 96
5.2 Create Structure . . . 97
5.3 Enter Data . . . 99

5.4 Clean Data 99
5.4.1 Interviewer Fraud 100
5.4.2 Suspicious Response Patterns 100
5.4.3 Data Entry Errors 102
5.4.4 Outliers 102
5.4.5 Missing Data 104
5.5 Describe Data 110
5.5.1 Univariate Graphs and Tables 110
5.5.2 Univariate Statistics 113
5.5.3 Bivariate Graphs and Tables 115
5.5.4 Bivariate Statistics 117
5.6 Transform Data (Optional) 120
5.6.1 Variable Respecification 120
5.6.2 Scale Transformation 121
5.7 Create a Codebook 123
5.8 The Oddjob Airways Case Study 124
5.8.1 Introduction to Stata 124
5.8.2 Finding Your Way in Stata 126
5.9 Data Management in Stata 134
5.9.1 Restrict Observations 134
5.9.2 Create a New Variable from Existing Variable(s) 135
5.9.3 Recode Variables 136
5.10 Example 137
5.10.1 Clean Data 138
5.10.2 Describe Data 139
5.11 Cadbury and the UK Chocolate Market (Case Study) 149
5.12 Review Questions 150
5.13 Further Readings 151
References 151

6 Hypothesis Testing & ANOVA 153
6.1 Introduction 153
6.2 Understanding Hypothesis Testing 154
6.3 Testing Hypotheses on One Mean 156
6.3.1 Step 1: Formulate the Hypothesis 156
6.3.2 Step 2: Choose the Significance Level 158
6.3.3 Step 3: Select an Appropriate Test 160
6.3.4 Step 4: Calculate the Test Statistic 168
6.3.5 Step 5: Make the Test Decision 171
6.3.6 Step 6: Interpret the Results 175
6.4 Two-Samples *t*-Test 175
6.4.1 Comparing Two Independent Samples 175
6.4.2 Comparing Two Paired Samples 177

6.5 Comparing More Than Two Means: Analysis of Variance (ANOVA) . . . 179
6.6 Understanding One-Way ANOVA . . . 180
6.6.1 Check the Assumptions . . . 181
6.6.2 Calculate the Test Statistic . . . 182
6.6.3 Make the Test Decision . . . 186
6.6.4 Carry Out Post Hoc Tests . . . 187
6.6.5 Measure the Strength of the Effects . . . 188
6.6.6 Interpret the Results and Conclude . . . 189
6.6.7 Plotting the Results (Optional) . . . 189
6.7 Going Beyond One-Way ANOVA: The Two-Way ANOVA . . . 190
6.8 Example . . . 198
6.8.1 Independent Samples *t*-Test . . . 198
6.8.2 One-way ANOVA . . . 202
6.8.3 Two-way ANOVA . . . 207
6.9 Customer Analysis at Crédit Samouel (Case Study) . . . 212
6.10 Review Questions . . . 213
6.11 Further Readings . . . 213
References . . . 214

7 Regression Analysis . . . 215
7.1 Introduction . . . 216
7.2 Understanding Regression Analysis . . . 216
7.3 Conducting a Regression Analysis . . . 219
7.3.1 Check the Regression Analysis Data Requirements . . . 219
7.3.2 Specify and Estimate the Regression Model . . . 222
7.3.3 Test the Regression Analysis Assumptions . . . 226
7.3.4 Interpret the Regression Results . . . 231
7.3.5 Validate the Regression Results . . . 237
7.3.6 Use the Regression Model . . . 239
7.4 Example . . . 243
7.4.1 Check the Regression Analysis Data Requirements . . . 244
7.4.2 Specify and Estimate the Regression Model . . . 248
7.4.3 Test the Regression Analysis Assumptions . . . 249
7.4.4 Interpret the Regression Results . . . 254
7.4.5 Validate the Regression Results . . . 258
7.5 Farming with AgriPro (Case Study) . . . 260
7.6 Review Questions . . . 262
7.7 Further Readings . . . 262
References . . . 263

8 Principal Component and Factor Analysis 265
8.1 Introduction 266
8.2 Understanding Principal Component and Factor Analysis 267
8.2.1 Why Use Principal Component and Factor Analysis? 267
8.2.2 Analysis Steps 269
8.3 Principal Component Analysis 270
8.3.1 Check Requirements and Conduct Preliminary Analyses 270
8.3.2 Extract the Factors 273
8.3.3 Determine the Number of Factors 278
8.3.4 Interpret the Factor Solution 280
8.3.5 Evaluate the Goodness-of-Fit of the Factor Solution 282
8.3.6 Compute the Factor Scores 283
8.4 Confirmatory Factor Analysis and Reliability Analysis 284
8.5 Structural Equation Modeling 289
8.6 Example 291
8.6.1 Principal Component Analysis 291
8.6.2 Reliability Analysis 304
8.7 Customer Satisfaction at Haver and Boecker (Case Study) . . . 306
8.8 Review Questions 308
8.9 Further Readings 309
References 309

9 Cluster Analysis 313
9.1 Introduction 314
9.2 Understanding Cluster Analysis 314
9.3 Conducting a Cluster Analysis 316
9.3.1 Select the Clustering Variables 316
9.3.2 Select the Clustering Procedure 321
9.3.3 Select a Measure of Similarity or Dissimilarity 333
9.3.4 Decide on the Number of Clusters 340
9.3.5 Validate and Interpret the Clustering Solution 344
9.4 Example 349
9.4.1 Select the Clustering Variables 350
9.4.2 Select the Clustering Procedure and Measure of Similarity or Dissimilarity 353
9.4.3 Decide on the Number of Clusters 354
9.4.4 Validate and Interpret the Clustering Solution 358
9.5 Oh, James! (Case Study) 362
9.6 Review Questions 363
9.7 Further Readings 364
References 365

10 Communicating the Results . . . 367
10.1 Introduction . . . 367
10.2 Identify the Audience . . . 368
10.3 Guidelines for Written Reports . . . 369
10.4 Structure the Written Report . . . 370
10.4.1 Title Page . . . 371
10.4.2 Executive Summary . . . 371
10.4.3 Table of Contents . . . 371
10.4.4 Introduction . . . 372
10.4.5 Methodology . . . 372
10.4.6 Results . . . 373
10.4.7 Conclusion and Recommendations . . . 383
10.4.8 Limitations . . . 384
10.4.9 Appendix . . . 384
10.5 Guidelines for Oral Presentations . . . 384
10.6 Visual Aids in Oral Presentations . . . 385
10.7 Structure the Oral Presentation . . . 386
10.8 Follow-Up . . . 387
10.9 Ethics in Research Reports . . . 388
10.10 Review Questions . . . 389
10.11 Further Readings . . . 389
References . . . 389

Glossary . . . 391

Index . . . 411

Introduction to Market Research 1

Keywords
American Marketing Association (AMA) • ESOMAR • Field service firms • Full service providers • Limited service providers • Segment specialists • Specialized service firms • Syndicated data

Learning Objectives
After reading this chapter, you should understand:

- What market and marketing research are and how they differ.
- How practitioner and academic market(ing) research differ.
- When market research should be conducted.
- Who provides market research and the importance of the market research industry.

1.1 Introduction

When Toyota developed the Prius—a highly fuel-efficient car using a hybrid petrol/ electric engine—it took a gamble on a grand scale. Honda and General Motors' previous attempts to develop frugal (electric) cars had not worked well. Just like Honda and General Motors, Toyota had also been working on developing a frugal car, but focused on a system integrating a petrol and electric engine. These development efforts led Toyota to start a project called Global Twenty-first Century aimed at developing a car with a fuel economy that was at least 50% better than similar-sized cars. This project nearly came to a halt in 1995 when Toyota encountered substantial technological problems. The company solved these problems, using nearly a thousand engineers, and launched the car, called the Prius, in Japan in 1997. Internal Toyota predictions suggested that the car was either going

E. Mooi et al., *Market Research*, Springer Texts in Business and Economics,
DOI 10.1007/978-981-10-5218-7_1

to be an instant hit, or that the product's acceptance would be slow, as it takes time to teach dealers and consumers about the technology. In 1999, Toyota decided to start working on launching the Prius in the US. Initial market research showed that it was going to be a difficult task. Some consumers thought it was too small for the US and some thought the positioning of the controls was poor for US drivers. There were other issues too, such as the design, which many thought was too strongly geared towards Japanese drivers.

While preparing for the launch, Toyota conducted further market research, which could, however, not reveal who the potential car buyers would be. Initially, Toyota thought the car might be tempting for people concerned with the environment, but market research dispelled this belief. Environmentalists dislike technology in general and money is a big issue for this group. A technologically complex and expensive car such as the Prius was therefore unlikely to appeal to them. Additional market research did little to identify any other good market segment. Despite the lack of conclusive findings, Toyota decided to sell the car anyway and to await the public's reaction. Before the launch, Toyota put a market research system in place to track the initial sales and identify where customers bought the car. After the formal launch in 2000, this system quickly found that celebrities were buying the car to demonstrate their concern for the environment. Somewhat later, Toyota noticed substantially increased sales figures when ordinary consumers became aware of the car's appeal to celebrities. It appeared that consumers were willing to purchase cars that celebrities endorse.

CNW Market Research, a market research company specializing in the automotive industry, attributed part of the Prius's success to its unique design, which clearly demonstrated that Prius owners were driving a different car. After substantial increases in the petrol price, and changes to the car (based on extensive market research) to increase its appeal, Toyota's total Prius sales reached about four million and the company is now the market leader in hybrid petrol/electric cars.

This example shows that while market research occasionally helps, sometimes it contributes little, or even fails. There are many reasons for market research's success varying. These reasons include the budget available for research, the support for market research in the organization, the implementation, and the market researchers' research skills. In this book, we will guide you step by step through the practicalities of the basic market research process. These discussions, explanations, facts, and methods will help you carry out successful market research.

1.2 What Is Market and Marketing Research?

Market research can mean several things. It can be the process by which we gain insight into how markets work. Market research is also a function in an organization, or it can refer to the outcomes of research, such as a database of customer purchases, or a report that offers recommendations. In this book, we focus on the market research process, starting by identifying and formulating the problem, continuing by determining the research design, determining the sample and method of data collection, collecting the data, analyzing the data, interpreting, discussing, and presenting the findings, and ending with the follow-up.

Some people consider marketing research and market research to be synonymous, whereas others regard these as different concepts. The **American Marketing Association (AMA)**, the largest marketing association in North America, defines marketing research as follows:

> The function that links the consumer, customer, and public to the marketer through information – information used to identify and define marketing opportunities and problems; generate, refine, and evaluate marketing actions; monitor marketing performance; and improve understanding of marketing as a process. Marketing research specifies the information required to address these issues, designs the method for collecting information, manages and implements the data collection process, analyzes the results, and communicates the findings and their implications (American Marketing Association 2004).

On the other hand, **ESOMAR**, the world organization for market, consumer and societal research, defines market research as:

> The systematic gathering and interpretation of information about individuals and organisations. It uses the statistical and analytical methods and techniques of the applied social, behavioural and data sciences to generate insights and support decision-making by providers of goods and services, governments, non-profit organisations and the general public. (ICC/ESOMAR international code on market, opinion, and social research and data analytics 2016).

Both definitions overlap substantially, but the AMA definition focuses on marketing research as a function (e.g., a department in an organization), whereas the ESOMAR definition focuses on the process. In this book, we focus on the process and, thus, on market research.

1.3 Market Research by Practitioners and Academics

Practitioners and academics are both involved in marketing and market research. Academic and practitioner views of market(ing) research differ in many ways, but also have many communalities.

There is, however, a key difference is their target groups. Academics almost exclusively undertake research with the goal of publishing in academic journals. Highly esteemed journals include the *Journal of Marketing*, *Journal of Marketing Research*, *Journal of the Academy of Marketing Science*, and the *International Journal of Research in Marketing*. On the other hand, practitioners' target group is the client, whose needs and standards include relevance, practicality, generalizability, and timeliness of insights. Journals, on the other hand, frequently emphasize methodological rigor and consistency. Academic journals are often difficult to read and understand, while practitioner reports should be easy to read.

Academics and practitioners differ greatly in their use of and focus on methods. Practitioners have adapted and refined some of the methods, such as cluster analysis and factor analysis, which academics developed originally.[1] Developing methods is often a goal in itself for academics. Practitioners are more concerned about the value of applying specific methods. Standards also differ. Clear principles and professional conduct as advocated by ESOMAR and the Australian Market & Social Research Society (AMSRS) (for examples, see https://www.esomar.org/uploads/public/knowledge-and-standards/codes-and-guidelines/ICCESOMAR-International-Code_English.pdf and http://www.amsrs.com.au/documents/item/194) mostly guide practitioners' methods. Universities and schools sometimes impose data collection and analysis standards on academics, but these tend not to have the level of detail advocated by ESOMAR or the AMSRS. Interestingly, many practitioners claim that their methods meet academic standards, but academics never claim that their methods are based on practitioner standards.

Besides these differences, there are also many similarities. For example, good measurement is paramount for academics and practitioners. Furthermore, academics and practitioners should be interested in each other's work; academics can learn much from the practical issues that practitioners faced, while practitioners can gain much from understanding the tools, techniques, and concepts that academics develop. Reibstein et al. (2009), who issued an urgent call for the academic marketing community to focus on relevant business problems, underlined the need to learn from each other. Several other researchers, such as Lee and Greenley (2010), Homburg et al. (2015), and Tellis (2017), have echoed this call.

1.4 When Should Market Research (Not) Be Conducted?

Market research serves several useful roles in organizations. Most importantly, market research can help organizations by providing answers to questions firms may have about their customers and competitors; answers that could help such firms improve their performance. Specific questions related to this include identifying market opportunities, measuring customer satisfaction, and assessing market shares. Some of these questions arise ad hoc, perhaps due to issues that the top management, or one of the departments or divisions, has identified. Much market research is, however, programmatic; it arises because firms systematically evaluate market elements. Subway, the restaurant chain, systematically measures customer satisfaction, which is an example of programmatic research. This type of research does not usually have a distinct beginning and end (contrary to ad hoc research), but is executed continuously over time and leads to daily, weekly, or monthly reports.

The decision to conduct market research may be taken when managers face an uncertain situation and when the costs of undertaking good research are (much) lower than good decisions' expected benefits. Researching trivial issues or issues that cannot be changed is not helpful.

[1]Roberts et al. (2014) and Hauser (2017) discuss the impact of marketing science tools on marketing practice.

Other issues to consider are the politics within the organization, because if the decision to go ahead has already been made (as in the Prius example in the introduction), market research is unnecessary. If market research is conducted and supports the decision, it is of little value—and those undertaking the research may have been biased in favor of the decision. On the other hand, market research is ignored if it rejects the decision.

Moreover, organizations often need to make very quick decisions, for example, when responding to competitive price changes, unexpected changes in regulation, or to the economic climate. In such situations, however, market research may only be included after decisions have already been made. Consequently, research should mostly not be undertaken when urgent decisions have to be made.

1.5 Who Provides Market Research?

Many organizations have people, departments, or other companies working for them to provide market research. In Fig. 1.1, we show who these providers of market research are.

Most market research is provided internally by specialized market research departments, or people tasked with this function. It appears that about 75% of organizations have at least one person tasked with carrying out market research. This percentage is similar across most industries, although it is much less in government sectors and, particularly, in health care (Iaccobucci and Churchill 2015).

In larger organizations, a sub department of the marketing department usually undertakes internally provided market research. Sometimes this sub department is not connected to a marketing department, but to other organizational functions, such as corporate planning or sales (Rouziès and Hulland 2014). Many large organizations even have a separate market research department. This system of having a separate market research department, or merging it with other

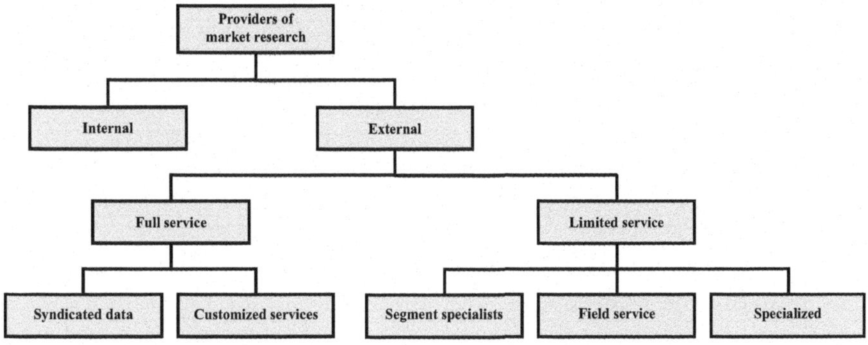

Fig. 1.1 The providers of market research

departments, seems to become more widespread, with the marketing function devolving increasingly into other functions within organizations (Sheth and Sisodia 2006).

The external providers of market research are a powerful economic force. In 2015, the Top 50 external providers had a collective turnover of about $21.78 billion (Honomichl 2016). The market research industry has also become a global field with companies such as The Nielsen Company (USA), Kantar (UK), GfK (Germany), and Ipsos (France), playing major roles outside their home markets. External providers of market research are either full service providers or limited ones.

Full service providers are large market research companies such as The Nielsen Company (http://www.nielsen.com), Kantar (http://www.kantar.com), and GfK (http://www.gfk.com). These large companies provide syndicated data and customized services. **Syndicated data** are data collected in a standard format and not specifically collected for a single client. These data, or analyses based on the data, are then sold to multiple clients. Large marketing research firms mostly collect syndicated data, as they have the resources to collect large amounts of data and can spread the costs of doing so over a number of clients. For example, The Nielsen Company collects syndicated data in several forms: Nielsen's Netratings, which collects information on digital media; Nielsen Ratings, which details the type of consumer who listens to the radio, watches TV, or reads print media; and Nielsen Homescan, which collects panel information on the purchases consumers make. These large firms also offer customized services by conducting studies for a specific client. These customized services can be very specific, such as helping a client carry out specific analyses.

Measuring TV audiences is critical for advertisers. But measuring the number of viewers per program has become more difficult as households currently have multiple TVs and may have different viewing platforms. In addition, "time shift" technologies, such as video-on-demand, have further complicated the tracking of viewer behavior. Nielsen has measured TV and other media use for more than 25 years, using a device called the (Portable) People Meter. This device measures usage of each TV viewing platform and instantly transmits the results back to Nielsen, allowing for instant measurement. Altogether, Nielsen captures about 40% of the world's viewing behavior.[2]

In the following seven videos, experts from The Nielsen Company discuss how the People Meter works.

(continued)

[2]See http://www.nielsen.com/eu/en/solutions/measurement/television.html for further detail.

Contrary to full service providers, which undertake nearly all market research activities, **limited service providers** specialize in one or more services and tend to be smaller companies. In fact, many of the specialized market research companies are one-man businesses and the owner—after (or besides) a practitioner or academic career—offers specialized services. Although there are many different types of limited service firms, we only discuss three of them: those focused on segmentation, field service, and specialized services.

Segment specialists concentrate on specific market segments. Skytrax, which focuses on market research in the airline and airport sector, is an example of such specialists. Other segment specialists do not focus on a particular industry, but on a type of customer; for example, Ethnic Focus (http://www.ethnicfocus.com), a UK-based market research firm, focuses on understanding ethnic minorities.

Field service firms, such as Survey Sampling International (http://www.surveysampling.com), focus on executing surveys, determining samples, sample sizes, and collecting data. Some of these firms also translate surveys, or provide addresses and contact details.

Specialized Service firms are a catch-all term for those firms with specific technical skills, thus only focusing on specific products, or aspects of products, such as market research on taste and smell. Specialized firms may also concentrate on a few highly specific market research techniques, or may focus on one or more highly specialized analysis techniques, such as time series analysis, panel data analysis, or quantitative text analysis. Envirosell (http://www.envirosell.com), a research and consultancy firm that analyzes consumer behavior in commercial environments, is a well-known example of a specialized service firm.

A choice between these full service and limited service market research firms boils down to a tradeoff between what they can provide (if this is highly specialized, you may not have much choice) and the price of doing so. In addition, if you have to combine several studies to gain further insight, full service firms may be better than multiple limited service firms. The fit and feel with the provider are obviously also highly important!

1.6 Review Questions

1. What is market research? Try to explain what market research is in your own words.
2. Imagine you are the head of a division of Procter & Gamble. You are just about ready to launch a new shampoo, but are uncertain about who might buy it. Is it useful to conduct a market research study? Should you delay the launch of the product?
3. Try to find the websites of a few market research firms. Look, for example, at the services provided by GfK and the Nielsen Company, and compare the extent of their offerings to those of specialized firms such as those listed on, for example, http://www.greenbook.org.
4. If you have a specialized research question, such as what market opportunities there are for selling music to ethnic minorities, would you use a full service or limited service firm (or both)? Please discuss the benefits and drawbacks.

1.7 Further Readings

American Marketing Association at http://www.marketingpower.com
Website of the American Marketing Association. Provides information on their activities and also links to two of the premier marketing journals, the Journal of Marketing and the Journal of Marketing Research.
Insights Association at http://www.insightsassociation.org/ *Launched in 2017, the Insights Association was formed through the merger of two organizations with long, respected histories of servicing the market research and analytics industry: CASRO (founded in 1975) and MRA (founded in 1957). The organization focuses on providing knowledge, advice, and standards to those working in the market research profession.*
The British Market Research Society at http://www.mrs.org.uk
The website of the British Market Research society contains a searchable directory of market research providers and useful information on market research careers and jobs.
Associação Brasileira de Empresas de Pesquisa (Brazilian Association of Research Companies) at http://www.abep.org/novo/default.aspx
The website of the Brazilian Association of Research Companies. It documents research ethics, standards, etc.
ESOMAR at http://www.esomar.org
The website of ESOMAR, the world organization for market, consumer and societal research. Amongst other activities, ESOMAR sets ethical and technical standards for market research and publishes books and reports on market research.
GreenBook: The guide for buyers of marketing research services at http://www.greenbook.org
This website provides an overview of many different types of limited service firms.

References

Hauser, J. R. (2017). Phenomena, theory, application, data, and methods all have impact. *Journal of the Academy of Marketing Science, 45*(1), 7–9.

Homburg, C., Vomberg, A., Enke, M., & Grimm, P. H. (2015). The loss of the marketing department's influence: Is it happening? And why worry? *Journal of the Academy of Marketing Science, 43*(1), 1–13.

Honomichl, J. (2016). 2016 Honomichl Gold Top 50. https://www.ama.org/publications/MarketingNews/Pages/2016-ama-gold-top-50-report.aspx

Iaccobucci, D., & Churchill, G. A. (2015). *Marketing research: Methodological foundations* (11th ed.). CreateSpace Independent Publishing Platform.

ICC/ESOMAR international code on market and social research. (2007). http://www.netcasearbitration.com/uploadedFiles/ICC/policy/marketing/Statements/ICCESOMAR_Code_English.pdf

Lee, N., & Greenley, G. (2010). The theory-practice divide: Thoughts from the editors and senior advisory board of EJM. *European Journal of Marketing, 44*(1/2), 5–20.

Reibstein, D. J., Day, G., & Wind, J. (2009). Guest editorial: Is marketing academia losing its way? *Journal of Marketing, 73*(4), 1–3.

Roberts, J. H., Kayand, U., & Stremersch, S. (2014). From academic research to marketing practice: Exploring the marketing science value chain. *International Journal of Research in Marketing, 31*(2), 128–140.

Rouziès, D., & Hulland, J. (2014). Does marketing and sales integration always pay off? Evidence from a social capital perspective. *Journal of the Academy of Marketing Science, 42*(5), 511–527.

Sheth, J. N., & Sisodia, R. S. (Eds.). (2006). *Does marketing need reform? In does marketing need reform? Fresh perspective on the future*. Armonk: M.E. Sharpe.

Tellis, G. J. (2017). Interesting and impactful research: On phenomena, theory, and writing. *Journal of the Academy of Marketing Science, 45*(1), 1–6.

The Market Research Process 2

Keywords
Causal research • Descriptive research • Ethnographies • Exploratory research • Field experiments • Focus groups • Hypotheses • In-depth interviews • Lab experiments • Market segments • Observational studies • Projective techniques • Research design • Scanner data • Test markets

Learning Objectives
After reading this chapter, you should understand:

- How to determine a research design.
- The differences between, and examples of, exploratory research, descriptive research, and causal research.
- What causality is.
- The market research process.

2.1 Introduction

How do organizations plan for market research processes? In this chapter, we explore the market research process and various types of research. We introduce the planning of market research projects, starting with identifying and formulating the problem and ending with presenting the findings and the follow-up (see Fig. 2.1). This chapter is also an outline of the chapters to come.

E. Mooi et al., *Market Research*, Springer Texts in Business and Economics,
DOI 10.1007/978-981-10-5218-7_2

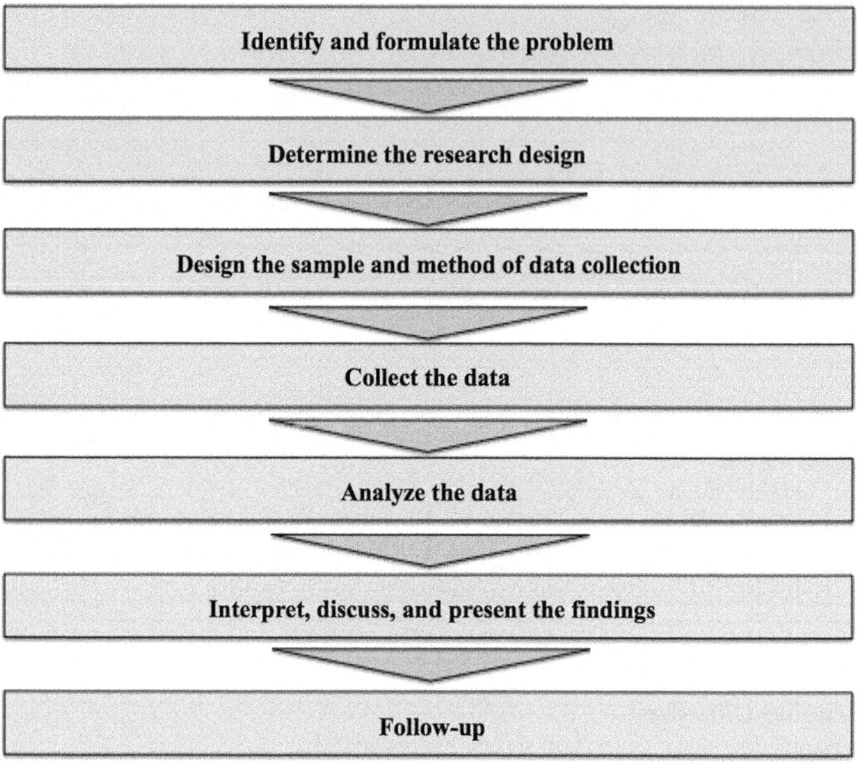

Fig. 2.1 The market research process

2.2 Identify and Formulate the Problem

The first step in setting up a market research process involves identifying and formulating the *research problem*. Identifying the research problem is valuable, but also difficult. To identify the "right" research problem, we should first identify the *marketing symptoms* or *marketing opportunities*. The marketing symptom is a problem that an organization faces. Examples of marketing symptoms include declining market shares, increasing numbers of complaints, or new products that consumers do not adopt. In some cases, there is no real problem, but instead a marketing opportunity, such as the potential benefits that new channels and products offer, or emerging market opportunities that need to be explored. Exploring marketing symptoms and marketing opportunities requires asking questions such as:

- Why is our market share declining?
- Why is the number of complaints increasing?
- Why are our new products not successful?
- How can we enter the market for 3D printers?
- How can we increase our online sales?

The research problems that result from such questions can come in different forms. Generally, we distinguish three *types of research problems*:

- ambiguous problems,
- somewhat defined problems, and
- clearly defined problems.

Ambiguous problems occur when we know very little about the issues that need to be solved. For example, ambiguity typically surrounds the introduction of radically new technologies or products. When Toyota planned to launch the Prius many years ago, critical, but little understood, issues arose, such as the features that were essential and even who the potential buyers of such a car were.

When we face somewhat defined problems, we know the issues (and variables) that are important for solving the problem, but not how they are related. For example, when an organization wants to export products, it is relatively easy to obtain all sorts of information on market sizes, economic development, and the political and legal system. However, how these variables impact the exporting success may be very uncertain.

When we face clearly defined problems, the important issues and variables, as well as their relationships, are clear. However, we do not know how to make the best possible choice. We therefore face the problem of how the situation should be optimized. A clearly defined problem may arise when organizations want to change their prices. While organizations know that increasing (or decreasing) prices generally leads to decreased (increased) demand, the precise relationship (i.e., how many units do we sell less when the price is increased by $1?) is unknown.

2.3 Determine the Research Design

The **research design** is related to the identification and formulation of the problem. Research problems and research designs are highly related. If we start working on an issue that has never been researched before, we seem to enter a funnel where we initially ask exploratory questions, because we as yet know little about the issues we face. These exploratory questions are best answered using an exploratory research design. Once we have a clearer picture of the research issue after our exploratory research, we move further into the funnel. Generally, we want to learn more by describing the research problem in terms of descriptive research. Once we have a reasonably complete picture of all the issues, it may be time to determine exactly how key variables are linked. We then move to the narrowest part of the funnel. We do this through causal (not *casual*!) research (see Fig. 2.2).

Each research design has different uses and requires the application of different analysis techniques. For example, whereas exploratory research can help formulate problems exactly or structure them, causal research provides exact insights into how variables relate. In Fig. 2.3, we provide several examples of different types of research, which we will discuss in the following paragraphs.

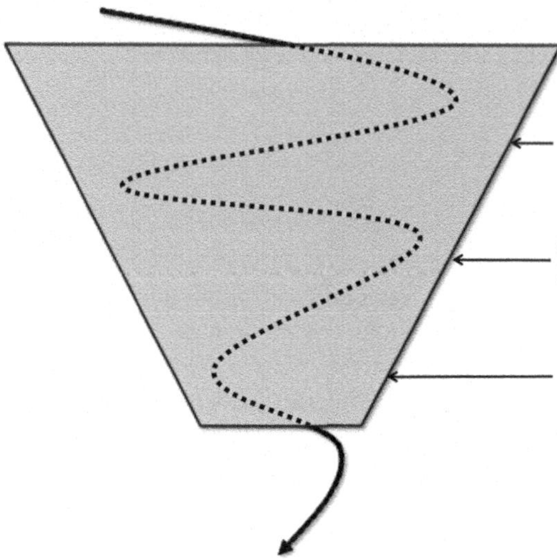

Fig. 2.2 The relationship between the marketing problem and the research design

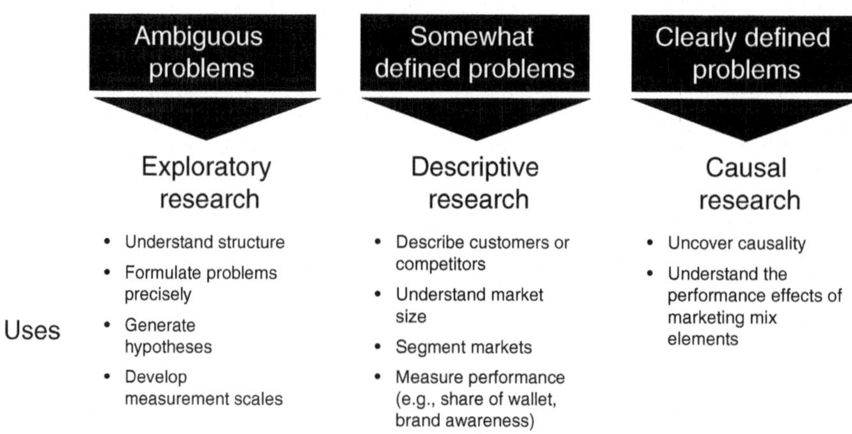

Fig. 2.3 Uses of exploratory, descriptive, and causal research

2.3.1 Exploratory Research

As its name suggests, the objective of **exploratory research** is to explore a problem or situation. As such, exploratory research has several key uses regarding the solving of ambiguous problems. It can help organizations formulate their problems exactly. Through initial research, such as interviewing potential customers, the opportunities and pitfalls may be identified that help determine or refine the research problem. It is crucial to discuss this information with the client to ensure that your findings are helpful. Such initial research also helps establish priorities (what is nice to know and what is important to know?) and eliminate impractical ideas. For example, market research helped Toyota dispel the belief that people concerned with the environment would buy the Prius, as this target group has an aversion to high technology and lacks spending power.

2.3.2 Uses of Exploratory Research

Exploratory research can be used to formulate problems precisely. For example, focus groups, in-depth interviews, projective techniques, observational studies, and ethnographies are often used to achieve this. In the following, we briefly introduce each technique, but provide more detailed descriptions in Chap. 4.

Focus groups usually have between 4 and 6 participants, who discuss a defined topic under the leadership of a moderator. The key difference between a depth interview and focus group is that focus group participants can interact with one another (e.g., "What do you mean by...?," "How does this differ from...."), thereby providing insight into group dynamics. **In-depth interviews** consist of an interviewer asking an interviewee several questions. Depth interviews allow probing on a one-to-one basis, which fosters interaction between the interviewer and the respondent. Depth interviews are required when the topic needs to be adjusted for each interviewee, for sensitive topics, and/or when the person interviewed has a very high status.

Projective techniques present people with pictures, words, or other stimuli to which they respond. For example, a researcher could ask what people think of BMW owners ("A BMW owner is someone who....") or could show them a picture of a BMW and ask them what they associate the picture with. Moreover, when designing new products, market researchers can use different pictures and words to create analogies to existing products and product categories, thus making the adoption of new products more attractive (Feiereisen et al. 2008).

Observational studies are frequently used to refine research questions and clarify issues. Observational studies require an observer to monitor and interpret participants' behavior. For example, someone could monitor how consumers spend their time in shops or how they walk through the aisles of a supermarket. These studies require a person, a camera or other tracking devices, such as radio frequency identification (RFID) chips, to monitor behavior. Other observational studies may comprise click stream data that track information on the web pages people have visited. Observational studies can also be useful to understand how people consume and/or use products. New technology is being developed in this area, for example, market research company Almax (also see Chap. 4) has developed the EyeSee Mannequin which helps observe who is attracted by store windows and reveals important details about customers, such as their age range, gender, ethnicity, and dwell time.

In the award-winning paper "An Exploratory Look at Supermarket Shopping Paths," Larson et al. (2005) analyze the paths individual shoppers take in a grocery store, which the RFID tags located on their shopping carts provide. The results provide new perspectives on many long-standing perceptions of shopper travel behavior within a supermarket, including ideas related to aisle traffic, special promotional displays, and perimeter shopping patterns. Before this study, most retailers believed that customers walked through the aisles systematically. Larson et al.'s (2005) research reveals this rarely happens.

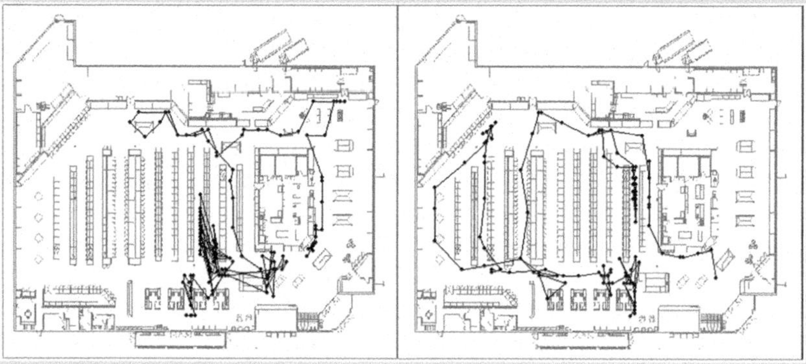

Ethnography (or ethnographic studies) originate from anthropology. In ethnographic research, a researcher interacts with consumers over a period to observe and ask questions. Such studies can consist of, for example, a researcher living with a family to observe how they buy, consume, and use products. For example, the market research company BBDO used ethnographies to understand consumers' rituals. The company found that many consumer rituals are ingrained in consumers in certain countries, but not in others. For example, women in Colombia, Brazil, and Japan are more than twice as likely to apply make-up when in their cars, than women in other countries. Miele, a German whitegoods producer, used ethnographies to understand how people with allergies do their washing and developed washing machines based on the insights gathered (Burrows 2014).

Exploratory research can also help establish research priorities. What is important to know and what is less important? For example, a literature search may reveal that there are useful previous studies and that new market research is not necessary. Exploratory research may also lead to the elimination of impractical ideas. Literature searches, just like interviews, may again help eliminate impractical ideas.

Another helpful aspect of exploratory research is the generation of **hypotheses**. A hypothesis is a claim made about a population, which can be tested by using sample results. For example, one could hypothesize that at least 10% of people in

France are aware of a certain product. Marketers frequently suggest hypotheses, because they help them structure and make decisions. In Chap. 6, we discuss hypotheses and how they can be tested in greater detail.

Another use of exploratory research is to develop measurement scales. For example, what questions can we use to measure customer satisfaction? What questions work best in our context? Do potential respondents understand the wording, or do we need to make changes? Exploratory research can help us answer such questions. For example, an exploratory literature search may use measurement scales that tell us how to measure important variables such as corporate reputation and service quality.

2.3.3 Descriptive Research

As its name implies, **descriptive research** is all about describing certain phenomena, characteristics or functions. It can focus on one variable (e.g., profitability) or on two or more variables at the same time ("what is the relationship between market share and profitability?" and "how does temperature relate to the sale of ice cream?"). Descriptive research often builds on previous exploratory research. After all, to describe something, we must have a good idea of what we need to measure and how we should measure it. Key ways in which descriptive research can help us include describing customers, competitors, market segments, and measuring performance.

2.3.4 Uses of Descriptive Research

Market researchers conduct descriptive research for many purposes. These include, for example, describing customers or competitors. For instance, how large is the UK market for pre-packed cookies? How large is the worldwide market for cruises priced $10,000 and more? How many new products did our competitors launch last year? Descriptive research helps us answer such questions. Much data are available for descriptive purposes, particularly on durable goods and fast moving consumer goods. One source of such data are **scanner data**, which are collected at the checkout of a supermarket where details about each product sold are entered into a vast database. By using scanner data, it is, for example, possible to describe the market for pre-packed cookies in the UK.

Descriptive research is frequently used to define **market segments**, or simply segments. Since companies can seldom connect with all their (potential) customers individually, they divide markets into groups of (potential) customers with similar needs and wants. Firms can then target each of these segments by positioning themselves in a unique segment (such as Ferrari in the high-end sports car market). Many market research companies specialize in market segmentation; an example is Claritas, which developed a segmentation scheme for the US market called *PRIZM* (*Potential Ratings Index by Zip Markets*). PRIZM segments consumers along a multitude of attitudinal, behavioral, and demographic characteristics; companies can use these segments to better target their customers. Segments have names, such as Up-and-Comers (young professionals with a college degree and a mid-level income) and Backcountry Folk (older, often retired people with a high school degree and low income).

Another important function of descriptive market research is to measure performance. Nearly all companies regularly track their sales across specific product categories to evaluate the performance of the firm, the managers, or specific employees. Such descriptive work overlaps with the finance or accounting departments' responsibilities. However, market researchers also frequently measure performance using measures that are quite specific to marketing, such as share of wallet (i.e., how much do people spend on a certain brand or company in a product category?) and brand awareness (i.e., do you know brand/company X?), or the Net Promotor Score, a customer loyalty metric for brands or firms (see Chap. 3 for more information).

2.3.5 Causal Research

Causal research is used to understand the relationships between two or more variables. For example, we may wish to estimate how changes in the wording of an advertisement impact recall. Causal research provides exact insights into how variables relate and may be useful as a test run to try out changes in the marketing mix. Market researchers undertake causal research less frequently than exploratory or descriptive research. Nevertheless, it is important to understand the delicate relationships between important marketing variables and the outcomes they help create. The key usage of causal research is to uncover *causality*. Causality is the relationship between an event (the cause) and a second event (the effect) when the second event is a consequence of the first. To claim causality, we need to meet the following four requirements:

- relationship between cause and effect,
- time order,
- controlling for other factors, and
- an explanatory theory.

First, the cause needs to be related to the effect. For example, if we want to determine whether price increases cause sales to drop, there should be a negative relationship or correlation between price increases and sales decreases (see Chap. 5). Note that people often confuse correlation and causality. Just because there is some type of relationship between two variables does not mean that the one caused the other (see Box 2.1).

Second, the cause needs to come before the effect. This is the time order's requirement. A price increase can obviously only have a causal effect on the sales if it occurred before the sales decrease.

Third, we need to control for other factors. If we increase the price, sales may go up, because competitors increase their prices even more. Controlling for other factors is difficult, but not impossible. In experiments, we design studies so that external factors' effect is nil, or as close to nil as possible. This is achieved by, for example, conducting experiments in labs where environmental factors, such as the conditions, are constant (controlled for). We can also use statistical tools that account for external influences to control for other factors. These statistical tools include an analysis of variance (see Chap. 6), regression analysis (see Chap. 7), and structural equation modeling (see end of Chap. 8).

Fourth, the need for a good explanatory theory is an important criterion. Without a theory, our effects may be due to chance and no "real" effect may be present. For example, we may observe that when we advertise, sales decrease. Without a good explanation of this effect (such as people disliking the advertisement), we cannot claim that there is a causal relationship.

Box 2.1 Correlation Does Not Automatically Imply Causation

Correlation does not automatically imply causality. For example, Fig. 2.4 plots US fatal motor vehicle crashes (per 100,000 people) against the harvested area of melons (in 1,000 acres) between 2000 and 2015.

Clearly, the picture shows a trend. If the harvested area of melons increases, the number of US fatal motor vehicle crashes increases. The resulting correlation of 0.839 is very high (we discuss how to interpret correlations in Chap. 5). While this correlation is the first requirement to determine causality, the story falls short when it comes to explanatory theory. What possible mechanism could explain the findings? This is likely a case of a *spurious correlation*, which is simply due to coincidence.

In the above situation, most people would be highly skeptical and would not interpret the correlation as describing a causal mechanism; in other instances, the situation is much less clear-cut. Think of claims that are part of everyday market research, such as "the new advertisement campaign caused a sharp increase in sales", "our company's sponsorship activities helped improve our company's reputation", or "declining sales figures are caused by competitors' aggressive price policies". Even if there is a correlation, the other requirements for causality may not be met. Causal research may help us determine if causality can be claimed.

(continued)

Box 2.1 (continued)

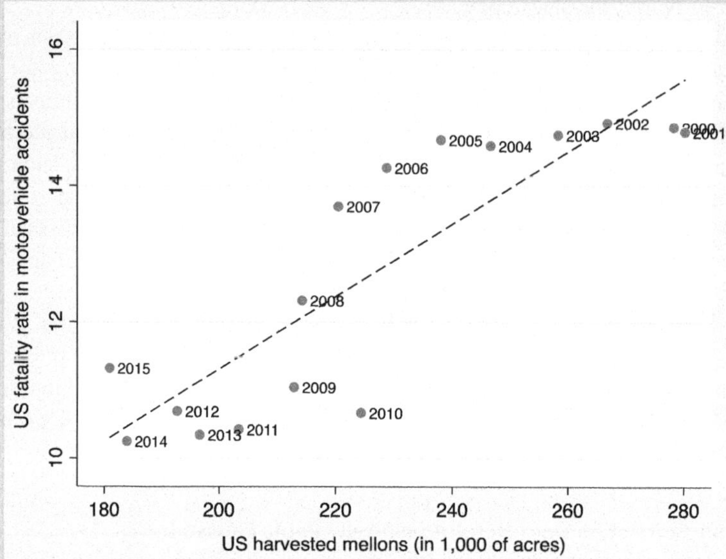

Fig. 2.4 Correlation and causation (the data were taken from the NHTSA Traffic Safety Facts, DOT HS 810780, and the United States Department of Agriculture, National Agricultural Statistics Service)

Some of these and other examples can be found in Huff (1993) or on Wikipedia. Furthermore, check http://www.tylervigen.com for more entertaining examples of spurious correlations—see also Vigen (2015).

http://en.wikipedia.org/wiki/Correlation_does_not_imply_causation

2.3.6 Uses of Causal Research

Experiments are a key type of causal research and come in the form of either lab or field experiments.

Lab experiments are performed in controlled environments (usually in a company or academic lab) to isolate the effects of one or more variables on a certain outcome. To do so, researchers impose a treatment (e.g., a new advertisement) that induces changes in one variable (e.g., the type of advertising appeal) and evaluate its impact on an outcome variable (e.g., product choice). **Field experiments** are like lab experiments in that they examine the impact of one or more variables on a certain outcome. However, field experiments are conducted in real-life settings (not set up in controlled environments), thus reducing (or even eliminating) plausible claims of causality (Gneezy 2017). On the plus side, their realism makes them attractive for market research purposes, as the observed effects can probably be generalized to similar settings. For example, isi (https://www.isi-goettingen.de/en), a German sensory market research company, regularly runs product acceptance tests in which consumers sequentially evaluate different products, interrupted by short breaks to neutralize their senses. These tests are traditionally run in sensory labs under controlled conditions. However, isi also runs field experiments in actual consumption environments. Figure 2.5 shows a photo of a field experiment the company ran in a coffeehouse to evaluate consumer ratings of different cappuccino products. We discuss experimental set-ups in more detail in Chap. 4.

Fig. 2.5 Field experiment

Field experiments are not always a good idea. In the city of Rotterdam, the local council tried to reduce bike accidents by turning the traffic lights at bike crossings at a very busy intersection green at the same time. While the idea was that bicyclists would pay more attention, it took less than a minute for two accidents to happen. Needless to say, the experiment was cancelled (see https://www.youtube.com/watch?v=QIsLSmbfaiQ, in Dutch only).

Test markets are a form of field experiment in which organizations in a geographically defined area introduce new products and services, or change the marketing mix to gauge consumer reactions. Test markets help marketers learn about consumer response, thus reducing the risks of a nationwide rollout of new products/services or changes in the marketing mix. For example, gps dataservice (http://www.gps-dataservice.de/en) runs several test markets for a series of major European retailers to evaluate the effect of treatments, such as new product launches, price changes, promotions, or product placements, on purchasing behavior. The company uses data from scanners, customer cards, and other sources (e.g., surveys, observations) to investigate their effects on sales. For example, *shelf tests* involve placing dummy packages in the usual shelves in selected stores, and determining the reactions to these new packages by means of shopper observations (e.g., eye or physical contact with the product; Fig. 2.6), surveys, and scanner data. In Chap. 4, we discuss test markets in more depth.

Fig. 2.6 Shelf test

2.4 Design the Sample and Method of Data Collection

Having determined the research design, we need to design a sampling plan and choose a data-collecting method. This involves deciding whether to use existing (secondary) data or to conduct primary research. We discuss this in more detail in Chap. 3.

2.5 Collect the Data

Collecting data is a practical, but sometimes difficult, part of the market research process. How do we design a survey? How do we measure attitudes toward a product, brand, or company if we cannot observe these attitudes directly? How do we get CEOs to respond? Dealing with such issues requires careful planning and knowledge of the marketing process. We discuss related key issues in Chap. 4.

2.6 Analyze the Data

Analyzing data requires technical skills. We discuss how to enter, clean, and describe data in Chap. 5. After this, we introduce key techniques, such as hypothesis testing and analysis of variance (ANOVA), regression analysis, principal component, factor analysis, and cluster analysis in Chaps. 6, 7, 8 and 9. In each of these chapters, we discuss the key theoretical choices and issues that market researchers face when using these techniques. We also illustrate how researchers can practically deal with these theoretical choices and issues by means of Stata.

2.7 Interpret, Discuss, and Present the Findings

When executing the market research process, researchers' immediate goals are interpreting, discussing, and presenting the findings. Consequently, researchers should provide detailed answers and actionable suggestions based on data and analysis techniques. The last step is to clearly communicate the findings and recommendations to help decision making and implementation. This is further discussed in Chap. 10.

2.8 Follow-Up

Market researchers often stop when the results have been interpreted, discussed, and presented. However, following up on the research findings is important too. Implementing market research findings sometimes requires further research, because suggestions or recommendations may not be feasible or practical and market conditions may have changed. From a market research firm's perspective, follow-up research on previously conducted research can be a good way of entering new deals for further research. Some market research never ends, for example, many firms track customer satisfaction continuously, but even such research can have follow-ups, for example, because the management may wish to know the causes of drops in customer satisfaction.

2.9 Review Questions

1. What is market research? Try to explain what market research is in your own words.
2. Why do we follow a structured process when conducting market research? Are there any shortcuts you can take? Compare, for example, Qualtrics' market research process (http://www.qualtrics.com/blog/marketing-research-process) with the process discussed above. What are the similarities and differences?
3. Describe what exploratory, descriptive, and causal research are and how they are related to one another. Provide an example of each type of research.
4. What are the four requirements for claiming causality? Do we meet these requirements in the following situations?
   - Good user design led to Google's Android becoming the market leader.
   - When Rolex charges a higher price, this increases sales.
   - More advertising causes greater sales.

2.10 Further Readings

Levitt, S. D., & Dubner, S. J. (2005). *Freakonomics. A rogue economist explores the hidden side of everything*. New York, NY: HarperCollins.

An entertaining book that discusses statistical (mis)conceptions and introduces cases of people confusing correlation and causation.

Levitt, S. D., & Dubner, S. J. (2009). *Superfreakonomics*. New York, NY: HarperCollins.

The follow-up book on Freakonomics. Also worth a read.

Nielsen Retail Measurement at http://www.nielsen.com/us/en/nielsen-solutions/nielsen-measurement/nielsen-retail-measurement.html

Pearl, J. (2009). *Causality, Models, reasoning, and inference*. New York, NY: Cambridge University Press.

This book provides a comprehensive exposition of the modern analysis of causation. Strongly recommended for readers with a sound background in statistics.

PRIZM by Claritas at http://www.claritas.com/MyBestSegments/Default.jsp?ID=20

This website allows looking up US lifestyle segments at the zip level.

References

Burrows, D. (2014). How to use ethnography for in-depth consumer insight. *Marketing Week*, May 9, 2014. https://www.marketingweek.com/2014/05/09/how-to-use-ethnography-for-in-depth-consumer-insight/. Accessed 21 Aug 2017.

Feiereisen, S., Wong, V., & Broderick, A. J. (2008). Analogies and mental simulations in learning for really new products: The role of visual attention. *Journal of Product Innovation Management, 25*(6), 593–607.

Gneezy, A. (2017). Field experimentation in marketing research. *Journal of Marketing Research, 54*(1), 140–143.

Huff, D. (1993). *How to lie with statistics*. New York: W. W. Norton & Company.

Larson, J. S., Bradlow, E. T., & Fader, P. S. (2005). An exploratory look at supermarket shopping paths. *International Journal of Research in Marketing, 22*(4), 395–414. http://papers.ssrn.com/sol3/papers.cfm?abstract_id=723821.

Vigen, T. (2015). *Spurious correlations*. New York: Hachette Books.

Data

3

Keywords

Armstrong and Overton procedure • Case • Census • Constant • Construct • Construct validity • Content validity • Criterion validity • Dependence of observations • Discriminant validity • Equidistance • Face validity • Formative constructs • Index • Index construction • Internal consistency reliability • Inter-rater reliability • Items • Latent concept • Latent variable • Measurement scaling • Multi-item constructs • Net Promoter Score (NPS) • Nomological validity • Non-probability sampling • Observation • Operationalization • Population • Predictive validity • Primary data • Probability sampling • Qualitative data • Quantitative data • Reflective constructs • Reliability • Sample size • Sampling • Sampling error • Scale development • Secondary data • Single-item constructs • Test-retest reliability • Unit of analysis • Validity • Variable

Learning Objectives

After reading this chapter, you should understand:

- How to explain what kind of data you use.
- The differences between primary and secondary data.
- The differences between quantitative and qualitative data.
- What the unit of analysis is.
- When observations are independent and when dependent.
- The difference between dependent and independent variables.
- Different measurement scales and equidistance.
- Validity and reliability from a conceptual viewpoint.
- How to set up different sampling designs.
- How to determine acceptable sample sizes.

E. Mooi et al., *Market Research*, Springer Texts in Business and Economics,
DOI 10.1007/978-981-10-5218-7_3

3.1 Introduction

Data are at the heart of market research. By data we mean a collection of facts that can be used as a basis for analysis, reasoning, or discussions. Think, for example, of people's answers to surveys, existing company records, or observations of shoppers' behaviors. "Good" data are vital, because they form the basis of useful market research. In this chapter, we discuss different types of data. This discussion will help you explain the data you use and why you do so. Subsequently, we introduce strategies for collecting data in Chap. 4.

3.2 Types of Data

Before we start discussing data, it is a good idea to introduce the terminology we will use. In the next sections, we will discuss the following four concepts:

- variables,
- constants,
- cases, and
- constructs.

A **variable** is an attribute whose value can change. For example, the price of a product is an attribute of that product and generally varies over time. If the price does not change, it is a **constant**. A **case** (or **observation**) consists of all the variables that belong to an object such as a customer, a company, or a country.

The relationship between variables and cases is that within one case we usually find multiple variables. Table 3.1 includes six variables: *type of car bought* and the customer's *age*, as well as *brand_1*, *brand_2*, and *brand_3*, which capture statements related to brand trust. In the lower rows, you see four observations.

Another important and frequently used term in market research is **construct**, which refers to a variable that is not directly observable (i.e., a **latent variable**). More precisely, a construct is used to represent latent concepts in statistical analyses. **Latent concepts** represent broad ideas or thoughts about certain phenomena that researchers have established and want to measure in their research (e.g., Bollen 2002). However, constructs cannot be measured directly, as respondents cannot articulate a single response that will completely and perfectly provide a measure of that concept. For example, constructs such as satisfaction, loyalty, and brand trust cannot be measured directly. However, we can measure satisfaction, loyalty, and brand trust by means of several **items**. The term items (or *indicators*) is normally used to indicate posed survey questions. Measuring constructs requires combining items to form a *multi-item scale*, an example of which appears in Table 3.1 in the form of three items *Brand_1* ("This brand's product claims are believable"), *Brand_2* ("This brand delivers what it promises"), and *Brand_3* ("This brand has a name that you can trust"). Bear in mind that not all items are

Table 3.1 Quantitative data

Variable name	*Type of car bought*	*Customer's Age*	*Brand_1*	*Brand_2*	*Brand_3*
Description	Name of car bought	Age in years	This brand's product claims are believable	This brand delivers what it promises	This brand has a name that you can trust
Customer 1	BMW 328i	29	6	5	7
Customer 2	Mercedes C180K	45	6	6	6
Customer 3	VW Passat 2.0 TFSI	35	7	5	5
Customer 4	BMW 525ix	61	5	4	5

Coding for *Brand_1*, *Brand_2*, and *Brand_3*: 1 = fully disagree, 7 = fully agree

constructs, for example, the *type of car bought* and *customer's age* in Table 3.1 are not a construct, as these are observable and a single response can measure it accurately and fully.

Like constructs, an **index** also consists of sets of variables. The difference is that an index is created by the variable's "causes." For example, we can create an index of information search activities, which is the sum of the information that customers require from dealers, the promotional materials, the Internet, and other sources. This measure of information search activities is also referred to as a *composite measure*, but, unlike a construct, the items in an index define what we want to measure. For example, the *Retail Price Index* consists of a "shopping" bag of common retail products multiplied by their price. Unlike a construct, each item in a scale captures a part of the index perfectly.

The procedure of combining several items is called **scale development**, **operationalization,** or, in the case of an index, **index construction**. These procedures involve a combination of theory and statistical analysis, such as factor analysis (discussed in Chap. 8) aimed at developing an appropriate construct measure. For example, in Table 3.1, *Brand_1*, *Brand_2*, and *Brand_3* are items that belong to a construct called *brand trust* (as defined by Erdem and Swait 2004). The construct is not an individual item that you see in the list, but it is captured by calculating the average of several related items. Thus, in terms of brand trust, the score of customer 1 is (6 + 5 + 7)/3 = 6.

But how do we decide which and how many items to use when measuring specific constructs? To answer these questions, market researchers make use of scale development procedures. These procedures follow an iterative process with several steps and feedback loops. For example, DeVellis (2017) provides a thorough introduction to scale development. Unfortunately, scale development requires much (technical) expertise. Describing each step is therefore beyond this book's scope. However, many scales do not require this procedure, as existing scales can be found in scale handbooks, such as the *Handbook of Marketing Scales* by Bearden et al. (2011). Furthermore, marketing and management journals frequently publish research articles that introduce new scales, such as for measuring the reputation of non-profit organizations (e.g., Sarstedt and Schloderer 2010) or for refining existing scales (e.g., Kuppelwieser and Sarstedt 2014). In Box 3.1, we introduce two distinctions that are often used to discuss constructs.

Box 3.1 Types of Constructs

In **reflective constructs**, the items are considered to be manifestations of an underlying construct (i.e., the items reflect the construct). Our brand trust example suggests a reflective construct, as the items reflect trust. Thus, if a respondent changes his assessment of brand trust (e.g., due to a negative brand experience), this is reflected in the answers to the three items. Reflective constructs typically use multiple items (3 or more) to increase the measurement stability and accuracy. If we have multiple items, we can use analysis techniques to inform us about the measurement quality, such as factor analysis and reliability analysis (discussed in Chap. 8). **Formative constructs** consist of several items that define a construct. A typical example is socioeconomic status, which is formed by a combination of education, income, occupation, and residence. If any of these measures increases, the socioeconomic status would increase (even if the other items did not change). Conversely, if a person's socioeconomic status increases, this would not necessarily go hand in hand with an increase in all four measures. The distinction between reflective and formative constructs is that they require different approaches to decide on the type and number of items. For example, reliability analyses (discussed in Chap. 8) cannot be run on formative measures. For an overview of this distinction, see Bollen and Diamantopoulos (2017), Diamantopoulos et al. (2008), or Sarstedt et al. (2016c).

Instead of using multiple items to measure constructs (i.e., **multi-item constructs**), researchers and practitioners frequently use single items (i.e., **single-item constructs**). For example, we may only use "This brand has a name that you can trust" to measure brand trust, instead of using three items. A popular single-item measure is the **Net Promoter Score (NPS)**, which aims to measure loyalty by using the single question: *"How likely are you to recommend our company/product/service to a friend or colleague?"* (Reichheld 2003). While this is a good way of making the questionnaire shorter, it also reduces the quality of your measures (e.g., Diamantopoulos et al. 2012, Sarstedt et al. 2016a, b). You should therefore avoid using single items to measure constructs unless you only need a rough proxy measure of a latent concept.

3.2.1 Primary and Secondary Data

Generally, we can distinguish between **primary data** and **secondary data**. While primary data are data that a researcher has collected for a specific purpose, another researcher collected the secondary data for another purpose.

The US Consumer Expenditure Survey (www.bls.gov/cex), which makes data available on what people in the US buy, such as insurance, personal care items, and food, is an example of secondary data. It also includes the prices people pay for these products and services. Since these data have already been collected, they are secondary data. If a researcher sends out a survey with various questions to find an answer to a specific issue, the collected data are primary data. If primary data are re-used to answer another research question, they become secondary data.

Secondary and primary data have their own specific advantages and disadvantages, which we illustrate in Table 3.2. The most important reasons for using secondary data are that they tend to be cheaper and quick to obtain access to (although lengthy processes may be involved). For example, if you want to have access to the US Consumer Expenditure Survey, all you have to do is to use your web browser to go to www.bls.gov/cex and to download the required files. However, the authority and competence of some research organizations could be a factor. For example, the claim that Europeans spend 9% of their annual income on health may be more believable if it comes from Eurostat (the statistical office of the European Community) rather than from a single, primary research survey.

However, important secondary data drawbacks are that they may not answer your research question. If you are, for example, interested in the sales of a specific

Table 3.2 The advantages and disadvantages of secondary and primary data

	Secondary data	Primary data
Advantages	– Tends to be cheaper	– Are recent
	– Sample sizes tend to be greater	– Are specific for the purpose
	– Tend to have more authority	– Are proprietary
	– Are usually quick to access	
	– Are easier to compare to other research using the same data	
	– Are sometimes more accurate (e.g., data on competitors)	
Disadvantages	– May be outdated	– Are usually more expensive
	– May not fully fit the problem	– Take longer to collect
	– There may be hidden errors in the data – difficult to assess the data quality	
	– Usually contain only factual data	
	– No control over data collection	
	– May not be reported in the required form (e.g., different units of measurement, definitions, aggregation levels of the data)	

product (and not in a product or service category), the US Expenditure Survey may not help much. In addition, if you are interested in the reasons for people buying products, this type of data may not help answer your question. Lastly, as you did not control the data collection, there may be errors in the data.

In contrast, primary data tend to be highly specific, because the researcher (you!) can influence what the research comprises. In addition, research to gather primary data can be carried out when and where required and competitors cannot access it. However, gathering primary data often requires much time and effort and is therefore usually expensive compared to secondary data.

As a rule, start looking for secondary data first. If they are available, and of acceptable quality, use them! We will discuss ways to gather primary and secondary data in Chap. 4.

3.2.2 Quantitative and Qualitative Data

Data can be quantitative or qualitative. **Quantitative data** are presented in values, whereas qualitative data are not. **Qualitative data** can take many forms, such as words, stories, observations, pictures, and audio. The distinction between qualitative and quantitative data is not as black-and-white as it seems, because quantitative data are based on qualitative judgments. For example, the questions on brand trust in Table 3.1 take the values of 1–7. There is no reason why we could not have used other values to code these answers, such as 0–6, but it is common practice to code the answers to a construct's items on a range of 1–7.

In addition, many of the sources that market researchers use, such as Twitter feeds or Facebook posts, produce qualitative data. Researchers can code attributes of the data, which describe a particular characteristic, thereby turning it into quantitative data. Think, for example, of how people respond to a new product in an interview. We can code the data by setting neutral responses to 0, somewhat positive responses to 1, positive responses to 2, and very positive responses to 3. We have therefore turned qualitative data into quantitative data. Box 3.2 shows an example of how to code qualitative data.

Box 3.2 Coding Qualitative Data

In 2016 Toyota launched new Prius, the Prius Prime (www.toyota.com/priusprime/). Not surprisingly, Facebook reactions were divided. Here are some examples of Facebook posts:

- "Love it! But why only 2 seats at the back? Is there any technical reason for that?"
- "Wondering if leather seats are offered? The shape of the seats looks super comfy!"

(continued)

Box 3.2 (continued)
- "Here's that big black grill on yet another Toyota. Will be very glad when this 'fashion faze' is over."

One way of structuring these responses is to consider the attributes mentioned in the posts. After reading them, you may find that, for example, the seat and styling are attributes. You can then categorize in respect of each post whether the response was negative, neutral, or positive. If you add the actual response, this can later help identify the aspect the posts liked (or disliked). As you can see, we have now turned qualitative data into quantitative data!

Attribute	Negative	Neutral	Positive
Seats	1-why only two seats?	2-are leather seats offered?	2-shape looks super comfy!
Styling	3-big black grill		1-love it!

Qualitative data's biggest strength is their richness, as they have the potential to offer detailed insights into respondents' perceptions, attitudes, and intentions. However, their downside is that qualitative data can be interpreted in many ways. Thus, the process of interpreting qualitative data is subjective. To reduce subjectivity, (multiple) trained researchers should code qualitative data. The distinction between quantitative and qualitative data is closely related to that between quantitative and qualitative research, which we discuss in Box 3.3. Most people think of quantitative data as being more factual and precise than qualitative data, but this is not necessarily true. Rather, how well qualitative data have been collected and/or coded into quantitative data is important.

3.3 Unit of Analysis

The **unit of analysis** is the level at which a variable is measured. Researchers often ignore this aspect, but it is crucial because it determines what we can learn from the data. Typical measurement levels include that of the respondents, customers, stores, companies, or countries. It is best to use data at the lowest possible level, because this provides more detail. If we need these data at another level, we can aggregate them. *Aggregating data* means that we sum up a variable at a lower level to create a variable at a higher level. For example, if we know how many cars all car dealers in a country sell, we can take the sum of all the dealer sales, to create a variable measuring countrywide car sales. Aggregation is not possible if we have incomplete or missing data at the lower levels.

Box 3.3 Quantitative Research and Qualitative Research

Market researchers often label themselves as either quantitative or qualitative researchers. The two types of researchers use different methodologies, different types of data, and focus on different research questions. Most people regard the difference between qualitative and quantitative as the difference between numbers and words, with quantitative researchers focusing on numbers and qualitative researchers on words. This distinction is not accurate, as many qualitative researchers use numbers in their analyses. The distinction should instead depend on when the information is quantified. If we know the values that may occur in the data even before the research starts, we conduct quantitative research. If we only know this after the data have been collected, we conduct qualitative research. Think of it in this way: If we ask survey questions and use a few closed questions, such as "Is this product of good quality?," and the respondents can either choose "Completely disagree," "Somewhat disagree," "Neutral," "Somewhat agree," and "Completely agree," we know that the data we will obtain from this will—at most—contain five different values. Because we know all possible values beforehand, the data are quantified beforehand. If, on the other hand, we ask someone "Is this product of good quality?," he or she could give many different answers, such as "Yes," "No," "Perhaps," "Last time yes, but lately...". This means we have no idea what the possible answer values are. Therefore, these data are qualitative. We can, however, recode these qualitative data, for example, as described in Box 3.2, and assign values to each response. Thus, we quantify the data, allowing further statistical analysis.

Qualitative research accounts for 17% of money spent in the market research industry, with quantitative research making up the rest.[1] Practically, market research is often hard to categorize as qualitative or quantitative, as it may include elements of both. Research that includes both elements is sometimes called *hybrid market research*, *fused market research*, or simply *mixed methodology*.

3.4 Dependence of Observations

A key issue for any data is the degree to which observations are related, or the **dependence of observations**. If we have exactly one observation from each individual, store, company, or country, we label the observations independent.

[1]See ESOMAR Global Market Research Report 2013.

That is, the observations are unrelated. If we have multiple observations of each individual, store, company, or country, we label them dependent. For example, we could ask respondents to rate a type of Cola, then show them an advertisement, and again ask them to rate the same type of Cola. Although the advertisement may influence the respondents, it is likely that the first response and second response will be related. That is, if the respondents first rated the Cola negatively, the chance is higher that they will continue to rate the Cola negative rather than positive after the advertisement. If the observations are dependent, this often impacts the type of analysis we should use. For example, in Chap. 6, we discuss the difference between the independent samples *t*-test (for *independent observations*) and the paired samples *t*-test (for *dependent observations*).

3.5 Dependent and Independent Variables

Dependent variables represent the outcome that market researchers study, while *independent variables* are those used to explain the dependent variable(s). For example, if we use the amount of advertising to explain sales, then advertising is the independent variable and sales the dependent.

This distinction is artificial, as all variables depend on other variables. For example, the amount of advertising depends on how important the product is for a company, the company's strategy, and other factors. However, this distinction is frequently used in the application of statistical methods. While researching relationships between variables, we, on the basis of theory and practical considerations, need to distinguish between the dependent and the independent variables beforehand.

3.6 Measurement Scaling

Not all data are equal! For example, we can calculate the respondents' average age in Table 3.1. However, if we would have coded the color of the car as black = 1, blue = 2, silver = 3 it would not make any sense to calculate the average. Why is this? The values that we have assigned 1, 2, and 3 are arbitrary; we could just as well have changed these value for any other. Therefore, choosing a different coding would lead to different results, which is meaningless. **Measurement scaling** refers to two things: the variables we use for measuring a certain construct (see discussion above) and the level at which a variable is measured, which we discuss in this section. This can be highly confusing!

There are four *levels of measurement*:

- nominal scale,
- ordinal scale,
- interval scale, and
- ratio scale.

Table 3.3 Measurement Scaling

	Label	Order	Differences	Origin is 0
Nominal scale	✓			
Ordinal scale	✓	✓		
Interval scale	✓	✓	✓	
Ratio scale	✓	✓	✓	✓

Fig. 3.1 Meaningless!

These scales relate to how we quantify what we measure. It is vital to know the scale on which something is measured, because, as the gender example above illustrates, the measurement scale determines the analysis techniques we can, or cannot, use. For example, as indicated above, it makes no sense to calculate. We will return to this issue in Chap. 5 and beyond. However, even when we know the scale, be aware that, as Fig. 3.1 shows, meaningful calculations are not always possible!

The *nominal scale* is the most basic level at which we can measure something. Essentially, if we use a nominal scale, we substitute a word for a numerical value. For example, we could code the color of each Prius sold: black = 1, blue = 2, silver = 3. In this example, the numerical values represent nothing more than a label.

The *ordinal scale* provides more information. If a variable is measured on an ordinal scale, increases or decreases in values give meaningful information. For example, if we code the Prius version people bought as the first generation = 1, second generation = 2, third generation = 3, and fourth generation = 4, we know whether the model is more recent. The ordinal scale provides information about the order of our observations. However, we do not know if the differences in the order are equally spaced. That is, we do not know if the difference between first generation and second generation is the same as between second and third generation, even though the difference in values (1–2 and 2–3) is equal.

If something is measured on an *interval scale*, we have precise information on the rank order at which something is measured and we can interpret the magnitude of the differences in values directly. For example, if the temperature in a car showroom is 23°C, we know that if it drops to 20°C, the difference is exactly 3°C. This difference of 3°C is the same as the increase from 23 to 26°C. This exact "spacing" is called **equidistance.** Equidistant scales are necessary for some analysis techniques, such as factor analysis (discussed in Chap. 8). What the interval scale does not give us, is an absolute zero point. If the temperature is 0°C it may feel cold, but the temperature can drop further. The value of 0 does not therefore mean that there is no temperature at all.

The *ratio scale* provides the most information. If something is measured on a ratio scale, we know that a value of 0 means that that the attribute of that particular variable is not present. For example, if a dealer sells zero Prius cars (value = 0) then he or she really sells none. Or, if we spend no money on advertising a Prius (value = 0), we really spend no money. Therefore, the origin of the variable is equal to 0.

While it is relatively easy to distinguish between the nominal and the interval scales, it is sometimes hard to see the difference between the interval and the ratio scales. The difference between the interval and the ratio scales can be ignored in most statistical methods. Table 3.3 shows the differences between these four scales.

3.7 Validity and Reliability

In any market research process, it is paramount to use "good" measures. Good measures are those that measure what they are supposed to measure and do so consistently. For example, if we are interested in knowing whether customers like a new TV commercial, we could show a commercial and ask the following two questions afterwards:

1. "Did you enjoy watching the commercial?," and
2. "Did the commercial provide the essential information required for a purchase decision?"

How do we know if these questions really measure whether or not the viewers liked the commercial? We can think of this as a measurement problem through which we relate what we want to measure—whether existing customers like a new TV commercial—with what we actually measure in terms of the questions we ask. If these relate perfectly, our actual measurement is equal to what we intend to measure and we have no measurement error. If these do not relate perfectly, we have *measurement error*.

This measurement error can be divided into a *systematic error* and a *random error*. We can express this as follows, where X_O stands for the observed score (i.e., what the customers indicated), X_T for the true score (i.e., what the customers' true liking of the commercial is), E_S for the systematic error, and E_R for the random error.

$$X_O = X_T + E_S + E_R$$

Systematic error is a measurement error through which we consistently measure higher, or lower, than we want to measure. If we were to ask customers, for example, to evaluate a TV commercial and offer them remuneration in return, they may provide more favorable information than they would otherwise have. This may cause us to think that the TV commercial is systematically more enjoyable than it is in reality. There may also be random errors. Some customers may be having a good day and indicate that they like a commercial, whereas others, who are having a bad day, may do the opposite.

Systematic errors cause the actual measurement to be consistently higher, or lower, than what it should be. On the other hand, random error causes (random) variation between what we actually measure and what we want to measure.

The systematic and random error concepts are important, because they relate to a measure's validity and reliability. **Validity** refers to whether we are measuring what we want to measure and, therefore, to a situation where the systematic error E_S is small. **Reliability** is the degree to which what we measure is free from random error and therefore relates to a situation where the E_R is zero. In Fig. 3.2, we illustrate the difference between reliability and validity by means of a target comparison. In this analogy, different measurements (e.g., of a customer's satisfaction with a specific service) are compared to arrows shot at a target. To measure each score, we have five measurements (indicated by the black circles), which correspond to, for example, questions asked in a survey. The cross indicates their average. Validity describes the cross's proximity to the bull's eye at the target center. The closer the average to the true score, the higher the validity. If several arrows are fired, reliability is the degree to which the arrows are apart. If all the arrows are close together, the measure is reliable, even though it is not necessarily near the bull's eye. This corresponds to the upper left box where we have a scenario in which the measure is reliable, but not valid. In the upper right box, both reliability and validity are given. In the lower left box, though, we have a situation in which the measure is neither reliable, nor valid. This is obviously because the repeated measurements are scattered around and the average does not match the true score. However, even if the latter were the case (i.e., if the cross were in the bull's eye), we would still not consider the measure valid. An unreliable measure can never be valid. If we repeated the measurement, say, five more times, the random error would probably shift the cross to a different position. Reliability is therefore a necessary condition for validity. This is also

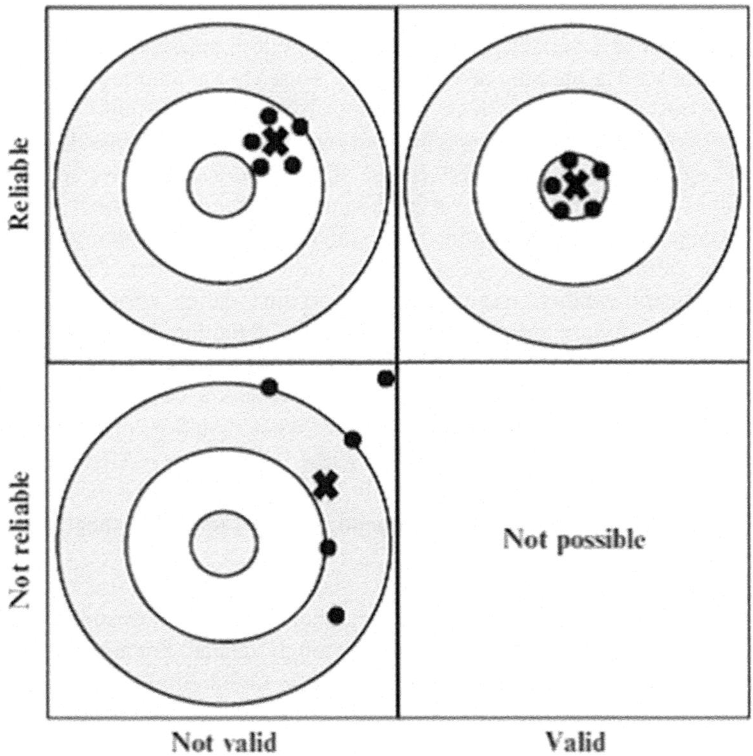

Fig. 3.2 Validity and reliability

why the scenario that is not reliable/valid (lower right box) is not included, as it is not possible for a measure to be valid, but not reliable.

3.7.1 Types of Validity

For some variables, such as length or income, we can objectively verify what the true score is. For constructs, such as satisfaction, loyalty and brand trust, this is impossible. From a philosophical point of view, one could even argue that there is no "true" score for a construct. So how do we know if a measure is valid? Because there is no objective way of verifying what we are measuring, several forms of validity have been developed, including face, content, predictive, criterion, discriminant, and nomological validity (Netemeyer et al. 2003). Researchers frequently summarize these validity types under the umbrella term **construct validity**, which relates to the correspondence between a measure at the conceptual level and a purported measure. The different types of validity help us understand the association between what we should measure and what we actually measure, thereby increasing the likelihood of adequately measuring the latent concept under consideration.

- **Face validity** is an absolute minimum requirement for a variable to be valid and refers to whether a variable reflects what you want to measure. Essentially, face validity exists if a measure seems to make sense. For example, if you want to measure trust, using items such as "this company is honest and truthful" makes a lot of sense, whereas "this company is not well known" makes little sense. Researchers should agree on the face validity before starting the actual measurement. Face validity is usually determined by using a sample of experts who discuss and agree on the degree of face validity (this is also referred to as *expert validity*).
- **Content validity** is strongly related to face validity, but is more formalized. To assess content validity, researchers need to first define what they want to measure and discuss what is included in the definition and what not. For example, trust between businesses is often defined as the extent to which a firm believes that its exchange partner is honest and/or benevolent (Geyskens et al. 1998). This definition clearly indicates what should be mentioned in the questions used to measure trust (honesty and benevolence). After researchers have defined what they want to measure, questions have to be developed that relate closely to the definition. Consequently, content validity is mostly achieved prior to the actual measurement.
- **Predictive validity** requires a measure to be highly correlated (see Chap. 5 for an introduction to correlations) with an outcome variable, measured at a later point in time, to which it is conceptually strongly related. For example, loyalty should lead to people purchasing a product in the future. Similarly, a measure of satisfaction should be predictive of people not complaining about a product or service. Assessing predictive validity requires collecting data at two points in time and therefore requires a greater effort. If both measures (i.e., the one to be evaluated and the outcome variable) are measured at the same point in time, we call this **criterion validity**.
- **Discriminant validity** ensures that a measure is empirically unique and represents phenomena of interest that other measures in a model do not capture. For example, customer satisfaction and customer loyalty are two distinct latent concepts. Discriminant validity requires the constructs used to measure these two concepts to also be empirically distinct (i.e., they should not correlate too highly).
- **Nomological validity** is the degree to which a construct behaves as it should in a system of related constructs. For example, customer expectations, perceived quality, and value have a significant influence on customer satisfaction. Similarly, satisfaction generally relates positively to customer loyalty. As such, you would expect the measure of satisfaction that you are evaluating to correlate with these measure.

3.7.2 Types of Reliability

How do we know if a measure is reliable? Three key factors are used to assess reliability: test-retest reliability, internal consistency reliability, and inter-rater reliability (Mitchell and Jolley 2013).

Test–retest reliability means that if we measure something twice (also called the *stability of the measurement*), we expect similar outcomes. The stability of measurement requires a market researcher to have collected two data samples, is therefore costly, and could prolong the research process. Operationally, researchers administer the same test to the same sample on two different occasions and evaluate how strongly the measurements are correlated. We would expect the two measurements to correlate highly if a measure is reliable. This approach is not without problems, as it is often hard, if not impossible, to survey the same people twice. Furthermore, the respondents may learn from past surveys, leading to *practice effects*. In addition, it may be easier to recall events the second time a survey is administered. Moreover, test–retest approaches do not work if a survey concentrates on specific time points. If we ask respondents to provide information on their last restaurant experience, the second test might relate to a different restaurant experience. Thus, test–retest reliability can only be assessed in terms of variables that are stable over time.

Internal consistency reliability is by far the most common way of assessing reliability. Internal consistency reliability requires researchers to simultaneously use multiple items to measure the same concept. Think, for example, of the set of questions commonly used to measure brand trust (i.e., "This brand's product claims are believable," "This brand delivers what it promises," and "This brand has a name that you can trust"). If these items relate strongly, there is a considerable degree of internal consistency. There are several ways to calculate indices of internal consistency, including split-half reliability and Cronbach's α (pronounced as alpha), which we discuss in Chap. 8.

Inter-rater reliability is used to assess the reliability of secondary data or qualitative data. If you want to identify, for example the most ethical organizations in an industry, you could ask several experts to provide a rating and then calculate the degree to which their answers relate.

3.8 Population and Sampling

A **population** is the group of units about which we want to make judgments. These units can be groups of individuals, customers, companies, products, or just about any subject in which you are interested. Populations can be defined very broadly, such as the people living in Canada, or very narrowly, such as the directors of large hospitals in Belgium. The research conducted and the research goal determine who or what the population will be.

Sampling is the process through which we select cases from a population. The most important aspect of sampling is that the selected sample is representative of the population. Representative means that the characteristics of the sample closely match those of the population. In Box 3.4, we discuss how to determine whether a sample is representative of the population.

Box 3.4 Do I Have a Representative Sample?

It is important for market researchers that their sample is representative of the population. How can we determine whether this is so?

- The best way to test whether the sample relates to the population is to use a database with information on the population (or to draw the sample from such databases). For example, the Amadeus database (www.bvdinfo.com) provides information on public and private companies around the world at the population level. We can (statistically) compare the information from these databases to the selected sample. However, this approach can only support the tested variables' representativeness; that is, the *specific representativeness*. Conversely, *global representativeness*—that is, matching the distribution of all the characteristics of interest to the research question, but which lie outside their scope—cannot be achieved without a census (Kaplan 1964).
- You can use (industry) experts to judge the quality of your sample. They may look at issues such as the type and proportion of organizations in your sample and population.
- To check whether the responses of people included in your research differ significantly from those of non-respondents (which would lead to your sample not being representative), you can use the **Armstrong and Overton procedure**. This procedure calls for comparing the first 50% of respondents to the last 50% in respect of key demographic variables. The idea behind this procedure is that later respondents more closely match the characteristics of non-respondents. If these differences are not significant (e.g., through hypothesis tests, discussed in Chap. 6), we find some support for there being little, or no, response bias (see Armstrong and Overton 1977). When the survey design includes multiple waves (e.g. the first wave of the survey is web-based and the second wave is by phone), this procedure is generally amended by comparing the last wave of respondents in a survey design to the earlier waves. There is some evidence that this procedure is better than Armstrong and Overton's original procedure (Lindner et al. 2001).
- Using follow-up procedures, a small sample of randomly chosen non-respondents can again be contacted to request their cooperation. This small sample can be compared against the responses obtained earlier to test for differences.

When we develop a sampling strategy, we have three key choices:

- census,
- probability sampling, and
- non-probability sampling.

Box 3.5 The US Census

https://www.youtube.com/user/uscensusbureau

If we are lucky and somehow manage to include every unit of the population in our study, we have conducted a **census**. Thus, strictly speaking, this is not a sampling strategy. Census studies are rare, because they are very costly and because missing just a small part of the population can have dramatic consequences. For example, if we were to conduct a census study of directors of Luxemburg banks, we may miss a few because they were too busy to participate. If these busy directors happen to be those of the very largest companies, any information we collect would underestimate the effects of variables that are more important at large banks. Census studies work best if the population is small, well-defined, and accessible. Sometimes census studies are also conducted for specific reasons. For example, the US Census Bureau is required to hold a census of all persons resident in the US every 10 years. Check out the US Census Bureau's YouTube channel using the mobile tag or URL in Box 3.5 to find out more about the US Census Bureau.

If we select part of the population, we can distinguish two types of approaches: probability sampling and non-probability sampling. Figure 3.3 provides an overview of the different sampling procedures, which we will discuss in the following sections.

3.8.1 Probability Sampling

Probability sampling approaches provide every individual in the population with a chance (not equal to zero) of being included in the sample (Cochran 1977, Levy and Lemeshow 2013). This is often achieved by using an accurate *sampling frame*, which is a list of individuals in the population. There are various sampling frames, such as Dun & Bradstreet's Selectory database (includes executives and

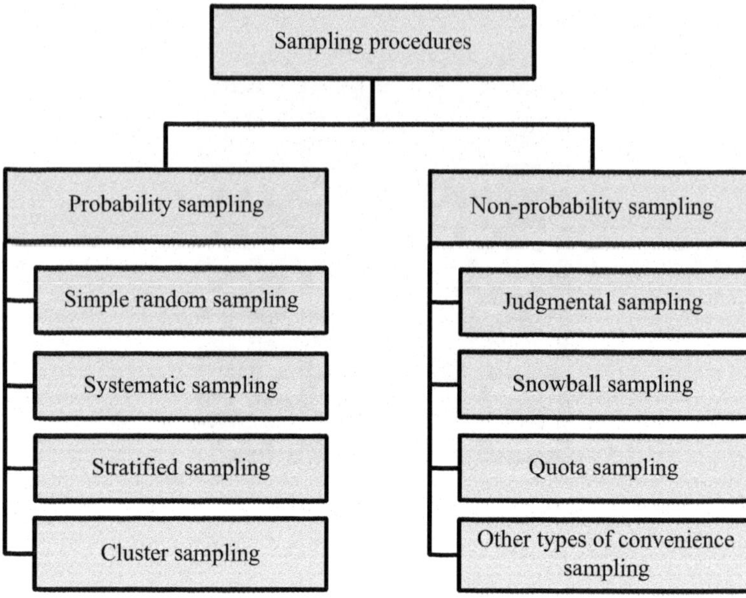

Fig. 3.3 Sampling procedures

companies), the Mint databases (includes companies in North and South America, Italy, Korea, the Netherlands, and the UK), and telephone directories. These sampling frames rarely cover the population of interest completely and often include outdated information, but are frequently used due to their ease of use and availability. If the sampling frame and population are very similar, we have little **sampling error**. Starting with a good-quality sampling frame, we can use several methods to select units from it (Sarstedt et al. 2017).

The easiest way is to use *simple random sampling*, which is achieved by randomly selecting the number of cases required. This can be achieved by using specialized software, or using Stata.[2]

Systematic sampling uses a different procedure. We first randomize the order of all the observations, number them and, finally, select every n^{th} observation. For example, if our sampling frame consists of 1000 firms and we wish to select just 100 firms, we could select the 1st observation, the 11th, the 21st, etc. until we reach the end of the sampling frame and have our 100 observations.

Stratified sampling and cluster sampling are more elaborate techniques of probability sampling requiring us to divide the sampling frame into different groups. When we use *stratified sampling*, we divide the population into several different homogenous groups called *strata*. These strata are based on key sample

[2]See www.stata.com/support/faqs/statistics/random-samples for details. Stata will be discussed in detail in Chap. 5 and beyond.

characteristics, such as different departments in organizations, or the areas in which consumers live. Subsequently, we draw a random number of observations from each stratum. While stratified sampling is more complex and requires accurate knowledge of the sampling frame and population, it also helps ensure that the sampling frame's characteristics are similar to those of the sample.

Cluster sampling requires dividing the population into different heterogeneous groups, with each group's characteristics similar to those of the population. For example, we can divide a country's consumers into different provinces, counties, and councils. Several of these groups could have key characteristics (e.g., income, political preference, household composition) in common, which are very similar (representative of) to those of the population. We can select one or more of these representative groups and use random sampling to select observations that represent this group. This technique requires knowledge of the sampling frame and population, but is convenient because gathering data from one group is cheaper and less time consuming.

Generally, all probability sampling methods allow for drawing representative samples from the target population. However, simple random sampling and stratified sampling are considered superior in terms of drawing representative samples. For a detailed discussion, see Sarstedt et al. (2017).

> Stata features several advanced methods to deal with sampling. A few are discussed on http://www.ats.ucla.edu/stat/stata/library/svy_survey.htm. For detail, please see http://www.stata.com/manuals13/svy.pdf

3.8.2 Non-probability Sampling

Non-probability sampling procedures do not give every individual in the population an equal chance of being included in the sample (Cochran 1977, Levy and Lemeshow 2013). This is a drawback, because the resulting sample is most certainly not representative of the population, which may bias the subsequent analyses' results. Nevertheless, non-probability sampling procedures are frequently used as they are easily executed, and are normally less costly than probability sampling methods. Popular non-probability procedures include judgmental sampling, snowball sampling, and quota sampling (Sarstedt et al. 2017).

Judgmental sampling is based on researchers taking an informed guess regarding which individuals should be included. For example, research companies often have panels of respondents who are continuously used in research. Asking these people to participate in a new study may provide useful information if we know, from experience, that the panel has little sampling frame error.

Snowball sampling involves existing study participants to recruit other individuals from among their acquaintances. Snowball sampling is predominantly used if access to individuals is difficult. People such as directors, doctors, and high-level managers often have little time and are, consequently, difficult to involve. If we can ask just a few of them to provide the names and the details of others in a similar position, we can expand our sample quickly and then access them. Similarly, if you post a link to an online questionnaire on your LinkedIn or Facebook page (or send out a link via email) and ask your friends to share it with others, this is snowball sampling.

Quota sampling occurs when we select observations for the sample that are based on pre-specified characteristics, resulting in the total sample having the same distribution of characteristics assumed to exist in the population being studied. In other words, the researcher aims to represent the major characteristics of the population by sampling a proportional amount of each (which makes the approach similar to stratified sampling). Let's say, for example, that you want to obtain a quota sample of 100 people based on gender. First you need to find what proportion of the population is men and what women. If you find that the larger population is 40% women and 60% men, you need a sample of 40 women and 60 men for a total of 100 respondents. You then start sampling and continue until you have reached exactly the same proportions and then stop. Consequently, if you already have 40 women for the sample, but not yet 60 men, you continue to sample men and discard any female respondents that come along. However, since the selection of the observations does not occur randomly, this makes quota sampling a non-probability technique. That is, once the quota has been fulfilled for a certain characteristic (e.g., females), you no longer allow any observations with this specific characteristic in the sample. This systematic component of the sampling approach can introduce a sampling error. Nevertheless, quota sampling is very effective and inexpensive, making it the most important sampling procedure in practitioner market research.

Convenience sampling is a catch-all term for methods (including the three non-probability sampling techniques just described) in which the researcher draws a sample from that part of the population that is close at hand. For example, we can use *mall intercepts* to ask people in a shopping mall if they want to fill out a survey. The researcher's control over who ends up in the sample is limited and influenced by situational factors.

3.8.3 Probability or Non-probability Sampling?

Probability sampling methods are recommended, as they result in representative samples. Nevertheless, judgmental and, especially, quota sampling might also lead to (specific) representativeness (e.g., Moser and Stuart 1953; Stephenson 1979). However, both methods' ability to be representative depends strongly on the researcher's knowledge (Kukull and Ganguli 2012). Only when the researcher considers all the factors that have a significant bearing on the effect under study, will these methods lead to a representative sample. However, snowball sampling never leads to a representative sample, as the entire process depends on the participants' referrals. Likewise, convenience sampling will almost never yield a representative sample, because observations are only selected if they can be accessed easily and conveniently. See Sarstedt et al. (2017) for further details.

3.9 Sample Sizes

After determining the sampling procedure, we have to determine the **sample size**. Larger sample sizes increase the precision of the research, but are also much more expensive to collect. The gains in precision decrease as the sample size increases (in Box 6.3 we discuss the question whether a sample size can be too large in the context of significance testing). It may seem surprising that relatively small sample sizes are precise, but the strength of samples comes from selecting samples accurately, rather their size. Furthermore, the required sample size has very little relation to the population size. That is, a sample of 100 employees from a company with 100,000 employees can be nearly as accurate as selecting 100 employees from a company with 1,000 employees.

There are some problems with selecting sample sizes. The first is that market research companies often push their clients to accept large sample sizes. Since the fee for market research services, such as those offered by Qualtrics or Toluna, is often directly dependent on the sample size, increasing the sample size increases the market research company's profit. Second, if we want to compare different groups, we need to multiply the required sample by the number of groups included. That is, if 150 observations are sufficient to measure how much people spend on organic food, 2 times 150 observations are necessary to compare singles and couples' expenditure on organic food.

The figures mentioned above are net sample sizes; that is, these are the actual (usable) number of observations we should have. Owing to non-response (discussed in Chaps. 4 and 5), a multiple of the initial sample size is normally necessary to obtain the desired sample size. Before collecting data, we should have an idea of the percentage of respondents we are likely to reach (often high), a percentage estimate of the respondents willing to help (often low), as well as a percentage estimate of the respondents likely to fill out the survey correctly (often high). For example, if we expect to reach 80% of the identifiable respondents, and if 25% are likely to help, and 75% of those who help are likely to fully fill out the questionnaire, only 15% ($0.80 \cdot 0.25 \cdot 0.75$) of identifiable respondents are likely to provide a usable response. Thus, if we wish to obtain a net sample size of 100, we need to send out $\left(\frac{\textit{desired sample size}}{\textit{likely usable responses}}\right) = 100/0.15 = 667$ surveys. In Chap. 4, we will discuss how we can increase response rates (the percentage of people willing to help).

3.10 Review Questions

1. Explain the difference between items and constructs.
2. What is the difference between reflective and formative constructs?
3. Explain the difference between quantitative and qualitative data and give examples of each type.
4. What is the scale on which the following variables are measured?
   - The amount of money a customer spends on shoes.

- A product's country-of-origin.
- The number of times an individual makes a complaint.
- A test's grades.
- The color of a mobile phone.

5. Try to find two websites offering secondary data and discuss the kind of data described. Are these qualitative or quantitative data? What is the unit of analysis and how are the data measured?
6. What are "good data"?
7. Discuss concepts reliability and validity. How do they relate to each other?
8. Please comment on the following statement: "Face and content validity are essentially the same."
9. What is the difference between predictive and criterion validity?
10. Imagine you have just been asked to execute a market research study to estimate the market for notebooks priced $300 or less. What sampling approach would you propose to the client?
11. Imagine that a university decides to evaluate their students' satisfaction. To do so, employees issue every 10th student at the student cafeteria on one weekday with a questionnaire. Which type of sampling is conducted in this situation? Can the resulting sample be representative of the student population?

3.11 Further Readings

Mitchell, M. L., & Jolley, J. M. (2013). *Research design explained* (8th ed.). Belmont, CA: Wadsworth.

The book offers an in-depth discussion of different types of reliability and validity, including how to assess them.

Churchill, G. A. (1979). A paradigm for developing better measures for marketing constructs. *Journal of Marketing Research, 16*(1), 64–73.

A landmark article that marked the start of the rethinking process on how to adequately measure constructs.

Cochran, W. G. (1977). *Sampling techniques* (3rd ed.). New York, NY: John Wiley and Sons.

This is a seminal text on sampling techniques, providing a thorough introduction to this topic. However, please note that most descriptions are rather technical and require a sound understanding of statistics.

Diamantopoulos A, Winklhofer HM (2001) Index construction with formative indicators: an alternative to scale development. *Journal of Marketing Research, 38*(2), 269–277.

In this seminal article the authors provide guidelines on how to operationalize formative constructs.

DeVellis, R. F. (2017). *Scale development: Theory and applications* (4th ed.). Thousand Oaks, CA: Sage.

This is a very accessible book which guides the reader through the classic way of developing multi-item scales. The text does not discuss how to operationalize formative constructs, though.

Marketing Scales Database at www.marketingscales.com/search/search.php

This website offers an easy-to-search database of marketing-related scales. A description is given of every scale; the scale origin, reliability, and validity are discussed and the items given.

Netemeyer, R. G., Bearden, W. O., & Sharma, S. (2003). *Scaling procedures: Issues and applications*. Thousand Oaks, CA: Sage.

Like DeVellis (2017), *this book presents an excellent introduction to the principles of the scale development of measurement in general.*

References

Armstrong, J. S., & Overton, T. S. (1977). Estimating nonresponse bias in mail surveys. *Journal of Marketing Research, 14*(3), 396–403.

Bearden, W. O., Netemeyer, R. G., & Haws, K. L. (2011). *Handbook of marketing scales. Multi-item measures for marketing and consumer behavior research* (3rd ed.). Thousand Oaks: Sage.

Bollen, K. A. (2002). Latent variables in psychology and the social sciences. *Annual Review of Psychology, 53*(1), 605–634.

Bollen, K. A., & Diamantopoulos, A. (2017). In defense of causal-formative indicators: A minority report. *Psychological Methods, 22*(3), 581–596.

Cochran, W. G. (1977). *Sampling techniques* (3rd ed.). New York: Wiley.

DeVellis, R. F. (2017). *Scale development: Theory and applications* (4th ed.). Thousand Oaks: Sage.

Diamantopoulos, A., Riefler, P., & Roth, K. P. (2008). Advancing formative measurement models. *Journal of Business Research, 61*(12), 1203–1218.

Diamantopoulos, A., Sarstedt, M., Fuchs, C., Wilczynski, P., & Kaiser, S. (2012). Guidelines for choosing between multi-item and single-item scales for construct measurement: A predictive validity perspective. *Journal of the Academy of Marketing Science, 40*(3), 434–449.

Erdem, T., & Swait, J. (2004). Brand credibility, brand consideration, and choice. *Journal of Consumer Research, 31*(1), 191–198.

Geyskens, I., Steenkamp, J.-B. E. M., & Kumar, N. (1998). Generalizations about trust in marketing channel relationships using meta-analysis. *International Journal of Research in Marketing, 15*(3), 223–248.

Kaplan, A. (1964). *The conduct of inquiry*. San Francisco: Chandler.

Kukull, W. A., & Ganguli, M. (2012). Generalizability. The trees, the forest, and the low-hanging fruit. *Neurology, 78*(23), 1886–1891.

Kuppelwieser, V., & Sarstedt, M. (2014). Confusion about the dimensionality and measurement specification of the future time perspective scale. *International Journal of Advertising, 33*(1), 113–136.

Levy, P. S., & Lemeshow, S. (2013). *Sampling of populations: Methods and applications* (5th ed.). Hoboken: Wiley.

Lindner, J. R., Murphy, T. H., & Briers, G. E. (2001). Handling nonresponse in social science research. *Journal of Agricultural Education, 42*(4), 43–53.

Mitchell, M. L., & Jolley, J. M. (2013). *Research design explained* (8th ed.). Belmont: Wadsworth.

Moser, C. A., & Stuart, A. (1953). An experimental study of quota sampling. *Journal of the Royal Statistical Society. Series A (General), 116*(4), 349–405.

Netemeyer, R. G., Bearden, W. O., & Sharma, S. (2003). *Scaling procedures: Issues and applications*. Thousand Oaks: Sage.

Reichheld, F. F. (2003). The one number you need to grow. *Harvard Business Review, 81*(12), 46–55.
Sarstedt, M., & Schloderer, M. P. (2010). Developing a measurement approach for reputation of nonprofit organizations. *International Journal of Nonprofit and Voluntary Sector Marketing, 15*(3), 276–299.
Sarstedt, M., Diamantopoulos, A., Salzberger, T., & Baumgartner, P. (2016a). Selecting single items to measure doubly-concrete constructs: A cautionary tale. *Journal of Business Research, 69*(8), 3159–3167.
Sarstedt, M., Diamantopoulos, A., & Salzberger, T. (2016b). Should we use single items? Better not. *Journal of Business Research, 69*(8), 3199–3203.
Sarstedt, M., Hair, J. F., Ringle, C. M., Thiele, K. O., & Gudergan, S. P. (2016c). Estimation issues with PLS and CBSEM: Where the bias lies! *Journal of Business Research, 69*(10), 3998–4010.
Sarstedt, M., Bengart, P., Shaltoni, A. M., & Lehmann, S. (2017, forthcoming). The use of sampling methods in advertising research: A gap between theory and practice. *International Journal of Advertising.*
Stephenson, C. B. (1979). Probability sampling with quotas: An experiment. *Public Opinion Quarterly, 43*(4), 477–496.

Getting Data

4

Keywords
Back-translation • Balanced scale • Big data • Closed-ended questions • Constant sum scale • Customer relationship management • Double-barreled questions • Equidistant scale • Ethnography • Experiments • Experimental design • External secondary data • External validity • Face-to-face interviews • Focus groups • Forced-choice scale • Free-choice scale • In-depth interviews • Internal secondary data • Internal validity • Laddering • Likert scale • Mail surveys • Manipulation checks • Means-end approach • Mixed mode • Mystery shopping • Observational studies • Open-ended questions • Personal interviews • Primary data • Projective techniques • Qualitative research • Rank order scales • Reverse-scaled items • Secondary data • Semantic differential scales • Sentence completion • Social desirability bias • Social media analytics • Social networking data • Surveys • Syndicated data • Telephone interviews • Test markets • Treatments • Unbalanced scale • Visual analogue scale • Web surveys • Verbatim items

Learning Objectives
After reading this chapter, you should understand:

- How to find secondary data and decide on their suitability.
- How to collect primary data.
- How to design a basic questionnaire.
- How to set up basic experiments.
- How to set up basic qualitative research.

4.1 Introduction

In the previous chapter, we discussed some of the key theoretical concepts and choices associated with collecting data. These concepts and choices included validity, reliability, sampling, and sample sizes. We also discussed different types

E. Mooi et al., *Market Research*, Springer Texts in Business and Economics,
DOI 10.1007/978-981-10-5218-7_4

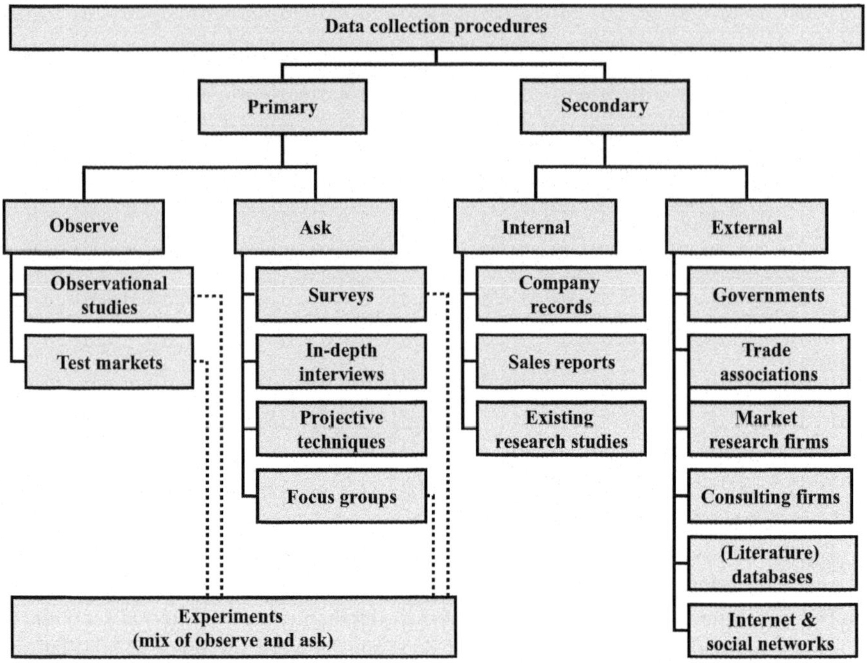

Fig. 4.1 Types of primary and secondary data sources

of data. Building on Chap. 3, this chapter discusses the practicalities of collecting data. First, we discuss how to work with secondary data. Before collecting primary data, market researchers should always consider secondary data, which are often available and do not depend on respondents' willingness to participate. Although secondary data have already been collected, you usually need to spend considerable effort preparing them for analysis, which we discuss first. If you find that the required secondary data are unavailable, outdated, or very costly, you may have to collect primary data. In the sections that follow, we discuss how to collect primary data through observations, surveys, and experiments. In Fig. 4.1, we provide an overview of some types of secondary and primary data.

4.2 Secondary Data

Secondary data are data that have already been gathered, often for a different research purpose and some time ago. Secondary data comprise internal secondary data, external secondary data, or a mix of both.

4.2.1 Internal Secondary Data

Internal secondary data are data that companies compiled for various reporting and analysis purposes. Much of these data have been collected and stored because "you can't manage what you don't measure."[1] Large companies have systems in place, such as Enterprise Resource Planning systems (usually abbreviated as ERP systems), in which vast amounts of customer, transaction, and performance data are stored. In general, internal secondary data comprise the following:

- company records,
- sales reports, and
- existing research studies.

Company records are a firm's repository of information. They may contain data from different business functions, such as finance, or **Customer Relationship Management** (**CRM**). The finance function may provide internal reviews of an organization's financial well-being and strategic advice, as it has access to the organization's financial and operational data. The term CRM refers to a system of databases and analysis software that tracks and predicts customer behavior. Firms such as IBM, Microsoft, and Oracle market the database systems that the analysis software utilizes. These database management systems often include information on, for example, purchasing behavior, (geo-)demographic customer data, and the after-sales service. This information is compiled to allow marketers to track individual customers over different sales channels and types of products in order to tailor their offerings to these customers' needs. Several information technology companies, such as SAP, Oracle, and Salesforce.com sell the analysis software that utilizes these databases. Companies use this software to, for example, identify customer trends, calculate their profitability per customer, or identify opportunities to sell new or different products. The CRM market is substantial, generating about $37 billion in 2017.[2]

Sales reports are created when products and services are sold to business-to-business clients. Many of these reports detail discussions held with clients, as well as the products and services sold. The reports therefore provide insights into customers' needs. Sales reports are also a means of retaining customers' suggestions regarding products and services, and can be a productive source of information. For example, DeMonaco et al. (2005) found that 59% of existing drugs had uses other than those that the producing company described. Because it is important to be aware of a drug's uses, sales discussions with doctors, hospitals, and research institutes can help this company market these drugs. When sales reports are available, they are often part of a CRM system.

[1]This quote has been attributed to Peter F. Drucker.

[2]www.gartner.com/DisplayDocument?id=2515815&ref=g_sitelink

Existing research studies are a good source of secondary data. You should, however, carefully consider whether existing research studies are still useful and what you can learn from them. Even if you believe their findings are outdated, their measures may be very useful. Consequently, if you wish to use existing research studies, it is important that you ascertain that enough of their details are available to make them useful.

4.2.2 External Secondary Data

External secondary data have been compiled outside a company for a variety of purposes. Important sources of secondary data, which we discuss next, include:

- governments,
- trade associations,
- market research firms,
- consulting firms,
- (literature) databases, and
- internet & social networks.

Governments often provide data that can be used for market research purposes. For example, The *CIA World Fact Book* provides information on the economy, politics, and other issues of nearly every country in the world. Eurostat (the statistics office of the European Union) provides detailed information on the economy and different market sectors of the European Union. Much of this information is free of charge and is an easy starting point for market research studies.

Trade associations are organizations representing different companies whose purpose is to promote their common interests. For example, the Auto Alliance—which consists of US automakers—provides information on the sector and lists the key issues it faces. The European Federation of Pharmaceutical Industries and Associations represents 1900 pharmaceutical companies operating in Europe. The federation provides a detailed list of key figures and facts, and regularly offers statistics on the industry. Most of the other trade associations also provide lists of their members' names and addresses. These can be used, for example, as a sampling frame (see Chap. 3). Most trade associations regard ascertaining their members' opinions a key task and therefore collect data regularly. These data are often included in reports that researchers can download from the organization's website. Such reports can be a short-cut to identifying key issues and challenges in specific industries. Sometimes, these reports are free of charge, but non-members usually need to pay a (mostly substantial) fee.

Market research firms are another source of secondary data. Especially large market research firms provide syndicated data that different clients can use (see Chap. 1). **Syndicated data** are standardized, processed information made available to multiple (potential) clients, usually for a substantial fee. Syndicated data often

Box 4.1 GfK Spex Retail
GfK is a large market research company. One of its databases, Spex Retail, provides resellers, distributors, manufacturers, and website portals with product data. In 20 languages, it offers aggregated data on more than seven million products from 20,000 manufacturers in 30 countries. This database provides details on IT, consumer electronics, household appliances, and other products. Spex Retail also provides insight into new products being launched by providing 70,000 new information sheets that describe new products or product changes every month. The data can also be used to map product categories. Such information helps its clients understand the market structure, or identify cross-selling or up-selling possibilities. See www.etilize.com/spex-plus-product-data.htm for more details, including a demo of its products.

allow the client's key measures (such as satisfaction or market share) to be compared against the rest of the market. Examples of syndicated data include the J.D. Power Initial Quality Study, which provides insights into the initial quality of cars in the US, and the J.D. Power Vehicle Ownership Satisfaction Study, which contains similar data on other markets, such as New Zealand and Germany. GfK Spex Retail, which we introduce in Box 4.1, is another important example of syndicated data.

Consulting firms are a rich source of secondary data. Most firms publish full reports or summaries of reports on their website. For example, McKinsey & Company publish the *McKinsey Quarterly*, a business journal that includes articles on current business trends, issues, problems, and solutions. Oliver Wyman publishes regular reports on trends and issues across many different industries. Other consulting firms, such as Gartner and Forrester, provide data on various topics. These data can be purchased and used for secondary analysis. For example, Forrester maintains databases on market segmentation, the allocation of budgets across firms, and the degree to which consumers adopt various innovations. Consulting firms provide general advice, information, and knowledge, while market research firms only focus on marketing-related applications. In practice, there is some overlap in the activities that consulting and market research firms undertake.

(Literature) databases comprise professional and academic journals, newspapers, and books. Two important literature databases are ProQuest (www.proquest.com) and JSTOR (www.jstor.org). ProQuest contains over 9,000 trade journals, business publications, and leading academic journals, including highly regarded publications such as the *Journal of Marketing* and the *Journal of Marketing Research*. A subscription is needed to gain access, although some papers are published as open access. Academic institutions often allow their students and, sometimes, their alumni to access these journals. JSTOR is like ProQuest, but is mostly aimed at academics. Consequently, it provides access to nearly all leading academic journals. A helpful feature of JSTOR is that the first page of academic articles (which contains the abstract) can be read free of charge. In addition, JSTOR's information is searchable via Google Scholar (discussed in Box 4.2). Certain database firms provide firm-level data, such as names and addresses. For example, Bureau van Dijk (www.bvdep.com), as well as Dun and Bradstreet (www.dnb.com), publish extensive lists of firm names, the industry in which they operate, their profitability, key activities, and address information. This information is often used as a sampling frame for surveys.

Internet data is a catch-all term that refers to data stored to track peoples' behavior on the Internet. Such data consist of *page requests* and *sessions*. A page request refers to people clicking on a link or entering a specific Internet address. A session is a series of these requests and is often identified by the IP number, a specific address that uniquely identifies the receiver for a period of time, or by means of a *tracking cookie*. With this information, researchers can calculate when and why people move from one page to another. The *conversion rate* is a specific type of information, namely the ratio of the number of purchases made on a website relative to the number of unique visitors, which often interests researchers. Facebook, Instagram, and LinkedIn, provide valuable information in the form of social networking profiles, which include personal details and information. These **social networking data** reflect how people would like others to perceive them and, thus, indicate consumers' intentions. Product or company-related user groups are of specific interest to market researchers. Take, for example, comments posted on a Facebook group site such as that of BMW or Heineken. An analysis of the postings helps provide an understanding of how people perceive these brands. Interpretations of such postings usually include analyzing five elements: the agent (who is posting?), the act (what happened, i.e., what aspect does the posting refer to?), the agency (what media is used to perform the action?), the scene (what is the background situation?), and the purpose (why do the agents act?). By analyzing this qualitative information, market researchers can gain insight into consumers' motives and actions. Casteleyn et al. (2009), for example, show that the Heineken Facebook posts reveal that the brand has a negative image in Belgium. The task of collecting, processing, analyzing, and storing social networking data is very challenging, due to the data's complexity and richness. To enable these tasks, researchers have combined theories and methods from a variety of disciplines (e.g., computer science, linguistics, statistics) in the emerging research field of **social media analytics** to develop new approaches to and method for analyzing social networking data. These include (1) *text mining* to derive high-quality information from text, (2) *social network analysis* to study the structure of the relationships between persons, organizations, or institutions in social networks, and (3) *trend analysis* to predict emerging topics in, for example, Twitter tweets or Facebook posts (Stieglitz et al. 2014). Several companies, such as Talkwaker (www.talkwalker.com), aggregate data from different websites (including blogs) and social media platforms, such as Twitter, Facebook, and Instagram, which they analyze. These sites also provide statistics, such as the number of mentions or complaints, thereby providing insight into people, brands, and products rated on various dimensions. Talkwalker, for example, conducted a sentiment analysis of a

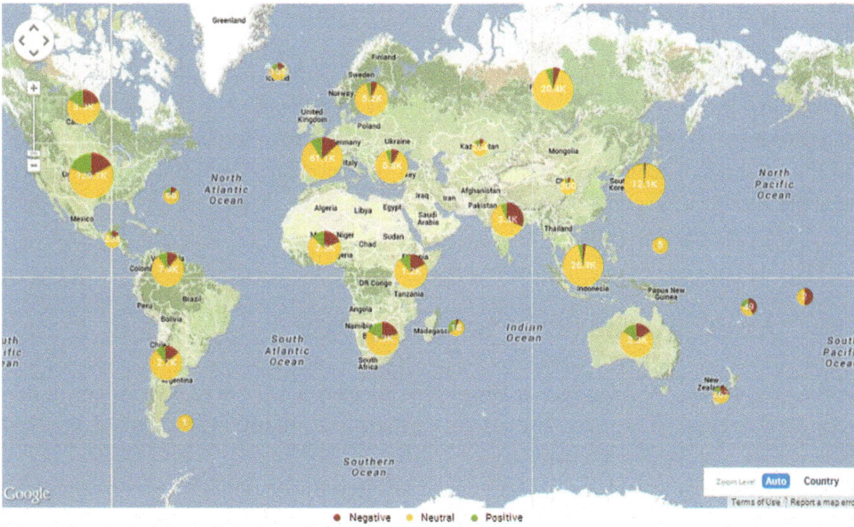

Fig. 4.2 Snapshot of sentiment analysis via www.talkwalker.com

major product recall by Toyota. Toyota found that its social media mentions increased sharply and were far more negative in the US and Europe than in Indonesia and Japan (see Fig. 4.2; https://www.talkwalker.com/blog/navigate-the-automotive-recall-storm-with-social-media-monitoring).

Social websites also provide quantitative information. For example, Facebook's Ad Manager provides information on the effectiveness of advertising on Facebook, including on measures, such as the click-through-rate, and on demographics, such as gender or location.

Big data is an important term in the context of Internet and social networking data. The term big data describes very large datasets, generally a mix of quantitative and qualitative data in very large volumes, which are automatically analyzed, often with the aim of making predictions. There is no commonly accepted definition of the term, but the use of big data has become very important very quickly. Big data's use is not unique to market research, but spans boundaries and often includes IT, operations, and other parts of organizations. Netflix, a provider of videos and movies, relies on big data. Netflix faces the challenge that it pays upfront for the videos and the movies it purchases, and therefore needs to understand which, and how many, of its customers will watch them. Netflix analyzes two billion hours of video each month in an endeavor to understand its customers' viewing behavior and to determine which videos and movies will become hits. Walmart, the largest retailer in the world, also uses big data. One of Walmart's challenges is to proactively suggest products and services to its customers. Using big data, Walmart connects information from many sources, including their location, and uses a product database to find related products. This helps Walmart make online recommendations.

4.3 Conducting Secondary Data Research

In Chap. 2, we discussed the market research process, starting with identifying and formulating the research question, followed by determining the research design. Once these two have been done, your attention should turn to designing the sample and the method of data collection. In respect of secondary data, this task involves the steps shown in Fig. 4.3.

4.3.1 Assess Availability of Secondary Data

Search engines (such as Google or Bing) provide easy access to many sources of the secondary data we have just discussed. Furthermore, many (specialist) databases also provide access to secondary data.

Search engines crawl through the Internet, regularly updating their contents. Algorithms, which include how websites are linked, and other peoples' searches evaluate this content and present a set of results. Undertaking searches by means of search engines requires careful thought. For example, the word order is important (put keywords first) and operators (such as +, −, and ~) may have to be added to restrict searches. In Box 4.2, we discuss the basics of using Google to search the Internet.

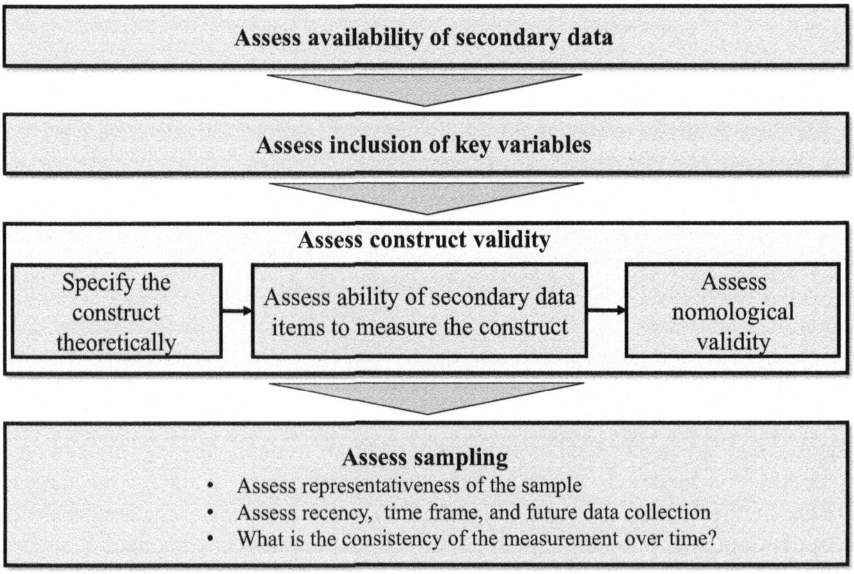

Fig. 4.3 Assessing secondary data

Box 4.2 Using Google to Searching for Secondary Data

Most people use Google daily, so why not use it for market research purposes too? Google has an easy interface, but if you use the standard search box, you may not find the data you are looking for. What must you then do to find useful data?

- You could use Google Scholar (https://scholar.google.com) if you are looking for scholarly information such as that found in academic journals. While you can *search* for any information, specific search items can usually only be accessed if you have an organizational, university, or library password.
- By using Google Books (https://books.google.com/books), you can enter several keywords to easily search through a very large catalogue of books. Google clearly indicates the books in which the search results are found and the pages on which the keywords occur. Ngram Viewer (https://books.google.com/ngrams/), which shows the relative frequency with which words are used, is a cool Google books tool.
- If you cannot find what you are looking for, Google allows you to refine your search. After you have searched you can typically select *All, Images, Videos, News, Maps, and More* to select the type of result you are interested in. Under Tools, you can select the country and time range from which results are to be shown.
- Google Public Data Explorer (https://www.google.com/publicdata/directory) facilitates exploration of a variety of public-interest datasets. These include the US Census Bureau, Eurostat, and the OECD datasets. This site is particularly useful if you want to obtain visualized data on economic indicators.
- Try using operators, which are signs that you use to restrict your research. For example, putting a minus symbol (−) (without a space) before a search word *excludes* this word from your findings. Putting a sequence of words, or an entire sentence in quotation marks (e.g., "a concise guide to market research") indicates that Google should only search for exact matches.

Databases contain existing data that can be used for market research purposes, which we discussed in the section "External Secondary Data." Lightspeed Research (www.lightspeedresearch.com), for example, maintains several databases, such as the Travel & Leisure Specialty Panel, providing details on selected market segments. GfK provides several databases that track retail sales. Nielsen maintains a large consumer panel of some 250,000 households in 27 countries. It is clearly not possible to provide an exhaustive list of the databases available, but an online search, a market research agency, or an expert should help you identify the options quickly.

Once a potential secondary data source has been located, the next task is to evaluate the available data. It is important to critically assess the (potential) data's fit with your needs. Figure 4.3 provides a set of criteria to help evaluate this fit.

4.3.2 Assess Inclusion of Key Variables

Measurement is the first element to assess. It consists of a set of criteria. You should first check whether the desired variables are included in the source. The key variables in which you are interested, or could use, should obviously be part of the data. Also check if these variables are included at the required level of analysis, which is called the aggregation level (see Chap. 3). For example, the American Customer Satisfaction Index (ACSI) satisfaction dataset reports on satisfaction at the company level;[3] therefore, if researchers need the measurement of satisfaction at a product, service, or store level, these data are inappropriate.

4.3.3 Assess Construct Validity

After checking that the desired variables are included in the source, the construct validity should be assessed (see Chap. 3 for a discussion of validity). Validity relates to whether variables measure what they should measure. *Construct validity* is a general term relating to how a variable is defined conceptually and its suggested (empirical) measure (see Chap. 3). Houston (2002) establishes a three-step method to assess the construct validity of secondary data measures.

- First, specify the theoretical definition of the construct in which you are interested. Satisfaction is, for example, often defined as the degree to which an experience conforms to expectations and the ideal.
- Second, compare your intended measure against this theoretical definition (or another acceptable definition of satisfaction). Conceptually, the items should fit closely.
- Third, assess if these items have *nomological validity* (see Chap. 3). For example, customer expectations, perceived quality, and value have a significant influence on customer satisfaction. Similarly, satisfaction generally relates positively to customer loyalty. As such, you would expect the measure of satisfaction that you are evaluating to correlate with these measures (if included in the database).

---

[3]See www.theacsi.org for a detailed description of how ACSI data are collected.

Taking these three steps is important, as the construct validity is often poor when secondary data are used.[4] See Raithel et al. (2012) for an application of this three-step process.

If there are multiple sources of secondary data, identical measures can be correlated to assess the construct validity. For example, the Thomson Reuter SDC Platinum and the Bioscan database of the American Health Consultants both include key descriptors of firms and financial information; they therefore have a considerable overlap regarding the measures included. This may raise questions regarding which databases are the most suitable, particularly if the measures do not correlate highly. Fortunately, databases have been compared; for example, Schilling (2009) compares several databases, including SDC Platinum.

4.3.4 Assess Sampling

Next, the sampling process of the collected secondary data should be assessed. First assess the population and the representativeness of the sample drawn from it. For example, the sampling process of Nielsen Homescan's data collection effort is based on probability sampling, which can lead to representative samples (see Chap. 3). Sellers of secondary data often emphasize the size of the data collected, but good sampling is more important than sample size! When sampling issues arise, these can be difficult to detect in secondary data, because the documents explaining the methodology behind secondary data (bases) rarely discuss the data collection's weaknesses. For example, in many commercial mailing lists 25% (or more) of the firms included routinely have outdated contact information, are bankrupt, or otherwise not accurately recorded. This means that the number of contactable firms is much lower than the number of firms listed in the database. In addition, many firms may not be listed in the database. Whether these issues are problematic depends on the research purpose. For example, descriptive statistics may be inaccurate if data are missing.

The recency of the data, the time period over which the data were collected, and future intentions to collect data should be assessed next. The data should be recent enough to allow decisions to be based on them. The data collection's timespan and the intervals at which the data were collected (in years, months, weeks, or in even more detail) should match the research question. For example, when introducing new products, market competitors' market share is an important variable, therefore such data must be recent. Also consider whether the data will be updated in the future and the frequency with which such updates will done. Spending considerable time on getting to know data that will not be refreshed can be frustrating! Other measurement issues include definitions that change over time. For example, many firms changed their definitions of loyalty from behavioral (actual purchases) to

---

[4]Issues related to construct validity in business marketing are discussed by, for example, Rindfleisch and Heide (1997). A more general discussion follows in Houston (2002).

attitudinal (commitment or intentions). Such changes make comparisons over time difficult, as measures can only be compared if the definitions are consistent.

4.4 Conducting Primary Data Research

Primary data are gathered for a specific research project or task. There are two ways of gathering primary data. You can observe consumers' behavior, for example, by means of observational studies, or test markets. Alternatively, you can ask consumers directly by means of surveys, in-depth interviews, predictive techniques, or focus groups. Experiments are a special type of research, which is normally a combination of observing and asking. We provide an overview of the various types of primary data collection methods in Fig. 4.1.

Next, we briefly introduce observing as a method of collecting primary data. We proceed by discussing how to conduct surveys. Since surveys are the main means of collecting primary data by asking, we discuss the process of undertaking survey research in enough detail to allow you to set up your own survey-based research project. We then discuss in-depth interviews, including a special type of test used in these interviews (projective techniques). Last, we discuss combinations of observing and asking – the basics of conducting experimental research.

4.4.1 Collecting Primary Data Through Observations

Observational studies can provide important insights that other market research techniques do not. Observational techniques shed light on consumers' and employees' behavior and can help answer questions such as: How do consumers walk through supermarkets?; how do they consume and dispose of products?; and how do employees spend their working day? Observational techniques are normally used to understand what people are doing rather than *why* they are doing it. They work well when people find it difficult to put what they are doing into words, such as shoppers from different ethnic backgrounds.

Most observational studies use video recording equipment, or trained researchers, who unobtrusively observe what people do (e.g., through one-way mirrors or by using recording equipment). Recently, researchers started using computer chips (called RFIDs) as observational equipment to trace consumers' shopping paths within a supermarket. Almax, an Italian company, has developed a special type of observational equipment. Their EyeSee product is an in-store mannequin equipped with a camera and audio recording equipment. The product also comprises software that analyze the camera recordings and provides statistical and contextual information, such as the shoppers' demographics. Such information is useful for developing targeted marketing strategies. For example, a retail company found that Chinese visitors prefer to shop in Spanish stores after 4 p.m., prompting these stores to increase their Chinese-speaking staff at these hours. Figure 4.4 shows what the EyeSee Mannequin looks like.

Fig. 4.4 The EyeSee Mannequin

Mystery shopping, when a trained researcher is asked to visit a store or restaurant and consume the products or services, is a specific type of observational study. For example, McDonalds and Selfridges, the latter a UK retail chain, both use mystery shoppers to ensure the quality of their services and products (see Box 4.3 for an MSNBC video on mystery shopping).

Sometimes observational studies are conducted in households, with researchers participating in them to see how the inhabitants buy, consume, and dispose of products or services. The type of study in which the researcher is a participant is called an **ethnography**. An example of an ethnography is Volkswagen's Moonraker project, in which several Volkswagen employees followed American drivers to gain an understanding of how their usage of and preferences for automobiles differ from those of European drivers (Kurylko 2005).

Test markets are a useful, but costly, type of market research in which a company introduces a new product or service to a specific geographic market. Test markets are sometimes also used to understand how consumers react to different marketing mix instruments, such as changes in pricing, distribution, or advertising and communication. Test marketing is therefore about changing a product or service offering in a real market and gauging consumers' reactions. While the results from such studies provide important insights into consumer behavior in a real-world setting, they are expensive and difficult to conduct. Some frequently used test markets include Hassloch in Germany, as well as Indianapolis and Nashville in the US.

Box 4.3 Using Mystery Shopping to Improve Customer Service

www.youtube.com/watch?v=JK2p6GMhs0I

4.4.2 Collecting Quantitative Data: Designing Surveys

There is little doubt that **surveys** are the mainstay of primary market research. While it may seem easy to conduct a survey (just ask what you want to know, right?), there are many issues that could turn good intentions into bad results. In this section, we discuss the key design choices for good surveys. A good survey requires at least seven steps. First, determine the survey goal. Next, determine the type of questionnaire required and the administration method. Thereafter, decide on the questions and the scale, as well as the design of the questionnaire. Conclude by pretesting and administering the questionnaire. We show these steps in Fig. 4.5.

4.4.2.1 Set the Survey Goal

Before you start designing the questionnaire, it is vital to consider the survey goal. Is it to collect quantitative data on customers' background, to assess customer satisfaction, or do you want to understand why and how customers complain? These different goals influence the type of questions asked (such as open-ended or closed-ended questions), the method of administration (e.g., by mail or on the Web), and other design issues discussed below. Two aspects are particularly relevant when designing surveys:

First, consider the information or advice you want to emerge from the study for which the survey is required. Say you are asked to help understand check-in waiting times at an airport. If the specific study question is to gain an understanding of how many minutes travelers are willing to wait before becoming dissatisfied, you should be able to provide an answer to the question: How much does travelers' satisfaction decrease with increased waiting time? If, on the other hand, the specific question is

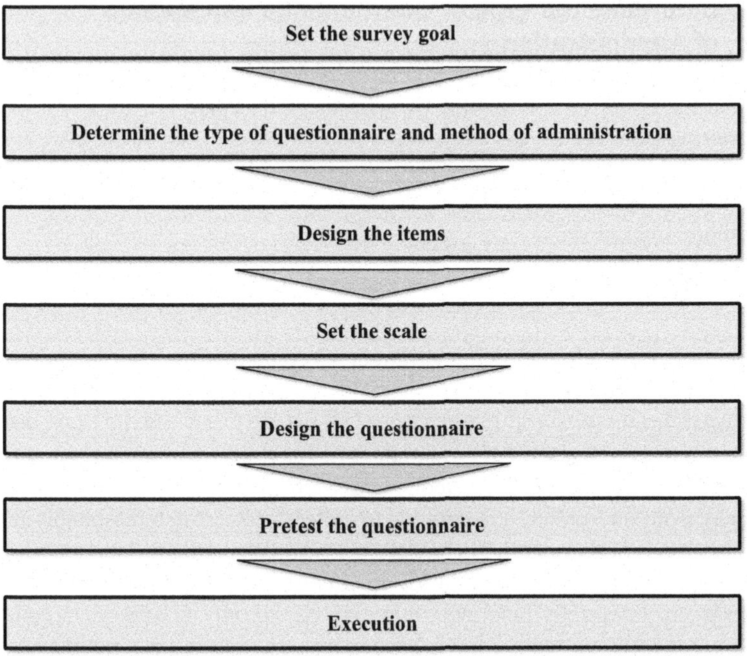

Fig. 4.5 Steps in designing surveys

to understand how people perceive waiting time (short or long), your questions should focus on how travelers perceive this time and, perhaps, what influences their perception. Thus, the information or advice you want to provide influences the questions that you should pose in a survey.

Second, consider the method required for the study early in the design process. For example, if a study's goal is to determine market segments, you should probably use cluster analysis (see Chap. 9). Similarly, if the study's goal is to develop a way to systematically measure customer satisfaction, you are likely to use factor analysis (see Chap. 8). This approach is crucial, as each method requires different types of data. Cluster analysis, for example, generally requires variables that are not too highly correlated, meaning that researchers need to use a type of questionnaire that can produce these data. On the other hand, factor analysis requires data that include different, but highly correlated, variables. If you use factor analysis to distinguish between the different aspects of consumer satisfaction, you need to design a survey that will produce data allowing you to conduct factor analysis.

4.4.2.2 Determine the Type of Questionnaire and Method of Administration

After determining the survey goal, you need to decide on the type of questionnaire you should use and how it should be administered. There are four key ways of administering a survey:

- personal interviews,
- telephone interviews,
- web surveys, and
- mail surveys.

In some cases, researchers combine different ways of administering surveys. This is called a mixed mode.

Personal interviews (or **face-to-face interviews**) can obtain high response rates, since engagement with the respondents is maximized, allowing rich information (visual expressions, etc.) to be collected. Moreover, since people find it hard to walk away from interviews, it is possible to collect answers to a reasonably lengthy set of questions. Consequently, personal interviews can support long surveys. It is also the best type of data collection for open-ended responses. In situations where the respondent is initially unknown, this may be the only feasible data collection type. Consequently, in-depth interviews may be highly preferable, but they are also the costliest per respondent. This is less of a concern if only small samples are required (in which case personal interviewing could be the most efficient). Other issues with personal interviews include the possibility of *interviewer bias* (i.e., a bias resulting from the interviewer's behavior e.g., in terms of his/her reactions or presentation of the questions), *respondent bias* to sensitive items, and the data collection usually takes more time. Researchers normally use personal interviewing when they require an in-depth exploration of opinions. Such interviewing may also help if drop out is a key concern. For example, if researchers collect data from executives around the globe, using methods other than face-to-face interviewing may lead to excessive non-response in countries such as Russia or China where face-to-face interviews are a sign of respect for and appreciation of the time taken. *CAPI*, which is the abbreviation of *computer-assisted personal interviews*, is a frequently used term in the context of in-depth interviewing. CAPI involves using computers during the interviewing process to, for example, route the interviewer through a series of questions, or to enter responses directly. Similarly, in *CASI* (*computer-assisted self-interviews*), the respondent uses a computer to complete the survey questionnaire without an interviewer administering it.

Telephone interviews allow researchers to collect data quickly. These interviews also support open-ended responses, although not as well as personal interviews. Moreover, interviewer bias can only be controlled moderately, since the interviewers follow predetermined protocols, and the respondent's interactions with others during the interview is strongly controlled. Telephone interviewing can be a good compromise between mailed interviews' low cost and the richness of in-depth interviews. *CATI* refers to *computer-assisted telephone interviews*, which are an important method of administering surveys. Until the 1990s, telephone interviews were generally conducted via landlines, but mobile phone usage has soared in the meantime. In many countries, mobile phone adoption rates are higher than landline adoption. This holds especially for African countries, India, and many European countries if younger consumers are the targeted interviewees (Vincente and Reis 2010). Consequently, mobile phone surveys have become dominant in market research.

A decade ago, the differences between landline and mobile phone surveys could be large, with younger and richer individuals being overrepresented in mobile phone surveys (Vincente et al. 2008). As the adoption of mobile phones increased, the use of landlines decreased, which introduced new problems in terms of sampling errors (Stern et al. 2014). For example, younger people are now less likely to have landlines. An additional issue is that landlines are fixed to a geographical area whereas mobile phones are not. As people move or travel, mobile phones are far less likely to give useful information about geographic areas. While recent research has shown that differences in response accuracy between mobile phone and landline surveys are small (e.g., with regard to social desirability bias), especially for questions that are not cognitively demanding (e.g., Kennedy and Everett 2011; Lynn and Kaminska 2013), the samples from which they are drawn can be very different in practice. There are some other differences as well. For example, the likelihood of full completion of surveys is higher for mobile calling, even though completion takes around 10–15% longer (Lynn and Kaminska 2013).

Web surveys (sometimes referred to as *CAWI*, or *computer-assisted web interviews*) are often the least expensive to administer and can be fast in terms of data collection, particularly since they can be set up very quickly. Researchers can administer web surveys to very large populations, even internationally, because, besides the fixed costs of setting up a survey, the marginal costs of administering additional web surveys are relatively low.

Many firms specializing in web surveys will ask $0.30 (or more) for each respondent, which is substantially lower than the costs of telephone interviews, in-depth interviews, and mail surveys. It is also easy to obtain precise quotes quickly. Qualtrics (http://www.qualtrics.com) is a leading web service provider that allows a specific type of respondent and a desired sample size to be chosen. For example, using Qualtrics's sample to survey 500 current owners of cars to measure their satisfaction costs $2,500 for a 10-min survey. This cost increases sharply if samples are hard to access and/or require compensation for their time. For example, surveying 500 purchasing managers by means of a 10-min survey costs approximately $19,500.

Web surveys also support complex survey designs with elaborate branching and skip patterns that depend on the response. For example, web surveys allow different surveys to be created for different types of products. Further, since web surveys reveal the questions progressively to the respondents, there is an option to channel them to the next question based on their earlier responses. This procedure is called *adaptive questioning*. In addition, web surveys can be created that allow respondents to automatically skip questions if they do not apply. For example, if a respondent has no experience with an iPad, researchers can create surveys that do not ask questions about this product. However, web surveys impose similar burdens on the respondents as mail surveys do (see below), which they may experience as an issue. This makes administering long web surveys difficult. Moreover, open-ended questions tend to be problematic, because few respondents are likely to provide answers, leading to a low item response. There is evidence that properly conducted web surveys lead to data as good as those obtained from mail surveys; in addition, the lack an interviewer and the resultant *interviewer bias* means they can provide better results than personal interviews (Bronner and Ton 2007; Deutskens et al. 2006). In web surveys, the respondents are also less exposed to evaluation apprehension and less inclined to respond with socially desirable behavior.[5] Web surveys are also used when a quick "straw poll" is needed on a subject.

It is important to distinguish between true web-based surveys used for collecting information on which marketing decisions will be based and polls, or very short surveys, on websites used to increase interactivity. These polls/short surveys are used to attract and keep people interested in websites and are not part of market research. For example, USA Today (http://www.usatoday.com), an American newspaper, regularly publishes short polls on their main website.

Mail surveys are paper-based surveys sent out to respondents. They are a more expensive type of survey research and are best used for sensitive items. Since no interviewer is present, there is no interviewer bias. However, mail surveys are a poor choice for complex survey designs, such as when respondents need to skip a large number of questions depending on previously asked questions, as this means that the respondent needs to correctly interpret the survey structure. Open-ended questions are also problematic, because few people are likely to provide answers to such questions if the survey is administered on paper. Other problems include a lack of control over the environment in which the respondent fills out the survey and that mail surveys take longer than telephone or web surveys. However, in some situations, mail surveys are the only way to gather data. For example, while executives rarely respond to web-based surveys, they are more likely to respond

[5]For a comparison of response behavior between CASI, CAPI, and CATI, see Bronner and Ton (2007).

to paper-based surveys. Moreover, if the participants cannot easily access the web (such as employees working in supermarkets, cashiers, etc.), handing out paper surveys is likely to be more successful.

The method of administration also has a significant bearing on a survey's maximum duration (Vesta Research 2016). As a rule of thumb, telephone interviews should be no longer than 20 min. When calling a on a mobile phone, however, the survey should not exceed 5 min. The maximum survey duration of web surveys is 20–10 min for social media-based surveys. Personal interviews and mail surveys can be much longer, depending on the context. For example, surveys comprising personal interviews on topics that respondents find important, could take up to 2 h. However, when topics are less important, mail surveys and personal interviews need to be considerably shorter.

Market researchers are increasingly using **mixed mode** approaches. An example of a mixed mode survey is when potential respondents are first approached by phone, asked to participate and confirm their email addresses, after which they are given access to a web survey. Mixed mode approaches are also used when people are first sent a paper survey and then called if they fail to respond.

Mixed mode approaches may help, because they signal that the survey is important. They may also help improve response rates, as people who are more visually oriented prefer mail and web surveys, whereas those who are aurally oriented prefer telephone surveys. However, there is only evidence of increased response rates when modes are offered sequentially and of mixed modes reducing response rates when offered simultaneously (Stern et al. 2014). By providing different modes, people can use the mode they most prefer. A downside of mixed mode surveys is that they are expensive and require a detailed address list (including a telephone number and matching email address). However, systematic (non) response is the most serious mixed mode survey issue. For example, when filling out mail surveys, the respondents have more time than when providing answers by telephone. If respondents need this time to think about their answers, the responses obtained from mail surveys may differ systematically from those obtained from telephone surveys.

4.4.2.3 Design the Items

Designing the items, whether for a personal interview, web survey, or mail survey, requires a great deal of thought. Take, for example, the survey item shown in Fig. 4.6.

It is unlikely that people can give meaningful answers to such an item. First, using a negation ("not") in sentences makes questions hard to understand. Second, the reader may not have an iPhone, or may not have experience using it. Third, the answer categories are unequally distributed. That is, there is one category above

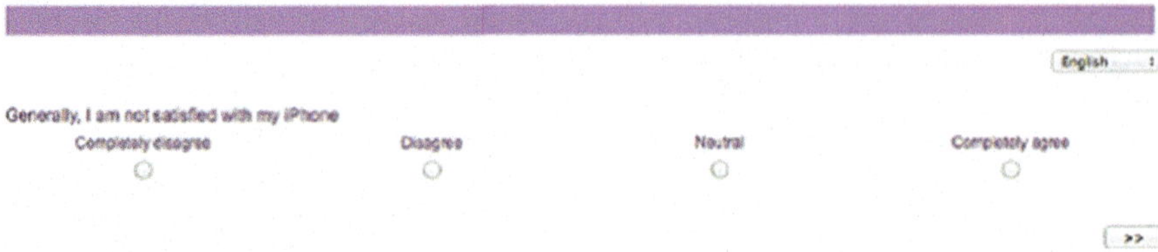

Fig. 4.6 Example of a bad survey item

neutral while there are two below. These issues are likely to create difficulties with understanding and answering questions, which may, in turn, cause validity and reliability issues. While the last aspect relates to the properties of the scale (see section *Set the Scale*), the first two issues refer to the item content and wording, which we will discuss next.

Item Content

When deciding on the item content, there are at least three essential rules you should keep in mind.

As a *first rule*, ask yourself whether everyone will be able to answer each item. If the item is, for example, about the quality of train transport and the respondent always travels by car, his or her answers will be meaningless. However, the framing of items is important, since, for example, questions about why the specific respondent does not use the train can yield important insights.

As a *second rule*, you should check whether respondents can construct or recall an answer. If you require details that possibly occurred a long time ago (e.g., "what information did the real estate agent provide when you bought/rented your current house?"), the respondents may have to "make up" an answer, which also leads to validity and reliability issues.

As a *third rule*, assess whether the respondents are willing to answer an item. If contents are considered sensitive (e.g., referring to sexuality, money, etc.), respondents may provide more socially desirable answers (e.g., by reporting higher or lower incomes than are actually true). Most notably, respondents might choose to select a position that they believe society favors (e.g., not to smoke or drink, or to exercise), inducing a **social desirability bias**. They may also not answer such questions at all. You should determine whether these items are necessary to attain the research objective. If they are not, omit them from the survey. The content that respondents may regard as sensitive is subjective and differs across cultures, age categories, and other variables. Use your common sense and, if necessary, use experts to decide whether the items are appropriate. In addition, make sure you pretest the survey and ask the pretest participants whether they were reluctant to provide certain answers. To reduce respondents' propensity to give socially desirable answers, use indirect questioning (e.g., "What do you believe other people think about. . .?"), frame questions as neutrally as possible, and suggest that many people exhibit behavior different from the norm (e.g., Brace 2004). Adjusting the response categories also helps when probing sensitive topics. For example, instead of directly asking about respondents' disposable income, you can provide various answering categories, which will probably increase their willingness to answer this question.

Designing the items also requires deciding whether to use open-ended or closed-ended questions. **Open-ended questions** provide little or no structure for respondents' answers. Generally, the researcher asks a question and the respondent writes down his or her answer in a box. Open-ended questions (also called **verbatim items**) are flexible and allow for explanation, making them particularly suitable for exploratory research. However, the drawback is that the respondents may be reluctant to provide detailed information. Furthermore, answering open-ended questions is more difficult and time consuming. In addition, their interpretation requires substantial coding. Coding issues arise when respondents provide many different answers (such as "sometimes," "maybe," "occasionally," or "once in a while") and the researcher has to divide these into categories (such as very infrequently, infrequently, frequently, and very frequently) for further statistical analysis. This coding is very time-consuming, subjective, and difficult. **Closed-ended questions** provide respondents with only a few categories from which to choose, which drastically reduces their burden, thereby inducing much higher response rates compared to open-ended questions. Closed-ended questions also facilitate immediate statistical analysis of the responses as no ex post coding is necessary. However, researchers need to identify answer categories in advance and limit this number to a manageable amount. Closed-ended questions are dominant in research and practice.

Item Wording

The golden rule of item wording is to keep it short and simple. That is, use simple words and avoid using jargon or slang as some respondents may misunderstand these. Similarly, keep grammatical complexities to a minimum by using active rather than passive voice, repeat the nouns instead of using pronouns, and avoiding possessive forms (Lietz 2010). Use short sentences (20 words max; Oppenheim 1992) to minimize the cognitive demands required from the respondents (Holbrook et al. 2006) and to avoid the risk of **double-barreled questions**, causing respondents to agree with one part of the question but not the other, or which they cannot answer without accepting a particular assumption. An example of such a double-barreled question is: "In general, are you satisfied with the products and services of the company?" A more subtle example is: "Do you have the time to read the newspaper every day?" This question also contains two aspects, namely "having time" and "reading the paper every day." The question "Do you read the newspaper every day?" followed by another about the reasons for a negative or a positive answer would be clearer (Lietz 2010).

Moreover, avoid using the word *not* or *no* where possible. This is particularly important when other words in the same sentence are negative, such as "unable," or "unhelpful," because sentences with two negatives (called a double negative) are hard to understand. For example, a question such as "I do not use the email function on my iPhone because it is unintuitive" is quite hard to follow. Also, avoid the use of *vague quantifiers*, such as "frequent" or "occasionally" (Dillman et al. 2014), which make it difficult for respondents to answer questions (what exactly is meant by "occasionally?"). They also make comparing responses difficult. After all, what one person considers "occasionally," may be "frequent" for another.[6] Instead, it is better to use frames that are precise ("once a week").

Never suggest an answer. For example, a question like "Company X has done very well, how do you rate it?" is highly suggestive and would shift the mean response to the positive end of the scale.

Many researchers recommend including **reverse-scaled items** in surveys. Reverse-scaled means the question, statement (if a Likert scale is used), or word pair (if a semantic differential scale is used) are reversed compared to the other items in the set. Reverse-scaled items act as cognitive "speed bumps" that alert inattentive respondents that the content varies (Weijters and Baumgartner 2012) and help reduce the effect of *acquiescence*, which relates to a respondent's tendency to agree with items regardless of the item content (Baumgartner and Steenkamp 2001; see also Chap. 5). Furthermore, reverse-scaled items help identify straight-lining, which occurs when a respondent marks the same response in almost all the items (Chap. 5). However, numerous researchers have shown that reverse-scaled items create problems, for example, by generating artificial factors, as respondents have greater difficulty verifying the item. This creates *misresponse rates* of 20% and higher (e.g., Swain et al. 2008; Weijters et al. 2013). Given these results, we suggest employing reverse-scaled items sparingly (Weijters and Baumgartner 2012), for example, only to identify straight-lining. If used, reverse-scaled items should not contain a particle negation such as "not" or "no."

Finally, when you undertake a survey in different countries (or adapt a scale from a different language), use professional translators, because translation is a complex process. Functionally, translating one language into another seems quite easy, as there are many websites, such as Google translate (translate.google.com), that do this. However, translating surveys requires preserving the conceptual equivalence of whole sentences and paragraphs; current software applications and websites cannot ensure this. In addition, cultural differences normally require changes to the instrument format or procedure. Back-translation is a technique to establish conceptual equivalence across languages. **Back-translation** requires translating a survey instrument into another language, after which another translator takes the translated survey instrument and translates it back into the original language (Brislin 1970). After the back-translation, the original and back-translated instruments are compared and points of divergence are noted. The translation is then corrected to more accurately reflect the intent of the wording in the original language.

[6]Foddy (1993) reported 445 interpretations of the word "usually," with the meaning assigned to the word varying, depending on, for example, the type of activity or who was asked about the activity.

4.4.2.4 Set the Scale

When deciding on scales, two separate decisions need to be made. First, you need to decide on the type of scale. Second, you need to set the properties of the scale you choose.

Type of Scale

Marketing research and practice have provided a variety of scale types. In the following, we discuss the most important (and useful) ones:

– likert scales,
– semantic differential scales, and
– rank order scales.

The most popular scale type used in questionnaires is the **Likert scale** (Liu et al. 2016). Likert scales are used to establish the degree of agreement with a specific statement. Such a statement could be "I am very committed to Oddjob Airlines." The degree of agreement is usually set by scale endpoints ranging from strongly disagree to strongly agree. Likert scales are used very frequently and are relatively easy to administer. Bear in mind that if the statement is too positive or negative, it is unlikely that the endings of the scale will be used, thereby reducing the number of answer categories actually used. We show an example of three Likert-scale-type items in Fig. 4.7, in which respondents assess the personality of the Oddjob Airways brand.

The semantic differential scale is another scale type that features prominently in market research. **Semantic differential scales** use an opposing pair of words, normally adjectives (e.g., young/old, masculine/feminine) constituting the endpoint of the scale. Respondents then indicate how well one of the word in each pair describes how he or she feels about the object to be rated (e.g., a company or brand). These scales are widely used in market research. As with Likert scales, 5 or 7 answer categories are commonly used (see the next section on the number of answer categories you should use). We provide an example of the semantic differential scale in Fig. 4.8, in which respondents provide their view of Oddjob Airways.

Rank order scales are a unique type of scale, as they force respondents to compare alternatives. In its basic form, a rank order scale allows respondents to indicate which alternative they rank highest, which alternative they rank second highest, etc. Figure 4.9 shows an example. The respondents therefore need to balance their answers instead of merely stating that everything is important. In a

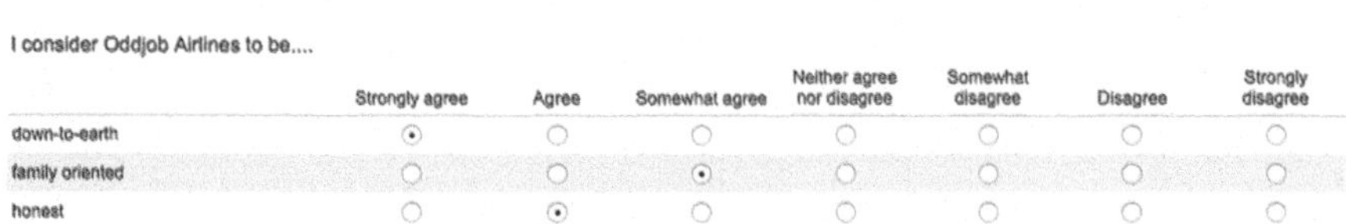

Fig. 4.7 Example of a 7-point Likert scale

Fig. 4.8 Example of a 7-point semantic differential scale

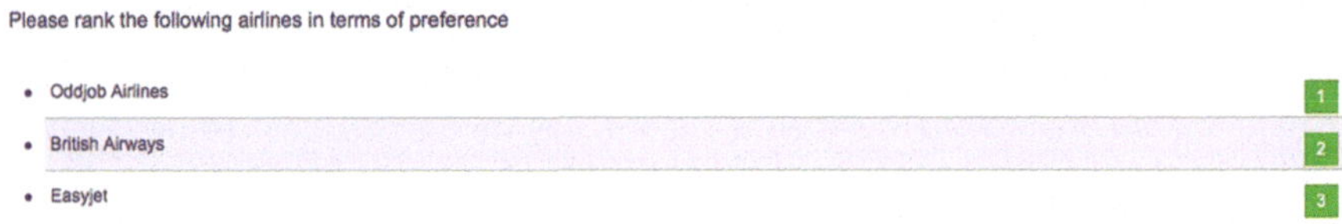

Fig. 4.9 Example of a rank order scale

more complicated form, rank order scales ask respondents to allocate a certain total number of points (often 100) to a number of alternatives. This is called the **constant sum scale**. Constant sum scales work well when a small number of answer categories are used (normally up to five). Generally, respondents find constant scales that have 6 or 7 answer categories somewhat challenging, while constant scales that have eight or more categories are very difficult to answer. The latter are thus best avoided.

In addition to these types of scaling, there are other types, such as graphic rating scales, which use pictures to indicate categories, and the *MaxDiff scale*, in which respondents indicate the most and least applicable items. We introduce the MaxDiff scale in the ⭳ Web Appendix (→ Downloads).

Properties of the Scale

After selecting the type of scale, we need to set the scale properties, which involves making several decisions:

- decide on the number of response categories,
- choose between forced-choice scale and free-choice scales,
- design the response categories,
- label the response categories,
- decide whether to use a "don't know" option, and
- choose between a balanced and unbalanced scale.

Decide on the number of response categories: When using closed-ended questions, the number of answer categories needs to be determined. In its simplest form, a survey could use just two answer categories (yes/no). Multiple categories (such as, "completely disagree," "disagree," "neutral," "agree," "completely agree") are used more frequently to allow for more nuances. When determining how many scale categories to use, we face a trade-off between having more variation and differentiation in the responses versus burdening the respondents too much, which can trigger different types of response biases (e.g., Weijters et al. 2010). Because 5-point scales and 7-point scales are assumed to achieve this trade-off well, their use has become common in marketing research (e.g., Fink 2003; Peterson 1997). Research on these two scale types in terms of their reliability and validity is inconclusive, but the differences between them are generally not very pronounced (e.g., Lietz 2010; Peng and Finn 2015; Weijters et al. 2010). Ten-point scales are often used in market research practice. However, scales with many answer categories often confuse respondents, because the wording differences between the scale points become trivial. For example, the difference between "tend to agree" and "somewhat agree" are subtle and respondents may not be able to differentiate between them. In such a case, respondents tend to choose categories in the middle of the scale (Rammstedt and Krebs 2007). Given this background, we recommend using 5-point or 7-point scales.

Finally, many web surveys now use levers that allow scaling on a continuum without providing response categories. This scale type is called a **visual analogue scale** and is especially recommended if small differences in response behavior need to be detected. Visual analogue scales are well known in paper-and-pencil-based research (especially in the medical sector) and have become increasingly popular in web surveys. A visual analogue scale consists of a line and two anchors, one at each end. The anchors often consist of verbal materials that mark opposite ends of a semantic dimension (e.g., good and bad). However, the anchors may also be pictures, or even sound files. Visual anchors, such as smileys, are often used with participants who may not fully grasp the meaning of verbal materials—for example, preschool children (Funke 2010). Compared to ordinal scales, measurement by means of visual analogue scales is more exact, leading to more narrow confidence intervals, and higher power statistical tests. This exactness helps detect smaller effect that may be unobservable with ordinal scales (Reips and Funke 2008). However, although not all web survey platforms offer visual analogue scales, you should use them if possible. Figure 4.10 shows an example of a visual analogue scale that measures customers' expectations of Oddjob Airlines.

Choose between a *forced-choice scale and a free-choice scale*: Sometimes, researchers and practitioners use 4-point or 6-point scales that omit the neutral

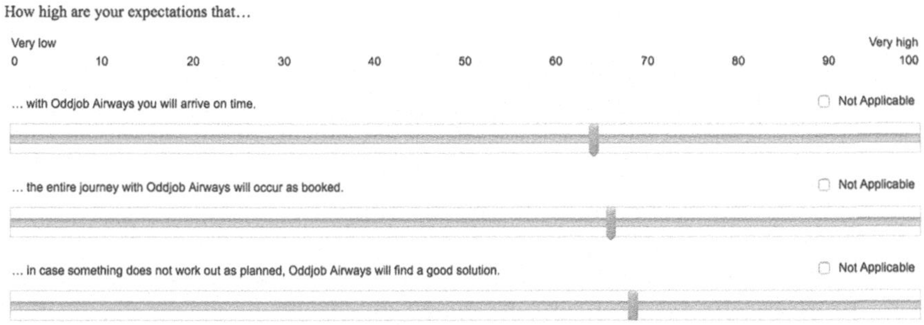

Fig. 4.10 Example of a visual analogue scale

category, thereby forcing the respondents to be positive or negative. Using such a **forced-choice scale** could bias the answers, leading to validity issues.[7] By providing a neutral category choice (i.e., a **free-choice scale**), the respondents are not forced to give a positive or negative answer. Many respondents feel more comfortable about participating in a survey using a free-choice scales (Nowlis et al. 2002). Furthermore, research has shown that including a neutral category minimizes response bias (Weijters et al. 2010). Therefore, we strongly suggest using free-choice scales and including a neutral category.

Design the response categories: When designing response categories for categorical variables, use response categories that are exclusive, so that answers do not overlap (e.g., age categories 0–4, 5–10, etc.). The question here is: How do we decide on the spacing between the categories? For example, should we divide US household income in the categories 0–$9,999, $10,000–$19,999, $20,000-higher, or use another way to set the categories? One suggestion is to use narrower categories if the respondent can recall the variable easily. A second suggestion is to space the categories so that we, as researchers, expect an approximately equal number of observations per category. In the example above, we may find that most households have an income of $20,000 or higher and that categories 0–$9,999 and $10,000–$19,999 are infrequently used. It is best to choose categories where equal percentages are expected such as 0–$24,999, $25,000–$44,999, $45,000–$69,999, $70,000–$109,999, $110,000–and higher. Although the range in each category differs, we can reasonably expect each category to hold about 20% of the responses if we sample US households randomly (https://en.wikipedia.org/wiki/Household_income_in_the_United_States#Household_income_over_time). Box 4.4 shows a real-life example of oddly chosen response categories (and there are a few other issues too!).

Box 4.4 Oddly Chosen Response Categories

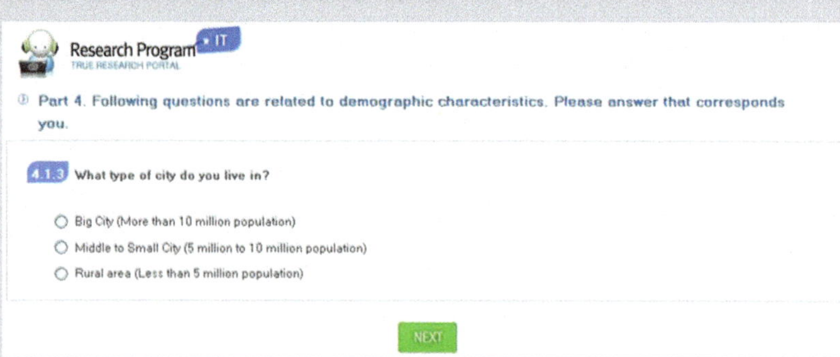

[7]Forced-choice scales can, of course, also be used for uneven response categories such as 5-point or 7-point Likert scales.

Note that the response categories also give the respondents the range of acceptable answers. Respondents generally view the middle of the scale as normal, or most common, and position themselves in relation to this (e.g., Revilla 2015). For example, Tourangeau and Smith (1996) found that the reported number of sexual partners in an open-ended question is more than twice as high when the answer category labels used in a previously asked closed-ended question on the same topic shifted to indicate higher numbers (from 0, 1, 2, 3, 4, 5 or more to 0, 1–4, 5–9, 10–49, 50–99, 100 or more).

Label the response categories: A common way of labeling response categories is to use endpoint labels only, omitting intermediary labels. For example, instead of labeling all five points of a Likert scale (e.g., "completely disagree," "disagree," "neutral," "agree," and "completely agree"), we only label the endpoints (e.g., "completely disagree" and "completely agree"). While this approach makes response category labeling easy, it also amplifies acquiescence, because the endpoints reinforce the respondents' agreement (Weijters et al. 2010). Conversely, if all the categories are labeled, this helps the respondents interpret and differentiate, making the midpoint more salient and accessible. Labeling the response categories fully also increases the scale reliability (Weng 2004), but reduces criterion validity. Drawing on Weijters et al. (2010), we generally recommend only labeling the endpoints. However, when using the items for prediction-type analyses, such as in correlation and regression analyses, it is beneficial to label all the categories.[8]

Decide whether or not to use a "don't know" option: Another important choice to make is whether or not to include a "don't know" option in the scaling (Lietz 2010). Using a "don't know" option allows the researcher to distinguish between those respondents who have a clear opinion and those who do not. Moreover, the respondents may find that answering the survey is slightly easier. While these are good reasons for including this category, the drawback is that there will then be missing observations. If many respondents choose not to answer, this will substantially reduce the number of surveys that can be used for analysis. Generally, when designing surveys, you should only include a "don't know" (or "undecided") option as an answer to questions that the respondents might genuinely not know, for example, when requiring answers to factual questions. If included, the option should appear at the end of the scale. The "don't know" option should not be included in other types of questions (such as on attitudes or preferences), as researchers are interested in the respondents' perceptions regardless of their knowledge of the subject matter.

Choose between a balanced and unbalanced scale: A **balanced scale** has an equal number of positive and negative scale categories. For example, in a 5-point Likert scale, we may have two negative categories ("completely disagree" and "disagree"), a neutral option, and two positive categories ("agree" and "completely

---

[8]Sometimes, researchers number the response options. For example, when numbering the response options of a 7-point scale you can use only positive numbers (1–7), or positive and negative numbers (−3 to +3). For recommendations, see Cabooter et al. (2016).

agree"). Besides this, the wording in a balanced scale should reflect equal distances between the scale items. This is called an **equidistant scale**, which is a requirement for analysis techniques such as factor analysis (Chap. 8). Consequently, we strongly recommend using a balanced scale instead of an **unbalanced scale**. A caveat of balanced scales is that many constructs cannot have negative values. For example, one can have some trust in a company or very little trust, but negative trust is highly unlikely. If a scale item cannot be negative, you will have to resort to an unbalanced scale in which the endpoints of the scales are unlikely to be exact opposites. Table 4.1 summarizes the key choices we have to make when designing surveys.

4.4.2.5 Design the Questionnaire

After determining the individual questions, you have to integrate these, together with other elements, to create the questionnaire. Questionnaire design involves the following elements:

- designing the starting pages of the questionnaire,
- choosing the order of the questions, and
- designing the layout and format.

Starting pages of the questionnaire: At the beginning of each questionnaire, the importance and goal are usually described to stress that the results will be treated confidentially, and to mention what they will be used for. This is usually followed by an example question (and answer), to demonstrate how the survey should be filled out. Keep this page very short when using mobile surveys.

If questions relate to a specific issue, moment, or transaction, you should indicate this clearly at the very beginning. For example, "Please provide answers to the following questions, keeping the purchase of product X in mind." If applicable, you should also point out that your survey is conducted in collaboration with a university, a recognized research institute, or a known charity, as this generally increases respondents' willingness to participate. Moreover, do not forget to provide a name and contact details for those participants who have questions, or if technical problems should arise. Consider including a photo of the research team, as this increases response rates. Lastly, you should thank the respondents for their time and describe how the questionnaire should be returned (for mail surveys).

Order of the questions: Choosing an appropriate question order is crucial, because it determines the questionnaire's logical flow and therefore contributes to high response rates. The order of questions is usually as follows:

1. *Screening questions* (typically simply referred to as *screeners*) come first. These questions determine what parts of the survey a respondent should fill out.
2. Next, ask questions relating to the study's key variables. This includes the dependent variables, followed by the independent variables.
3. Use a *funnel approach*. That is, ask questions that are more general first and then move on to details. This makes answering the questions easier as the order helps the respondents recall. Make sure that sensitive questions are put at the very end of this section.

Table 4.1 A summary of some of the key choices when designing survey items

Aspect	Recommendation
Item content	
Can all the respondents answer the question asked?	Ensure that all potential respondents can answer all items. If they cannot, ask screener questions to direct them. If the respondents cannot answer questions, they should be able to skip them.
Can the respondents construct or recall the answers?	If the answer is no, you should use other methods to obtain information (e.g., secondary data or observations). Moreover, you should ask the respondents about major aspects before zooming in on details to help them recall answers.
Do the respondents want to answer each question?	If the questions concern "sensitive" subjects, check whether they can be omitted. If not, stress the confidentiality of the answers and mention why these answers are useful for the researcher, the respondent, or society before introducing them.
Open-ended or closed-ended questions?	Keep the subsequent coding in mind. If easy coding is possible beforehand, design a set of exhaustive answer categories. Further, remember that open-ended scale items have a much lower response rate than closed-ended items.
Item wording	
Grammar and sentences	Use simple wording and grammar. Keep sentences short. Avoid negations, vague quantifiers, and double-barreled questions. Employ reverse-scaled items sparingly. Use back-translation if necessary.
Type of scale	
Number of response categories	Use visual analogue scales. If not available, use 5-point or 7-point scales.
Forced-choice or free-choice scale?	Use a free-choice scale by including a neutral category.
Design of the response categories	Ensure that the response categories are exclusive and that each category has approximately the same percentage of responses.
What scaling categories should you use (closed-ended questions only)?	Use visual analogue scales if possible, otherwise Likert scales. If the question requires this, use semantic differential scales, or rank order scales.
Labeling of response categories	Label endpoints only. When using the items for prediction-type analyses, label all categories.
Inclusion of a "Don't know" option	Include a "Don't know" option only for items that the respondent might genuinely not know. If included, place this at the end of the scale.
Balanced or unbalanced scale?	Always use a balanced scale. There should be an exact number of positive and negative wordings in the scale items. The words at the ends of the scale should be exact opposites.

4. Questions related to demographics are placed last if they are not part of the screening questions. If you ask demographic questions, always check whether they are relevant to the research goal and if they are not already known.[9] In addition, check if these demographics are likely to lead to non-response. Asking about demographics, like income, educational attainment, or health, may result in a substantial number of respondents refusing to answer. If such sensitive demographics are not necessary, omit them from the survey. Note that in certain countries asking about a respondent's demographic characteristics means you must adhere to specific laws, such as the Data Protection Act 1998 in the UK.

If your questionnaire comprises several sections (e.g., in the first section, you ask about the respondents' buying attitudes and, in the following section, about their satisfaction with the company's services), you should make the changing context clear to the respondents.

Layout and format of the survey: The layout of both mail and web-based surveys should be concise and should conserve space where possible, particularly in respect of mobile surveys. Avoid using small and colored fonts, which reduce readability. Booklets work well for mail-based surveys, since postage is cheaper if surveys fit in standard envelopes. If this is not possible, single-sided stapled paper can also work. When using web-based surveys, it is good to have a counter that tells the respondents what percentage of the questions they have already filled out. This gives them some indication of how much time they are likely to need to complete the survey. Make sure the layout is simple and compatible with mobile devices and tablets. Qualtrics and other survey tools offer mobile-friendly display options, which should always be ticked.

4.4.2.6 Pretest the Questionnaire and Execution

We have already mentioned the importance of pretesting the survey several times. Before any survey is sent out, you should *pretest* the questionnaire to enhance its clarity and to ensure the client's acceptance of the survey. Once the questionnaire is in the field, there is no way back! You can pretest questionnaires in two ways. In its simplest form, you can use a few experts (say 3–6) to read the survey, fill it out, and comment on it. Many web-based survey tools allow researchers to create a pretested version of their survey, in which there is a text box for comments behind every question. Experienced market researchers can spot most issues right away and should be employed to pretest surveys. If you aim for a very high quality survey, you should also send out a set of preliminary (but proofread) questionnaires to a small sample of 50–100 respondents. The responses (or lack thereof) usually indicate possible problems and the preliminary data can be analyzed to determine the potential results. Never skip pretesting due to time issues, since you are likely to run into problems later!

[9] The demographics of panels (see Sect. 4.4.2.2) are usually known.

Box 4.5 Dillman et al.'s (2014) Recommendations on How to Increase Response Rates

It is becoming increasingly difficult to get people to fill out surveys. This may be due to over-surveying, dishonest firms that disguise sales as research, and a lack of time. Dillman et al. (2014) discuss four steps to increase response rates:

1. Send out a pre-notice letter indicating the importance of the study and announcing that a survey will be sent out shortly.
2. Send out the survey with a sponsor letter, again indicating the importance of the study.
3. Follow up after 3–4 weeks with a thank you note (for those who responded) and the same survey, plus a reminder (for those who did not respond).
4. Call or email those who have still not responded and send out a thank you note to those who replied during the second round.

Motivating potential respondents to participate is an increasingly important aspect of survey research. In addition to Dillman et al.'s (2014) recommendations on how to increase response rates (Box 4.5), incentives are used. A simple example of such an incentive is to provide potential respondents with a cash reward. In the US, one-dollar bills are often used for this purpose. Respondents who participate in (online) research panels often receive points that can be exchanged for products and services. For example, Research Now, a market research company, gives its Canadian panel members AirMiles that can be exchanged for, amongst others, free flights. A special type of incentive is to indicate that some money will be donated to a charity for every returned survey. ESOMAR, the world organization for market and social research (see Chap. 10), suggests that incentives for interviews or surveys should "be kept to a minimum level proportionate to the amount of their time involved, and should not be more than the normal hourly fee charged by that person for their professional consultancy or advice."

Another incentive is to give the participants a chance to win a product or service. For example, you could randomly give a number of participants gifts. The participants then need to disclose their name and address in exchange for a chance to win a gift so that they can be reached. While this is not part of the research itself, some respondents may feel uncomfortable providing their contact details, which could potentially reduce the response rate.

Finally, reporting the findings to the participants is an incentive that may help participation (particularly in professional settings). This can be done by providing a general report of the study and its findings, or by providing a customized report detailing the participant's responses and comparing them with all the other responses (Winkler et al. 2015). Obviously, anonymity needs to be assured so that the participants cannot compare their answers to those of other individual responses.

4.5 Basic Qualitative Research

Qualitative research is mostly used to gain an understanding of *why* certain things happen. It can be used in an exploratory context by defining problems in more detail, or by developing hypotheses to be tested in subsequent research. Qualitative research also allows researchers to learn about consumers' perspectives and vocabulary, especially if they are not familiar with the context (e.g., the industry). As such, qualitative research offers importance guidance when little is known about consumers' attitudes and perceptions or the market (Barnham 2015).

As discussed in Chap. 3, qualitative research leads to the collection of qualitative data. One can collect *qualitative data* by explicitly informing the participants that you are doing research (directly observed qualitative data), or you can simple observe the participants' behavior without them being explicitly aware of the research goals (indirectly observed qualitative data). There are ethical issues associated with conducting research if the participants are not aware of the research purpose. Always check the regulations regarding what is allowed in your context and what not. It is, in any case, good practice to brief the participants on their role and the goal of the research once the data have been collected.

The two key forms of directly observed qualitative data are in-depth interviews and focus groups. Together, they comprise most of the conducted qualitative market research. First, we will discuss in-depth interviews, which are—as the terms suggests—interviews conducted with one participant at a time, allowing for high levels of personal interaction between the interviewer and respondent. Next, we will discuss projective techniques, a frequently used type of testing procedure in in-depth interviews. Lastly, we will introduce focus group discussions, which are conducted with multiple participants.

4.5.1 In-Depth Interviews

In-depth interviews are qualitative conversations with participants on a specific topic. These participants are often consumers, but, in a market research study, they may also be the decision-makers, who are interviewed to gain an understanding of their clients' needs. They may also be government or company representatives. The structure levels of interviews vary. In their simplest form, interviews are unstructured and the participants talk about a topic in general. This works well if you want to obtain insight into a topic, or as an initial step in a research process. Interviews can also be fully structured, meaning all the questions and possible answer categories are decided beforehand. This way allows you to collect quantitative data. However, most in-depth interviews for gathering qualitative data are semi-structured and contain a series of questions that need to be addressed, but have no specific format regarding what the answers should look like. The person interviewed can make additional remarks, or discuss somewhat related issues, but is not allowed to wander off too far. In these types of interviews, the interviewer often asks questions like "that's interesting, could you explain?," or "how come. . .?" to gain more insight into the issue. In highly structured interviews, the interviewer has a fixed set of questions and often a fixed amount of time for each person's response. The goal of structured interviews is to maximize the comparability of the answers. Consequently, the set-up of the questions and the structure of the answers need to be similar.

In-depth interviews are unique in that they allow probing on a one-to-one basis, thus fostering an interaction between the interviewer and interviewee. In-depth interviews also work well when those being interviewed have very little time and when they do not want the information to be shared with the other study participants. This is, for example, probably the case when you discuss marketing strategy decisions with CEOs. The drawbacks of in-depth interviews include the amount of time the researcher needs to spend on the interview itself and on traveling (if the interview is conducted face-to-face and not via the telephone), as well as on transcribing the interview.

When conducting in-depth interviews, a set format is usually followed. First, the interview details are discussed, such as confidentiality issues, the interview topic, the structure, and the duration. Moreover, the interviewer should disclose whether the interview is being recorded and inform the interviewee that there is no right or wrong answer, just opinions on the subject. The interviewer should also try to be open and maintain eye contact with the interviewee. Interviewers can end an interview by informing their respondents that they have reached the last question and thanking them for their time.

Interviews are often used to investigate means-end issues, in which researchers try to understand what ends consumers aim to satisfy and which means (consumption) they use to do so. A **means-end approach** involves first determining a product's attributes. These are the functional product features, such as the speed a car can reach or its acceleration. Subsequently, researchers look at the functional consequences that follow from the product benefits. This could be driving fast. The psychosocial consequences, or personal benefits, are derived from the functional benefits and, in this example, could include an enhanced status, or being regarded as successful. Finally, the psychosocial benefits are linked to people's personal values or life goals, such as a desire for success or acceptance. Analyzing and identifying the relationships between these steps is called **laddering**.

4.5.2 Projective Techniques

Projective techniques describe a special type of testing procedure, usually used as part of in-depth interviews. These techniques provide the participants with a stimulus and then gauge their responses. Although participants in projective techniques know that they are participating in a market research study, they may not be aware of the research's specific purpose. The stimuli that the projective techniques provide are ambiguous and require a response from the participants. Sentence completion is a key form of projective techniques, for example:

An iPhone user is someone who:	....
The Apple brand makes me think of:	....
iPhones are most liked by:	....

In this example, the respondents are asked to express their feelings, ideas, and opinions in a free format.

Projective techniques' advantage is that they allow for responses when people are unlikely to respond if they were to know the study's exact purpose. Thus, projective techniques can overcome self-censoring and allow expression and fantasy. In addition, they can change a participant's perspective. Think of the previous example. If the participants are iPhone users, the sentence completion example asks them how they think other people regard them, not what they think of the iPhone. A drawback is that projective techniques require the responses to be interpreted and coded, which can be difficult.

4.5.3 Focus Groups

Focus groups are interviews conducted with a number of respondents at the same time and led by a moderator. This moderator leads the interview, structures it, and often plays a central role in transcribing the interview later. Focus groups are usually semi or highly structured. The group usually comprises between 4 and 6 people to allow for interaction between the participants and to ensure that all the participants have a say. The duration of a focus group interview varies, but is often between 30 and 90 min for company employee focus groups and between 60 and 120 min for consumers. When focus groups are held with company employees, moderators usually travel to the company and conduct their focus group in a room. When consumers are involved, moderators often travel to a market research company, or hotel, where the focus group meets in a conference room. Market research companies often have specially equipped conference rooms with, for example, one-way mirrors, built-in microphones, and video recording devices.

How are focus groups structured? They usually start with the moderator introducing the topic and discussing the background. Everyone in the group is introduced to all the others to establish rapport. Subsequently, the moderator encourages the members of the focus group to speak to one another, instead of asking the moderator for confirmation. Once the focus group members start discussing topics with one another, the moderator tries to stay in the background, merely ensuring that the discussions stay on-topic. Afterwards, the participants are briefed and the discussions are transcribed for further analysis.

Focus groups have distinct advantages: they are relatively cheap compared to in-depth interviews, work well with issues that are socially important, or which require spontaneity. They are also useful for developing new ideas. On the downside, focus groups do not offer the same opportunity for probing as interviews do, and also run a greater risk of going off-topic. Moreover, a few focus group members may dominate the discussion and, especially in larger focus groups, "voting" behavior may occur, hindering real discussions and the development of new ideas. Table 4.2 summarizes the key differences between focus groups and in-depth interviews.

Table 4.2 Comparing focus groups and in-depth interviews

	Focus groups	In-depth interviews
Group interactions	Group interaction, which may stimulate the respondents to produce new thoughts.	There is no group interaction. The interviewer is responsible for stimulating the respondents to produce new ideas.
Group/peer pressure	Group pressure and stimulation may clarify and challenge thinking.	In the absence of group pressure, the respondents' thinking is not challenged.
	Peer pressure and role playing.	With just one respondent, role playing is minimized and there is no peer pressure.
Respondent competition	Respondents compete with one another for time to talk. There is less time to obtain in-depth details from each participant.	Individuals are alone with the interviewer and can express their thoughts in a non-competitive environment. There is more time to obtain detailed information.
Peer influence	Responses in a group may be biased by other group members' opinions.	With one respondent, there is no potential of other respondents influencing this person.
Subject sensitivity	If the subject is sensitive, respondents may be hesitant to talk freely in the presence of other people.	If the subject is sensitive, respondents may be more likely to talk.
Stimuli	The volume of stimulus materials that can be used is somewhat limited.	A fairly large amount of stimulus material can be used.
Interviewer schedule	It may be difficult to assemble 8 or 10 respondents if they are a difficult type to recruit (such as busy executives).	Individual interviews are easier to schedule.

4.6 Collecting Primary Data Through Experimental Research

In Chap. 2, we discussed causal research and briefly introduced experiments as a means of conducting research. The goal of experiments is to avoid unintended influences through the use of randomization. Experiments are typically conducted by manipulating one variable, or a few, at a time. For example, we can change the price of a product, the type of product, or the package size to determine whether these changes affect important outcomes such as attitudes, satisfaction, or intentions. Simple field observations are often unable to establish these relationships, as inferring causality is problematic. Imagine a company that wants to introduce a new type of soft drink aimed at health-conscious consumers. If the product were to fail, the managers would probably conclude that the consumers did not like the product. However, many (usually unobserved) variables, such as competitors' price cuts, changing health concerns, and a lack of availability, can also influence new products' success.

4.6.1 Principles of Experimental Research

Experiments deliberately impose one or more **treatments** and then observe the outcome of a specific treatment (Mitchell and Jolley 2013). This way, experiments attempt to isolate how one change affects an outcome. The outcome(s) is (are) the dependent variable(s) and the independent variable(s) (also referred to as factors) are used to explain the outcomes. To examine the influence of the independent variable(s) on the dependent variable(s), the participants are subjected to treatments. These are supposed to manipulate the participants by putting them in different situations. A simple form of treatment could be an advertisement with and without humor. In this case, the humor is the independent variable, which can take two levels (i.e., with or without humor). If we manipulate, for example, the price between low, medium, and high, we have three levels. When selecting independent variables, we normally include those that marketers care about and which are related to the marketing and design of the products and services. To assure that the participants do indeed perceive differences, **manipulation checks** are conducted. For example, if two different messages with and without humor are used as stimuli, we could ask the respondents which of the two is more humorous. Such manipulation checks help establish an experiment's validity.

Care should be taken not to include too many of these variables in order to keep the experiment manageable. An experiment that includes four independent variables, each of which has three levels and includes every possible combination (called a *full factorial design*) requires $4^3 = 64$ treatments. Large numbers of levels (five or more) will dramatically increase the complexity and cost of the research. Finally, *extraneous variables*, such as the age or income of the experiment participant, are not changed as part of the experiment. However, it might be important to control for their influence when setting up the experimental design.

Experiments can run in either a lab or field environment. *Lab experiments* are performed in controlled environments (usually in a company or academic lab), thereby allowing for isolating the effects of one or more variables on a certain outcome. Extraneous variables' influence can also be controlled for. However, field experiments' often lack **internal validity**, which is the extent to which we can make causal claims based on the study results. Lab experiments often take place in highly stylized experimental settings, which typically ignore real-world business conditions, they therefore usually lack **external validity**, which is the extent to which the study results can be generalized to similar settings. *Field experiments*, on the other hand, are performed in natural environments (e.g., in a supermarket or in-home) and, and such, generally exhibit high degrees of external validity. However, since controlling for extraneous variables' impact is difficult in natural environments, field experiments usually lack internal validity (Gneezy 2017).

4.6.2 Experimental Designs

Experimental design refers to an experiment's structure. There are various types of experimental designs. To clearly separate the different experimental designs, researchers have developed the following notation:

O:	A formal observation or measurement of the dependent variable. Subscripts below an observation O such as $_1$ or $_2$, indicate measurements at different points in time.
X:	The test participants' exposure to an experimental treatment.
R:	The random assignment of participants. Randomization ensures control over extraneous variables and increases the experiment's reliability and validity.

In the following, we will discuss the most prominent experimental designs:

- one-shot case study,
- before-after design,
- before-after design with a control group, and
- solomon four-group design.

The simplest form of experiment is the *one-shot case study*. This type of experiment is structured as follows:[10]

X	O_1

This means we have one treatment (indicated by X), such as a new advertising campaign. After the treatment, we await reactions to it and then measure the outcome of the manipulation (indicated by O_1), such as the participants' willingness

[10]If one symbol precedes another, it means that the first symbol precedes the next one in time.

to purchase the advertised product. This type of experimental set-up is common, but does not tell us if the effect is causal. One reason for this is that we did not measure anything before the treatment and therefore cannot assess what the relationship between the treatment and outcome is. The participants' willingness to purchase the product was perhaps higher before they were shown the advertisement, but since we did not measure their willingness to purchase before the treatment, this cannot be ruled out. Consequently, this design does not allow us to establish causality.

The simplest type of experiment that allows us to make causal inferences—within certain limits—is the *before-after design* used for one group. The notation for this design is:

O_1	X	O_2

We have one measurement before (O_1) and one after a treatment (O_2). Thus, this type of design can be used to determine whether an advertisement has a positive, negative, or no effect on the participants' willingness to purchase a product.

A problem with this type of design is that we do not have a standard of comparison with which to contrast the change between O_1 and O_2. While the advertisement may have increased the participants' willingness to purchase, this might have been even higher if they had not seen the advertisement. The reason for this is that the before-after-design does not control for influences occurring between the two measurements. For example, negative publicity after the initial measurement could influence the subsequent measurement. These issues make the "real" effect of the advertisement hard to assess.

If we want to have a much better chance of identifying the "real" effect of a treatment, we need a more complex setup, called the *before-after design with a control group*. In this design, we add a control group who is not subjected to the treatment X. The notation of this type of experiment is:

Experimental group (R)	O_1	X	O_2
Control group (R)	O_3		O_4

The effect attributed to the experiment is the difference between O_1 and O_2 minus the difference between O_3 and O_4. For example, if the participants' willingness to purchase increases much stronger in the experimental group (i.e., O_2 is much higher than O_1) than in the control group (i.e., O_4 is slightly higher than or equal to O_3), the advertisement has had an impact on the participants.

The random assignment of the participants to the experimental and the control groups (indicated by R), is an important element of this experimental design. This means that, for any given treatment, every participant has an equal probability of being chosen for one of the two groups. This ensures that participants with different characteristics are spread randomly (and, it is hoped, equally) between the treatment(s), which will neutralize self-selection. Self-selection occurs when participants can select themselves into either the experimental or the control group. For example, if participants who like sweets participate in a cookie tasting test, they will certainly try to give the treatment (cookie) a try! See Mooi and Gilliland (2013) for an example of self-selection and an analysis method.

However, the before-after experiment with a control group does have limitations. The initial measurement O_1 may alert the participants that they are being studied, which may bias the post measurement O_2. This effect is also referred to as the *before measurement effect*, or the *testing effect* (Campbell and Stanley 1966). Likewise, the initial measurement O_1 may incite the participants to drop out of the experiment, leading to no recorded response for O_2. If there is a systematic reason for the participants to drop out, this will threaten the experiment's validity.

The *Solomon four-group design* is an experimental design accounting for before measurement effects and is structured as follows (Campbell and Stanley 1966):

Experimental group 1 (R)	O_1	X	O_2
Control group 1 (R)	O_3		O_4
Experimental group 2 (R)		X	O_5
Control group 2 (R)			O_6

The design is much more complex, because we need to measure the effects six times and administer two treatments. This method provides an opportunity to control for the before measurement effect of O_1 on O_2. The design also provides several measures of the treatment's effect (i.e., (O_2–O_4), (O_2–O_1)–(O_4–O_3), and (O_6–O_5)). If these measures agree, the inferences about the treatment's effect are much stronger.

In Chap. 6, we discuss how to analyze experimental data using ANOVA and various other tests.

4.7 Review Questions

1. What is the difference between primary and secondary data? Can primary data become secondary data?
2. Please search for and download two examples of secondary data sources found on the Internet. Discuss two potential market research questions that each dataset can answer.
3. Imagine you are asked to understand what consumer characteristics make these consumers likely to buy a BMW i3 (http://www.bmw.com). How would you collect the data? Would you start with secondary data, or would you start collecting primary data directly? Do you think it is appropriate to collect qualitative data? If so, at what stage of the process should you do so?
4. What are the different reasons for choosing interviews rather than focus groups? What choice would you make if you want to understand CEOs' perceptions of the economy, and what would seem appropriate when you want to understand how consumers feel about a newly introduced TV program?

	Strongly disagree	Somewhat disagree	Somewhat agree	Agree	Completely agree
I am satisfied with the performance and reliability of my iPhone.	□	□	□	□	□
	Strongly disagree	Somewhat disagree	Neutral	Somewhat agree	Completely agree
I am satisfied with the after-sales service of my iPhone.	□	□	□	□	□
	Not at all important	Not important	Neutral	Important	Very important
Which of the following iPhone features do you find most important?					
Camera	□	□	□	□	□
Music player	□	□	□	□	□
App store	□	□	□	□	□
Web browser	□	□	□	□	□
Mail	□	□	□	□	□

5. Consider the following examples of survey items relating to how satisfied iPhone users are with their performance, reliability, and after-service. Please assess their adequacy and make suggestions on how to revise the items, if necessary.
6. Make recommendations regarding the following aspects of scale design: the number of response categories, the design and labeling of response categories, and the inclusion of a "don't know" option.
7. Describe the Solomon four-group design and explain which of the simpler experimental design problems it controls for.
8. If you were to set up an experiment to ascertain what type of product package (new or existing) customers prefer, what type of experiment would you choose? Please discuss.

4.8 Further Readings

Barnham, C. (2015). Quantitative and qualitative research: Perceptual foundations. *International Journal of Market Research*, *57*(6), 837–854.

This article reflects on the classic distinction between quantitative and qualitative research and the underlying assumptions.

Campbell, D. T., & Stanley, J. C. (1966). *Experimental and quasi-experimental designs for research*. Chicago: Wadsworth Publishing.

Cook, T. D., & Campbell, D. T. (1979). *Quasi-experimentation: Design and analysis issues for field settings*. Chicago: Wadsworth Publishing.

These are the two great books on experimental research.

Dillman, D. A., Smyth, J. D., & Christian, L. M. (2014). *Internet, phone, mail, and mixed-mode surveys: The tailored design method* (4th ed.). Hoboken: Wiley.

This book gives an excellent overview of how to create questionnaires and how to execute them. Mixed-mode surveys and web surveys are specifically addressed.

FocusGroupTips.com.

This website provides a thorough explanation of how to set-up focus groups from planning to reporting the results.

Lietz, P. (2010). Current state of the art in questionnaire design: A review of the literature. *International Journal of Market Research, 52*(2), 249–272.

Reviews survey design choices from an academic perspective.

Mystery Shopping Providers Association (www.mysteryshop.org).

This website discusses the mystery shopping industry in Asia, Europe, North America, and South America.

Veludo-de-Oliveira, T. M, Ikeda, A. A., & Campomar, M. C. (2006). Laddering in the practice of marketing research: Barriers and solutions. *Qualitative Market Research: An International Journal, 9*(3), 297–306.

This article provides an overview of laddering, the various forms of laddering, and biases that may result and how these should be overcome.

References

Barnham, C. (2015). Quantitative and qualitative research: Perceptual foundations. *International Journal of Market Research, 57*(6), 837–854.

Baumgartner, H., & Steenkamp, J. B. E. M. (2001). Response styles in marketing research: A cross-national investigation. *Journal of Marketing Research, 38*(2), 143–156.

Brace, I. (2004). *Questionnaire design. How to plan, structure and write survey material for effective market research*. London: Kogan Page.

Brislin, R. W. (1970). Back-translation for cross-cultural research. *Journal of Cross-Cultural Psychology, 1*(3), 185–216.

Bronner, F., & Ton, K. (2007). The live or digital interviewer. A comparison between CASI, CAPI and CATI with respect to differences in response behaviour. *International Journal of Market Research, 49*(2), 167–190.

Cabooter, E., Weijters, B., Geuens, M., & Vermeir, E. (2016). Scale format effects on response option interpretation and use. *Journal of Business Research, 69*(7), 2574–2584.

Casteleyn, J., André, M., & Kris, R. (2009). How to use facebook in your market research. *International Journal of Market Research, 51*(4), 439–447.

DeMonaco, H. J., Ayfer, A., & Hippel, E. V. (2005). The major role of clinicians in the discovery of off-label drug therapies. *Pharmacotherapy, 26*(3), 323–332.

Deutskens, E., de Jong, A., de Ruyter, K., & Martin, W. (2006). Comparing the generalizability of online and mail surveys in cross-national service quality research. *Marketing Letters, 17*(2), 119–136.

Dillman, D. A., Smyth, J. D., & Christian, L. M. (2014). *Internet, phone, mail, and mixed-mode surveys: The tailored design method* (4th ed.). Hoboken: Wiley.

Fink, A. (2003). *How to ask survey questions*. Thousand Oaks: Sage.

Foddy, W. (1993). *Constructing questions for interviews and questionnaires. Theory and practice in social science research.* Cambridge: Cambridge University Press.

Funke, F. (2010). *Internet-based measurement with visual analogue scales. An experimental investigation.* Dissertation, Eberhard Karls Universität Tübingen. Available online at: https://publikationen.uni-tuebingen.de/xmlui/handle/10900/49480

Holbrook, A., Cho, Y. I. K., & Johnson, T. (2006). The impact of question and respondent characteristics on comprehension and mapping difficulties. *Public Opinion Quarterly, 70*(4), 565–595.

Houston, M. B. (2002). Assessing the validity of secondary data proxies for marketing constructs. *Journal of Business Research, 57*(2), 154–161.

Kennedy, C., & Everett, S. E. (2011). Use of cognitive shortcuts in landline and cell phone surveys. *Public Opinion Quarterly, 75*(2), 336–348.

Kurylko, D. T. (2005). Moonraker project seeks marketing savvy for VW. *Automotive News Europe, 10*(17), 22.

Lietz, P. (2010). Research into questionnaire design. A summary of the literature. *International Journal of Market Research, 52*(2), 249–272.

Liu, M., Lee, S., & Conrad, F. G. (2016). Comparing extreme response styles between agree-disagree and item-specific scales. *Public Opinion Quarterly, 79*(4), 952–975.

Lynn, P., & Kaminska, O. (2013). The impact of mobile phones on survey measurement error. *Public Opinion Quarterly, 77*(2), 586–605.

Mooi, E., & Gilliland, D. I. (2013). How contracts and enforcement explain transaction outcomes. *International Journal of Research in Marketing, 30*(4), 395–405.

Mitchell, M. L., & Jolley, J. M. (2013). *Research design explained* (8th ed.). Belmont: Wadsworth.

Nowlis, S. M., Kahn, B. E., & Dhar, R. (2002). Coping with ambivalence: The effect of removing a neutral option on consumer attitude and preference judgments. *Journal of Consumer Research, 29*(3), 319–334.

Oppenheim, A. N. (1992). *Questionnaire design, interviewing and attitude measurement.* London: Pinter.

Peng, L., & Finn, A. (2015). Assessing response format effects on the scaling of marketing stimuli. *International Journal of Market Research, 58*(4), 595–619.

Peterson, R. A. (1997). A quantitative analysis of rating-scale response variability. *Marketing Letters, 8*(1), 9–21.

Raithel, S., Sarstedt, M., Scharf, S., & Schwaiger, M. (2012). On the value relevance of customer satisfaction. Multiple drivers in multiple markets. *Journal of the Academy of Marketing Science, 40*(4), 509–525.

Rammstedt, B., & Krebs, D. (2007). Does response scale format affect the answering of personality scales? *European Journal of Psychological Assessment, 23*(1), 32–38.

Reips, U.-D., & Funke, F. (2008). Interval-level measurement with visual analogue scales in Internet-based research: VAS generator. *Behavior Research Methods, 40*(3), 699–704.

Revilla, M. (2015). Effect of using different labels for the scales in a web survey. *International Journal of Market Research, 57*(2), 225–238.

Rindfleisch, A., & Heide, J. B. (1997). Transaction cost analysis: Past, present, and future applications. *Journal of Marketing, 61*(4), 30–54.

Schilling, M. A. (2009). Understanding the alliance data. *Strategic Management Journal, 30*(3), 233–260.

Stern, M. J., Bilgen, I., & Dillman, D. A. (2014). The state of survey methodology challenges, dilemmas, and new frontiers in the era of the tailored design. *Field Methods, 26*(3), 284–301.

Stieglitz, S., Dang-Xuan, L., Bruns, A., & Neuberger, C. (2014). Social media analytics, An interdisciplinary approach and its implications for information systems. *Business & Information Systems Engineering, 6*(2), 89–96.

Swain, S. D., Weathers, D., & Niedrich, R. W. (2008). Assessing three sources of misresponse to reversed Likert items. *Journal of Marketing Research, 45*(1), 116–131.

Tourangeau, R., & Smith, T. W. (1996). Asking sensitive questions: The impact of data collection mode, question format, and question context. *Public Opinion Quarterly, 60*(2), 275–304.

Vesta Research. (2016). Rules of thumb for survey length. *Vesta Research Blog*. Available at: http://www.verstaresearch.com/blog/rules-of-thumb-for-survey-length/

Vicente, P., & Reis, E. (2010). Marketing research with telephone surveys: Is it time to change? *Journal of Global Marketing, 23*(4), 321−332.

Vincente, P., Reis, E., & Santos, M. (2008). Using mobile phones for survey research. *International Journal of Market Research, 51*(5), 613–633.

Weijters, B., & Baumgartner, H. (2012). Misresponse to reversed and negated items in surveys: A review. *Journal of Marketing Research, 49*(5), 737–747.

Weijters, B., Cabooter, E., & Schillewaert, N. (2010). The effect of rating scale format on response styles: The number of response categories and response category labels. *International Journal of Research in Marketing, 27*(3), 236–247.

Weijters, B., Baumgartner, H., & Schillewaert, N. (2013). Reversed item bias: An integrative model. *Psychological Methods, 18*(3), 320–334.

Weng, L.-J. (2004). Impact of the number of response categories and anchor labels on coefficient alpha and test-retest reliability. *Educational and Psychological Measurement, 64*(6), 956–972.

Winkler, T. J., Sarstedt, M., Keil, M., & Rost, P. (2015). Selfsurvey.org: A platform for prediction-based benchmarking and feedback-enabled survey research. In *Proceedings of the European conference on information systems*, Paper 204, Münster, Germany.

Descriptive Statistics 5

Keywords
Acquiescence • Aggregation • Bar chart • Bivariate statistics • Box plot • Codebook • Construct score • Correlation • Covariance • Crosstabs • Data entry errors • Dummy variables • Extreme response styles • Frequency table • Histogram • Inconsistent answers • Index • Interquartile range • Interviewer fraud • Item non-response • Line chart • Listwise deletion • Little's MCAR test • Log transformation • Mean • Measures of centrality • Measures of dispersion • Median • Middle response styles • Missing (completely) at random • Missing data • Multiple imputation • Non-random missing • Outliers • Pie chart • Range • Range standardization • Reverse-scaled items • Scale transformation • Scatter plot • Skewed data • Stata • Standard deviation • Standardizing variables • Straight-lining • Survey non-response • Suspicious response patterns • Transforming data • Univariate statistics • Variable respecification • Variance • Workflow • *z*-standardization

Learning Objectives
After reading this chapter, you should understand:

- The workflow involved in a market research study.
- Univariate and bivariate descriptive graphs and statistics.
- How to deal with missing values.
- How to transform data (*z*-transformation, log transformation, creating dummies, aggregating variables).
- How to identify and deal with outliers.
- What a codebook is.
- The basics of using Stata.

E. Mooi et al., *Market Research*, Springer Texts in Business and Economics,
DOI 10.1007/978-981-10-5218-7_5

5.1 The Workflow of Data

Market research projects involving data become more efficient and effective if they have a proper **workflow of data**, which is a strategy to keep track of the entering, cleaning, describing, and transforming of data. These data may have been collected through surveys or may be secondary data (Chap. 3). Entering, cleaning, and analyzing bits of data haphazardly is not a good strategy, since it increases the likelihood of making mistakes and makes it hard to replicate results. Moreover, without a good data workflow, it becomes hard to document the research process and to cooperate on projects. For example, how can you outsource the data analysis if you cannot indicate what the data are about or what specific values mean? Finally, a lack of a good workflow increases the risk of duplicating work or even losing data. In Fig. 5.1, we show the steps required to create and describe a dataset after the data have been collected. We subsequently discuss each step in greater detail.

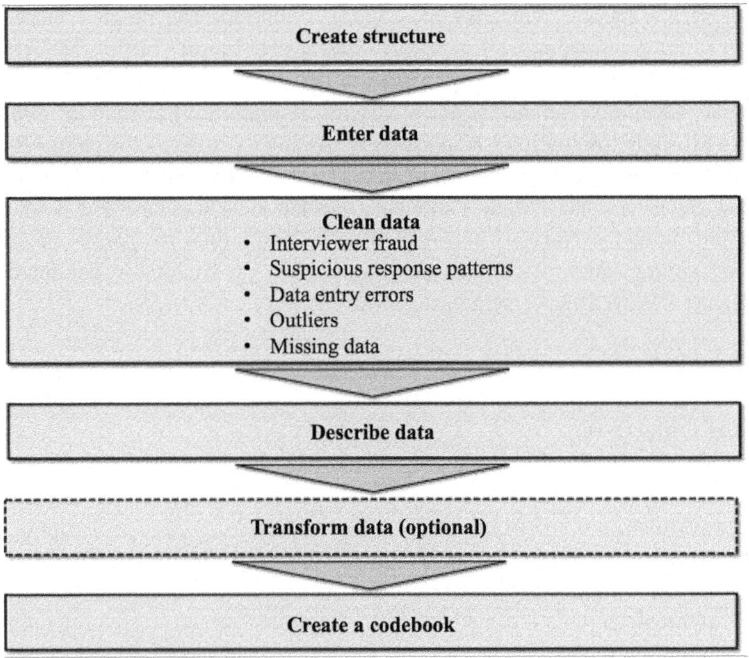

Fig. 5.1 The workflow of data

5.2 Create Structure

The basic idea of setting up a good workflow is that good planning saves the researcher time and allows other researchers to do their share of the analysis and/or replicate the research. After the data collection phase, the first step is to save the available data. We recommend keeping track of the dataset by providing data and data-related files in separate directories by means of Windows Explorer or macOS Finder. This directory should have subdirectories for at least: (1) the data files, (2) commands, (3) a temporary directory, and (4) related files; that is, a directory with files that are directly related to a project, such as the survey used to collect the data.[1]

In Table 5.1, we show an example of a directory structure. Within the main directory, there are four subdirectories, each with distinct files. Notice that in the **Data files** subdirectory, we have the original dataset, two modified datasets (one without missing data and one which includes several transformations of the data), as well as a zip file that contains the original dataset. If the data file is contained in a zip or other archive file, it is stored and unlikely to be modified, but can be easily opened if the working file is accidentally overwritten or deleted. In the **Data files** subdirectories, we distinguish between two files with the suffix **rev1** and **rev2**. We use **rev** (abbreviation of revision), but you however, can choose another file name

Table 5.1 Example of a directory structure for saving market-research-related files

Directory name	Subdirectory name	Example file names
Oddjob	Data files	oddjob.dta
		oddjob.zip
		oddjob rev1.dta
		oddjob rev2.dta
	Command files	Missing data analysis.do
		Descriptives.do
		Factor analysis.do
		Regression analysis.do
	Temporary files	Missing data analysis rev1.smcl
		Descriptives rev1.smcl
		Factor analysis rev1.smcl
		Regression analysis rev1.smcl
	Related files	Codebook.docx
		Survey.pdf
		Initial findings–presentation to client.pptx
		Findings–presentation to client.pptx
		Recommendations rev1.docx
		Recommendations rev2.docx

[1]Alternatively, you could also choose one of the many control system versions, including Subversion, Git, and Mecurial, which enable simple branching and project management. These systems work well with version control in centralized and in distributed environments.

as long as it clearly indicates the revision on which you are working. In the **Command files** subdirectory we store all commands that were used to manage our data. These commands may relate to different project phases, including a missing data analysis, descriptives, factor analysis, and other methods used over the course of the project. As the name indicates, **Temporary files** serve as intermediary files that are kept until the final data or command files are established, after which they are removed. Finally, in the **Related Files** subdirectory, we have a codebook (more on this later), the survey, two presentations, and two documents containing recommendations.

Another aspect of creating a structure is setting up the variables for your study properly. This involves making decisions on the following elements:

- variable names,
- variablc labcls,
- data type, and
- coding of variables.

The variable names should be clear and short so that they can be read in the dialog boxes. For example, if you have three questions on product satisfaction, three on loyalty, and several descriptors (age and gender), you could code these variables as *satisfaction1*, *satisfaction2*, *satisfaction3*, *loyalty1*, *loyalty2*, *loyalty3*, *age*, and *gender*.

In Stata, and most other statistical software programs, you can include *variable labels* that describe what each variable denotes. The description generally includes the original question if the data were collected by means of surveys. Another point to consider is *variable coding*. Coding means assigning values to a variable. When collecting quantitative data, the task is relatively easy; we use values that correspond with the answers for Likert and semantic differential scales (see Chap. 4). For example, when using a 7-point Likert scale, responses can be coded as 1–7 or as 0–6 (with 0 being the most negative and 6 being the most positive response). Open-ended questions (qualitative data) require more effort, usually involving a three-step process. First, we collect all the responses. In the second step, we group these responses. Determining the number of groups and the group to which a response belongs is the major challenge in this step. Two or three market researchers usually code the responses independently to prevent the process from becoming too subjective and thereafter discuss the differences that may arise. The third step is providing a value for each group. Stata can perform such analyses with the help of the additional software package *Wordstat* (https://provalisresearch.com/products/content-analysis-software/wordstat-for-stata/). Please see Krippendorff (2012) for more details about coding qualitative variables.

Once a system has been set up to keep track of your progress, you need to consider safeguarding your files. Large companies usually have systems for creating backups (extra copies as a safeguard). If you are working alone or for a small company, you are probably responsible for this. You should save your most recent and second most recent version of your file on a separate drive and have multiple copies of your entire drive! Always keep at least two copies and never keep both backups in the same place, because you could still lose all your work through theft, fire, or an accident! You can use cloud storage services, such as Dropbox, Google Drive, or Microsoft's OneDrive for small projects to prevent loss. Always read the terms of the cloud storage services carefully to determine whether your data's privacy is guaranteed.

5.3 Enter Data

How do we enter survey or experimental data into a dataset? Specialized software is often used for large datasets, or datasets created by professional firms. For example, Epidata (http://www.epidata.dk, freely downloadable) is frequently used to enter data from paper-based surveys, Entryware's mobile survey (http://www.techneos.com) to enter data from personal intercepts or face-to-face interviewing, and Voxco's Interviewer CATI for telephone interviewing. Stata has no dedicated data entry platform. It does, however, facilitate the use of StatTransfer (http://www.stattransfer.com), a software program designed to simplify the transfer of statistical data between many software packages, including Stata, SPSS, Excel, and SAS.

Such software may not be available for smaller projects, in which case data should be entered directly into Stata. A significant drawback of direct data entry is the risk of typing errors, for which Stata cannot check. Professional software, such as Epidata, can directly check if values are admissible. For example, if a survey question has only two answer categories, such as gender (coded 0/1), Epidata (and other packages) can directly check if the value entered is 0 or 1, and not any other value. The software also allows for using multiple typists when very large amounts of data need to be entered simultaneously (note that Epidata can export directly to Stata).

5.4 Clean Data

Cleaning data is the next step in the workflow. It requires checking for:

- interviewer fraud,
- suspicious response patterns,
- data entry errors,
- outliers, and
- missing data.

These issues require researchers to make decisions very carefully. In the following, we discuss each issue in greater detail.

5.4.1 Interviewer Fraud

Interviewer fraud is a difficult and serious issue. It ranges from interviewers "helping" respondents provide answers to entire surveys being falsified. Interviewer fraud often leads to incorrect results. Fortunately, we can avoid and detect interviewer fraud in various ways. First, never base interviewers' compensation on the number of completed responses they submit. Second, check and control for discrepancies in respondent selection and responses. If multiple interviewers were used, each of whom collected a reasonably large number of responses ($n > 100$), a selection of the respondents should be similar. This means that the average responses obtained should also be similar. In Chap. 6 we will discuss techniques to test this. Third, if possible, contact a random number of respondents afterwards for their feedback on the survey. If a substantial number of people claim they were not interviewed, interviewer fraud is likely. Furthermore, if people were previously interviewed on a similar subject, the factual variables collected, such as their gender, should not change (or no more than a trivial percentage), while variables such as a respondent's age and highest education level should only move up. We can check this by means of descriptive statistics. If substantial interviewer fraud is suspected, the data should be discarded. You should check for interviewer fraud during the data collection process to safeguard the quality of data collection and minimize the risk of having to discard the data in the end.

5.4.2 Suspicious Response Patterns

Before analyzing data, we need to identify **suspicious response patterns**. There are two types of response patterns we need to look for:

- straight-lining, and
- inconsistent answers.

Straight-lining occurs when a respondent marks the same response in almost all the items. For example, if a 7-point scale is used to obtain answers and the response pattern is 4 (the middle response), or if the respondent selects only 1s, or only 7s in all the items. A common way of identifying straight-lining is by including one or more **reverse-scaled items** in a survey (see Chap. 4). Reverse-scaled means that the way the question, statement (when using a Likert scale), or word pair (when using a semantic differential scale) is reversed compared to the other items in the set. Box 5.1 shows an example of a four-item scale for measuring consumers' attitude toward the brand (e.g., Sarstedt et al. 2016) with one reverse-scaled item printed in bold. By evaluating the response patterns, we can differentiate between those respondents who are not consistent for the sake of consistency and those who are merely mindlessly consistent. Note, however, that this only applies if respondents do not tick the middle option. Straight-lining is very common, especially in web surveys where respondents generally pay less attention to the answers. Likewise,

Box 5.1 An Example of a Scale with Reverse-Scaled Items (in Bold)

Please rate the *brand* in the advertisement on the following dimensions:

Dislike	1	2	3	4	5	6	7	Like
Unpleasant	1	2	3	4	5	6	7	Pleasant
Good	**1**	**2**	**3**	**4**	**5**	**6**	**7**	**Bad**
Expensive	1	2	3	4	5	6	7	Inexpensive
Useless	1	2	3	4	5	6	7	Useful

long surveys and those with many similarly worded items trigger straight-lining (Drolet and Morrison 2001). An alternative is to note potential straight-lined responses and include this as a separate category in the subsequent statistical analyses. This step avoids the need to reduce the sample and indicates the size and direction of any bias.

However, straight-lining can also be the result of *culture-specific response styles*. For example, respondents from different cultures have different tendencies regarding selecting the mid points (**middle response styles**) or the end points of a response scale (**extreme response styles**). Similarly, respondents from different cultures have different tendencies regarding agreeing with statements, regardless of the item content; this tendency is also referred to as **acquiescence** (Baumgartner and Steenkamp 2001). For example, respondents from Spanish-speaking countries tend to show higher extreme response styles and high acquiescence, while East Asian (Japanese and Chinese) respondents show a relatively high level of middle response style. Within Europe, the Greeks stand out as having the highest level of acquiescence and a tendency towards an extreme response style. Harzing (2005) and Johnson et al. (2005) provide reviews of culture effects on response behavior.

Inconsistent answers also need to be addressed before analyzing your data. Many surveys start with one or more screening questions. The purpose of a screening question is to ensure that only individuals who meet the prescribed criteria complete the survey. For example, a survey of mobile phone users may screen for individuals who own an iPhone. If an individual indicates that he/she does not have an iPhone, this respondent should be removed from the dataset.

Surveys often ask the same question with slight variations, especially when reflective measures (see Box 3.1 in Chap. 3) are used. If a respondent gives a different answer to very similar questions, this may raise a red flag and could suggest that the respondent did not read the questions closely, or simply marked answers randomly to complete the survey as quickly as possible.

5.4.3 Data Entry Errors

When data are entered manually, **data entry errors** occur routinely. Fortunately, such errors are easy to spot if they happen outside the variable's range. That is, if an item is measured using a 7-point scale, the lowest value should be 1 (or 0) and the highest 7 (or 6). We can check if this is true by using descriptive statistics (minimum, maximum, and range; see next section). Data entry errors should always be corrected by going back to the original survey. If we cannot go back (e.g., because the data were collected using face-to-face interviews), we need to delete this specific observation for this specific variable.

More subtle errors—for example, incorrectly entering a score of 4 as 3—are difficult to detect using statistics. One way to check for these data entry errors is to randomly select observations and compare the entered responses with the original survey. We do, of course, expect a small number of errors (below 1%). If many data entry errors occur, the dataset should be entered again.

Manual double data entry is another method to detect data entry errors. That is, once the data has been entered manually, a second data checker enters the same data a second time and the two separate entries are compared to ensure they match. Entries that deviate from one another or values that fall outside the expected range of the scales (e.g., 7-point Likert scales should have values that fall within this range) are then indicative of data entry errors (Barchard and Verenikina 2013). Various studies reveal that—although double data entry is more laborious and expensive—it still detects errors better than single data entry (Barchard and Pace 2011; Paulsen et al. 2012).

5.4.4 Outliers

Data often contain **outliers**, which are values situated far from all the other observations that may influence results substantially. For example, if we compare the average income of 20 households, we may find that the incomes range between $20,000 and $100,000, with the average being $45,000. If we considered an additional household with an income of, say, $1 million, this would increase the average substantially.

> Malcolm Gladwell's (2008) book "Outliers: The Story of Success" is an entertaining study of how some people became exceptionally successful (outliers).

5.4.4.1 Types of Outliers

Outliers must be interpreted in the context of the study and this interpretation should be based on the types of information they provide. Depending on the source of their uniqueness, outliers can be classified into three categories:

- The first type of outlier is produced by data collection or entry errors. For example, if we ask people to indicate their household income in thousands of US dollars, some respondents may just indicate theirs in US dollars (not thousands). Obviously, there is a substantial difference between $30 and $30,000! Moreover, (as discussed before) data entry errors occur frequently. Outliers produced by data collection or entry errors should be deleted, or we need to determine the correct values by, for example, returning to the respondents.
- A second type of outlier occurs because exceptionally high or low values are a part of reality. While such observations can influence results significantly, they are sometimes highly important for researchers, because the characteristics of outliers can be insightful. Think, for example, of extremely successful companies, or users with specific needs long before most of the relevant marketplace also needs them (i.e., lead users). Deleting such outliers is not appropriate, but the impact that they have on the results must be discussed.
- A third type of outlier occurs when *combinations* of values are exceptionally rare. For example, if we look at income and expenditure on holidays, we may find someone who earns $1,000,000 and spends $500,000 of his/her income on holidays. Such combinations are unique and have a very strong impact on the results (particularly the correlations that we discuss later in this chapter). In such situations, the outlier should be retained, unless specific evidence suggests that it is not a valid member of the population under study. It is very useful to flag such outliers and discuss their impact on the results.

5.4.4.2 Detecting Outliers

In a simple form, outliers can be detected using univariate or bivariate graphs and statistics.[2] When searching for outliers, we need to use multiple approaches to ensure that we detect all the observations that can be classified as outliers. In the following, we discuss both routes to outlier detection:

Univariate Detection

The univariate detection of outliers examines the distribution of observations of each variable with the aim of identifying those cases falling outside the range of the "usual" values. In other words, finding outliers means finding observations with very low or very high variable values. This can be achieved by calculating the

[2]There are multivariate techniques that consider three, or more, variables simultaneously in order to detect outliers. See Hair et al. (2010) for an introduction, and Agarwal (2013) for a more detailed methodological discussion.

minimum and maximum value of each variable, as well as the range. Another useful option for detecting outliers is by means of box plots, which are a means of visualizing the distribution of a variable and pinpointing those observations that fall outside the range of the "usual" values. We introduce the above statistics and box plots in greater detail in the *Describe Data* section.

It is important to recognize that there will always be observations with exceptional values in one or more variables. However, we should strive to identify outliers that impact the presented results.

Bivariate Detection

We can also examine pairs of variables to identify observations whose combinations of variables are exceptionally rare. This is done by using a *scatter plot*, which plots all observations in a graph where the x-axis represents the first variable and the y-axis the second (usually *dependent*) variable (see the *Describe Data* section). Observations that fall markedly outside the range of the other observations will show as isolated points in the scatter plot.

A drawback of this approach is the number of scatter plots that we need to draw. For example, with 10 variables, we need to draw 45 scatter plots to map all possible combinations of variables! Consequently, we should limit the analysis to only a few relationships, such as those between a dependent and independent variable in a regression. Scatterplots with large numbers of observations are often problematic when we wish to detect outliers, as there is usually not just one dot, or a few isolated dots, just a cloud of observations where it is difficult to determine a cutoff point.

5.4.4.3 Dealing with Outliers

In a final step, we need to decide whether to delete or retain outliers, which should be based on whether we have an explanation for their occurrence. If there is an explanation (e.g., because some exceptionally wealthy people were included in the sample), outliers are typically retained, because they are part of the population. However, their impact on the analysis results should be carefully evaluated. That is, one should run an analysis with and without the outliers to assess if they influence the results. If the outliers are due to a data collection or entry error, they should be deleted. If there is no clear explanation, outliers should be retained.

5.4.5 Missing Data

Market researchers often have to deal with **missing data**. There are two levels at which missing data occur:

- Entire surveys are missing (survey non-response).
- Respondents have not answered all the items (item non-response).

Survey non-response (also referred to as *unit non-response*) occurs when entire surveys are missing. Survey non-response is very common and regularly only

5–25% of respondents fill out surveys. Although higher percentages are possible, they are not the norm in one-shot surveys. Issues such as inaccurate address lists, a lack of interest and time, people confusing market research with selling, privacy issues, and respondent fatigue also lead to dropping response rates. The issue of survey response is best solved by designing proper surveys and survey procedures (see Box 4.5 in Chap. 4 for suggestions).

Item non-response occurs when respondents do not provide answers to certain questions. There are different forms of missingness, including people not filling out or refusing to answer questions. Item non-response is common and 2–10% of questions usually remain unanswered. However, this number greatly depends on factors, such as the subject matter, the length of the questionnaire, and the method of administration. Non-response can be much higher in respect of questions that people consider sensitive and varies from country to country. In some countries, for instance, reporting incomes is a sensitive issue.

The key issue with item non-response is the type of pattern that the missing data follow. Do the missing values occur randomly, or is there some type of underlying system?[3] Once we have identified the type of missing data, we need to decide how to treat them. Figure 5.2 illustrates the process of missing data treatment, which we will discuss next.

5.4.5.1 The Three Types of Missing Data: Paradise, Purgatory, and Hell

We generally distinguish between three types of missing data:

- missing completely at random ("paradise"),
- missing at random ("purgatory"), and
- non-random missing ("hell").

Data are **missing completely at random** (**MCAR**) when the probability of data being missing is unrelated to any other measured variable and is unrelated to the variable with missing values. MCAR data thus occurs when there is no systematic reason for certain data points being missing. For example, MCAR may happen if the Internet server hosting the web survey broke down temporarily for a random reason. Why is MCAR paradise? When data are MCAR, observations with missing data are indistinguishable from those with complete data. If this is the case and little data are missing (typically less than 10% in each variable) listwise deletion can be used. Listwise deletion means that we only analyze complete cases; in most statistical software, such as Stata, this is a default option. Note that this default option in Stata only works when estimating models and only applies to the variables included in the model. When more than 10% of the data are missing, we can use multiple imputation (Eekhout et al. 2014), a more complex approach to missing data treatment that we discuss in the section *Dealing with Missing Data*.

[3]For more information on missing data, see https://www.iriseekhout.com

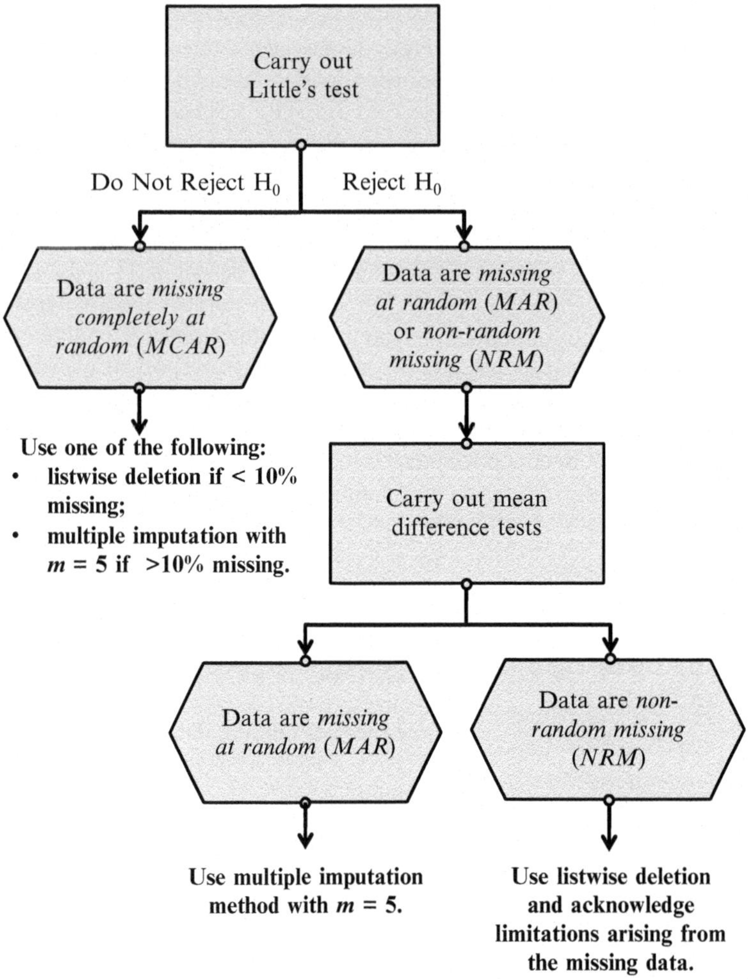

Fig. 5.2 Treating missing data

Unfortunately, data are rarely MCAR. If a missing data point (e.g., x_i) is unrelated to the observed value of x_i, but depends on another observed variable, we consider the data **missing at random** (**MAR**). In this case, the probability that the data point is missing varies from respondent to respondent. The term MAR is unfortunate, because many people confuse it with MCAR; however, the label has stuck. An example of MAR is when women are less likely to reveal their income. That is, the probability of missing data depends on the gender and not on the income. Why is MAR purgatory? When data are MAR, the missing value pattern is not random, but this can be handled by more sophisticated missing data techniques such as multiple imputation techniques. In the ⭳ Web Appendix (→ Downloads), we will illustrate how to impute a dataset with missing observations.

Lastly, data are **non-random missing** (**NRM**) when the probability that a data point (e.g., x_i) is missing depends on the variable x and on other unobserved factors. For example, very affluent and poor people are less likely to indicate their income. Thus, the missing income values depend on the income variable, but also on other unobserved factors that inhibit the respondents from reporting their incomes. This is the most severe type of missing data ("hell"), as even sophisticated missing data techniques do not provide satisfactory solutions. Thus, any result based on NRM data should be considered with caution. NRM data can best be prevented by extensive pretesting and consultations with experts to avoid surveys that cause problematic response behavior. For example, we could use income categories instead of querying the respondents' income directly, or we could simply omit the income variable.

A visualization of these three missingness mechanisms can be found under https://iriseekhout.shinyapps.io/MissingMechanisms/

5.4.5.2 Testing for the Type of Missing Data

When dealing with missing data, we must ascertain the missing data's type. If the dataset is small, we can browse through the data for obvious nonresponse patterns. However, missing data patterns become more difficult to identify with an increasing sample size and number of variables. Similarly, when we have few observations, patterns should be difficult to spot. In these cases, we should use one (or both) of the following diagnostic tests to identify missing data patterns:

- Little's MCAR test, and
- mean difference tests.

Little's MCAR test (Little 1998) analyzes the pattern of the missing data by comparing the observed data with the pattern expected if the data were randomly missing. If the test indicates no significant differences between the two patterns, the missing data can be classified as MCAR. Put differently, the null hypothesis is that the data are MCAR. Thus,

- if we do **not** reject the null hypothesis, we assume that the data are MCAR, and
- if we reject the null hypothesis, the data are either MAR or NRM.

If the data cannot be assumed to be MCAR, we need to test whether the missing pattern is caused by another variable in the dataset by using the procedures discussed in Chap. 6.

Looking at group means and their differences can also reveal missing data problems. For example, we can run a two independent samples *t*-test to explore whether there is a significant difference in the mean of a continuous variable (e.g., income) between the group with missing values and the group without missing

Table 5.2 Example of response issues

	Low income	Medium income	High income
Response	65	95	70
Non-response	35	5	30
$N = 300$			

values. In respect of nominal or ordinal variables, we could tabulate the occurrence of non-responses against different groups' responses. If we put the (categorical) variable about which we have concerns in one column of a table (e.g., income category), and the number of (non-)responses in another, we obtain a table similar to Table 5.2.

Using the χ^2*-test* (pronounced as *chi-square*), which we discuss under nonparametric tests in the ⭳ Web Appendix (→ Downloads), we can test if there is a significant relationship between the respondents' (non-)responses in respect of a certain variable and their income. In this example, the test indicates that there is a significant relationship between the respondents' income and the (non-)response behavior in respect of another variable, supporting the assumption that the data are MAR. We illustrate Little's MCAR test, together with the missing data analysis and imputation procedures in this chapter's appendix ⭳ Web Appendix (→ Downloads).

5.4.5.3 Dealing with Missing Data

Research has suggested a broad range of approaches for dealing with missing data. We discuss the listwise deletion and the multiple imputation method.

Listwise deletion uses only those cases with complete responses in respect of all the variables considered in the analysis. If any of the variables used have missing values, the observation is omitted from the computation. If many observations have some missing responses, this decreases the usable sample size substantially and hypotheses are tested with less power (the power of a statistical test is discussed in Chap. 6).

Multiple imputation is a more complex approach to missing data treatment (Rubin 1987; Carpenter and Kenward 2013). It is a simulation-based statistical technique that facilitates inference by replacing missing observations with a set of possible values (as opposed to a single value) representing the uncertainty about the missing data's true value (Schafer 1997). The technique involves three steps. First, the missing values are replaced by a set of plausible values not once, but *m* times (e.g., five times). This procedure yields *m* imputed datasets, each of which reflects the uncertainty about the missing data's correct value (Schafer 1997). Second, each of the imputed *m* datasets are analyzed separately by means of standard data methods. Third and finally, the imputed results from all *m* datasets (with imputed values) are combined into a single multiple-imputation dataset to produce statistical inferences with valid confidence intervals. This is necessary to reflect the uncertainty related to the missing values. According to the literature, deciding on the number of imputations, *m*, can be very challenging, especially when the patterns of the missing data are unclear. As a rule of thumb, an *m* of at least 5 should be

Table 5.3 Data cleaning issues and how to deal with them

Problem	Action
Interviewer fraud	– Check with respondents whether they were interviewed and correlate with previous data if available.
Suspicious response patterns	– Check for straight lining. – Include reverse-scaled items. – Consider removing the cases with straight-lined responses. – Consider cultural differences in response behavior (middle and extreme response styles, acquiescence). – Check for inconsistencies in response behavior.
Data entry errors	– Use descriptive statistics (minimum, maximum, range) to check for obvious data entry errors. – Compare a subset of surveys to the dataset to check for inconsistencies.
Outliers	– Identify outliers by means of univariate descriptive statistics (minimum, maximum, range), box plots, and scatter plots. – Outliers are usually retained unless they: . . . are a result of data entry errors, . . . do not fit the objective of the research, or . . . influence the results severely (but report results with and without outliers for transparency).
Missing data	– Check the type of missing data by running Little's MCAR test and, if necessary, mean differences tests. – When the data are MCAR, use either listwise deletion or the multiple imputation method with an *m* of 5. – When the data are MAR, use the multiple imputation method with an *m* of 5. – When the data are NRM, use listwise deletion and acknowledge the limitations arising from the missing data.

sufficient to obtain valid inferences (Rubin 1987; White et al. 2011). Additional information about the multiple imputation techniques available when using Stata can be found in the ⤓ Web Appendix (→ Downloads).

Now that we have briefly reviewed the most common approaches for handling missing data, there is still one unanswered question: Which one should you use? As shown in Fig. 5.2, if the data are MCAR, listwise deletion is recommended (Graham 2012) when the missingness is less than 10% and multiple imputation when this is greater than 10%. When the data are not MCAR but MAR, listwise deletion yields biased results. You should therefore use the multiple imputation method with an *m* of 5 (White et al. 2011). Finally, when the data are NRM, the multiple imputation method provides inaccurate results. Consequently, you should choose listwise deletion and acknowledge the limitations arising from the missing data. Table 5.3 summarizes the data cleaning issues discussed in this section.

5.5 Describe Data

Once we have performed the previous steps, we can turn to the task of describing the data. Data can be described one variable at a time (univariate descriptives) or in terms of the relationship between two variables (bivariate descriptives). We further divide univariate and bivariate descriptives into graphs and tables, as well as statistics.

The choice between the two depends on the information we want to convey. Graphs and tables can often tell a non-technical person a great deal. On the other hand, statistics require some background knowledge, but have the advantage that they take up little space and are exact. We summarize the different types of descriptive statistics in Fig. 5.3.

5.5.1 Univariate Graphs and Tables

In this section, we discuss the most common *univariate graphs* and *univariate tables*:

- bar chart,
- histogram,
- box plot,
- pie chart, and the
- frequency table.

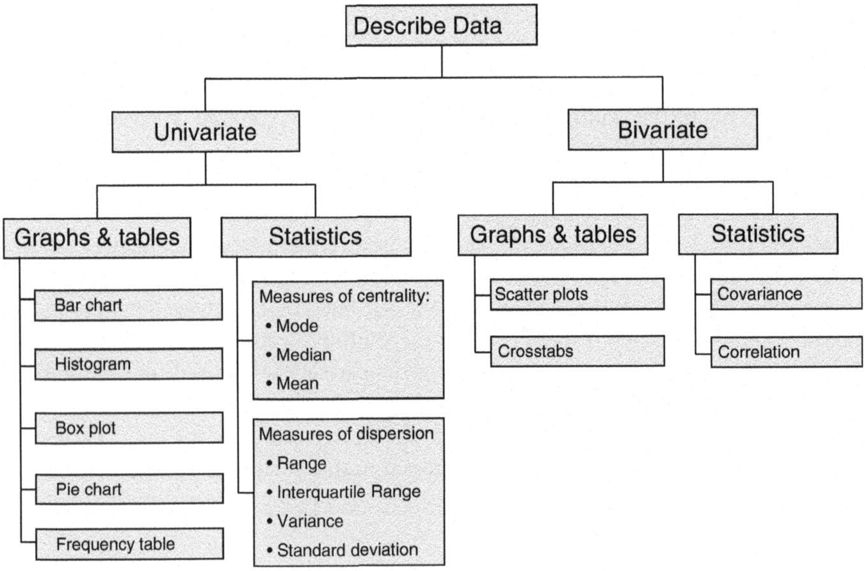

Fig. 5.3 The different types of descriptive statistics

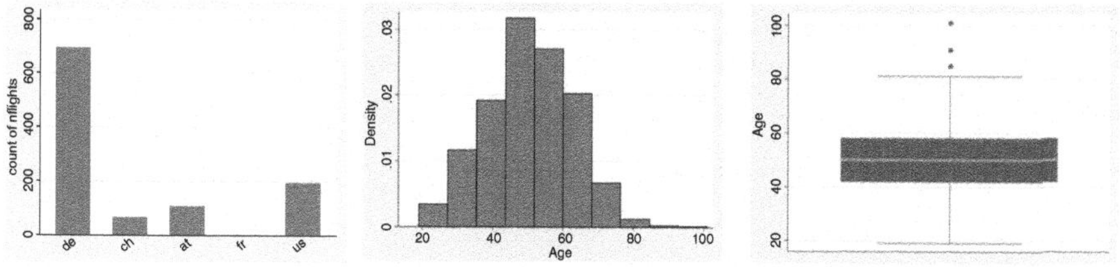

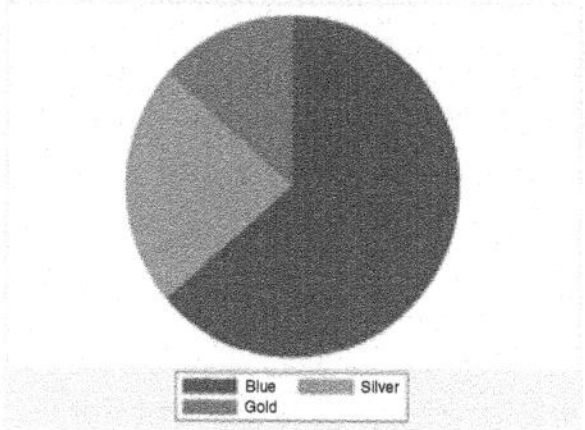

. ta country

Home country	Freq.	Percent	Cum.
de	695	65.26	65.26
ch	66	6.20	71.46
at	108	10.14	81.60
fr	1	0.09	81.69
us	195	18.31	100.00
Total	1,065	100.00	

Fig. 5.4 From *top left* to *bottom right*; the bar chart, histogram, box plot, pie chart, and frequency table

Figure 5.4 draws on these different types of charts and tables to provide information on the characteristics of travelers taken from the Oddjob Airways dataset that we use throughout this book.

A **bar chart** (Fig. 5.4 top left) is a graphical representation of a single categorical variable indicating each category's frequency of occurrence. However, each bar's height can also represent other indices, such as centrality measures or the dispersion of different data groups (see next section). Bar charts are primarily useful for describing nominal or ordinal variables. Histograms should be used for interval or ratio-scaled variables.

A **histogram** (Fig. 5.4 top middle) is a graph that shows how frequently categories made from a continuous variable occur. Differing from the bar chart, the variable categories on the x-axis are divided into (non-overlapping) classes of equal width. For example, if you create a histogram for the variable *age*, you can use classes of 21–30, 31–40, etc. A histogram is commonly used to examine the distribution of a variable. For this purpose, a curve following a specific distribution (e.g., normal) is superimposed on the bars to assess the correspondence of the actual distribution to the desired (e.g., normal) distribution. Given that overlaying a normal curve makes most symmetric distributions look more normal then they are, you should be cautious when assessing normality by means of histograms. In Chap. 6 we will discuss several options for checking the normality of data.

Histograms plot continuous variables with ranges of the variables grouped into intervals (bins), while bar charts plot nominal and ordinal variables.

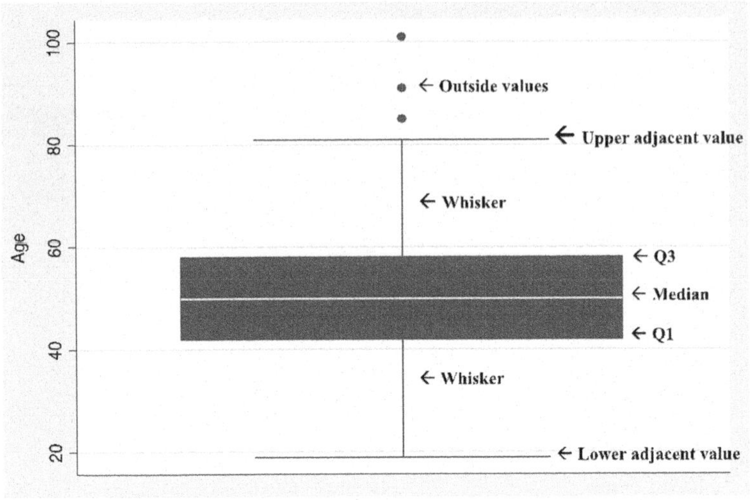

Fig. 5.5 Elements of the box plot

Another way of displaying the distribution of a (continuous) variable is the **box plot** (Fig. 5.4 top right) (also referred to as a **box-and-whisker plot**). The box plot is a graph representing a variable's distribution and consists of elements expressing the dispersion of the data. Note that several elements refer to terminologies discussed in the *Univariate Statistics* section. Figure 5.5 shows a box plot for the variable age based on the Oddjob Airways dataset.

- *Outside values* are observations that fall above the 3rd quartile +1.5 interquartile range.
- The *upper adjacent value* represents observations with the highest value that fall within the 3rd quartile +1.5 interquartile range.
- The upper line extending the box (*whisker*) represents the distance to observations with the highest values that fall within the following range: 3rd quartile + interquartile range. If there are no observations within this range, the line is equal to the maximum value.
- The top and bottom of the box describe the 3rd quartile (top) and 1st quartile (bottom); that is, the box contains the middle 50% of the data, which is equivalent to the interquartile range.
- The solid line inside the box represents the *median*.
- The lower line extending the box (*whisker*) represents the distance to the smallest observation that is within the following range: 1st quartile − interquartile range. If there are no observations within this range, the line is equal to the minimum value.
- The *lower adjacent value* represents observations with lowest values that fall inside the 3rd quartile −1.5 interquartile range.

We can make statements about the dispersion of the data with a box plot. The larger the box, the greater the observations' variability. Furthermore, the box plot helps us identify outliers in the data.

The **pie chart** (i.e., Fig. 5.4 bottom left) visualizes how a variable's different values are distributed. Pie charts are particularly useful for displaying percentages of variables, because people interpret the entire pie as being 100%, and can easily see how often values occur. The limitation of the pie chart is, however, that it is difficult to determine the size of segments that are very similar.

A **frequency table** (i.e., Fig. 5.4 bottom right) is a table that includes all possible values of a variable in absolute terms (i.e., frequency), how often they occur relatively (i.e., percentage), and the percentage of the cumulative frequency, which is the sum of all the frequencies from the minimum value to the category's upper bound (i.e., cumulative frequency). It is similar to the histogram and pie chart in that it shows the distribution of a variable's possible values. However, in a frequency table, all values are indicated exactly. Like pie charts, frequency tables are primarily useful if variables are measured on a nominal or ordinal scale.

5.5.2 Univariate Statistics

Univariate statistics fall into two groups: those describing centrality and those describing the dispersion of variables. Box 5.2 at the end of this section shows sample calculation of the statistics used on a small set of values.

5.5.2.1 Measures of Centrality

Measures of centrality (sometimes referred to as *measures of central tendency*) are statistical indices of a "typical" or "average" score. There are two main types of measures of centrality, the median and the mean.[4]

The **median** is the value that occurs in the middle of the set of scores if they are ranked from the smallest to the largest, and it therefore separates the lowest 50% of cases from the highest 50% of cases. For example, if 50% of the products in a market cost less than $1,000, then this is the median price. Identifying the median requires at least ordinal data (i.e., it cannot be used with nominal data).

The most commonly used measure of centrality is the **mean** (also called the *arithmetic mean* or, simply, the *average*). The mean (abbreviated as $\bar{x}$) is the sum of each observation's value divided by the number of observations:

$$\bar{x} = \frac{\text{Sum}(x)}{n} = \frac{1}{n}\sum_{i=1}^{n} x_i$$

[4]The mode is another measure. However, unlike the median and mean, it is ill-defined, because it can take on multiple values. Consequently, we do not discuss the mode.

In the above formula, x_i refers to the value of observation i of variable x and n refers to the total number of observations. The mean is only useful for interval or ratio-scaled variables.

Each measure of centrality has its own use. The mean is most frequently used, but is sensitive to very small or large values. Conversely, the median is not sensitive to outliers. Consequently, the relationship between the mean and the median provides us with valuable information about a variable's distribution. If the mean and the median are about the same, the variable is likely to be symmetrically distributed (i.e., the left side of the distribution mirrors the right side). If the mean differs from the median, this suggests that the variable is asymmetrically distributed and/or contains outliers. This is the case when we examine the prices of a set of products valued $500, $530, $530, and $10,000; the median is $530, while the mean is $2,890. This example illustrates why a single measure of centrality can be misleading. We also need to consider the variable's dispersion to gain a more complete picture.

5.5.2.2 Measures of Dispersion

Measures of dispersion provide researchers with information about the variability of the data; that is, how far the values are spread out. We differentiate between four types of measures of dispersion:

- range,
- interquartile range,
- variance, and
- standard deviation.

The **range** is the simplest measure of dispersion. It is the difference between the highest and the lowest value in a dataset and can be used on data measured at least on an ordinal scale. The range is of limited use as a measure of dispersion, because it provides information about extreme values and not necessarily about "typical" values. However, the range is valuable when screening data, as it allows for identifying data entry errors. For example, a range of more than 6 on a 7-point Likert scale would indicate an incorrect data entry.

The **interquartile range** is the difference between the 3rd and 1st quartile. The 1st *quartile* corresponds to the value separating the 25% lowest values from the 75% largest values if the values are ordered sequentially. Correspondingly, the 3rd quartile separates the 75% lowest from the 25% highest values. The interquartile range is particularly important for drawing box plots.

The **variance** (generally abbreviated as s^2) is a common measure of dispersion. The variance is the sum of the squared differences of each value and a variable's mean, divided by the sample size minus 1. The variance is only useful if the data are interval or ratio-scaled:

$$s^2 = \frac{\sum_{i=1}^{n} (x_i - \bar{x})^2}{n - 1}$$

The variance tells us how strongly observations vary around the mean. A low variance indicates that the observations tend to be very close to the mean; a high variance indicates that the observations are spread out. Values far from the mean increase the variance more than those close to the mean.

The most commonly used measure of dispersion is the **standard deviation** (usually abbreviated as s). It is the square root of—and, therefore, a variant of—the variance:

$$s = \sqrt{s^2} = \sqrt{\frac{\sum_{i=1}^{n} (x_i - \bar{x})^2}{n - 1}}$$

The variance and standard deviation provide similar information, but while the variance is expressed on the same scale as the original variable, the standard deviation is standardized. Consequently, the following holds for normally distributed variables (this will be discussed in the following chapters in more detail):

- 66% of all observations are between plus and minus one standard deviation units from the mean,
- 95% of all observations are between plus and minus two standard deviation units from the mean, and
- 99% of all observations are between plus and minus three standard deviation units from the mean.

Thus, if the mean price is \$1,000 and the standard deviation is \$150, then 66% of all the prices fall between \$850 and \$1150, 95% fall between \$700 and \$1300, and 99% of all the observations fall between \$550 and \$1,450.

5.5.3 Bivariate Graphs and Tables

There are several *bivariate graphs* and *tables*, of which the scatter plot and the crosstab are the most important. Furthermore, several of the graphs, charts, and tables discussed in the context of univariate analysis can be used for bivariate analysis. For example, box plots can be used to display the distribution of a variable in each group (category) of nominal variables.

A **scatter plot** (see Fig. 5.6) uses both the y and x-axis to show how two variables relate to one another. If the observations almost form a straight diagonal line in a

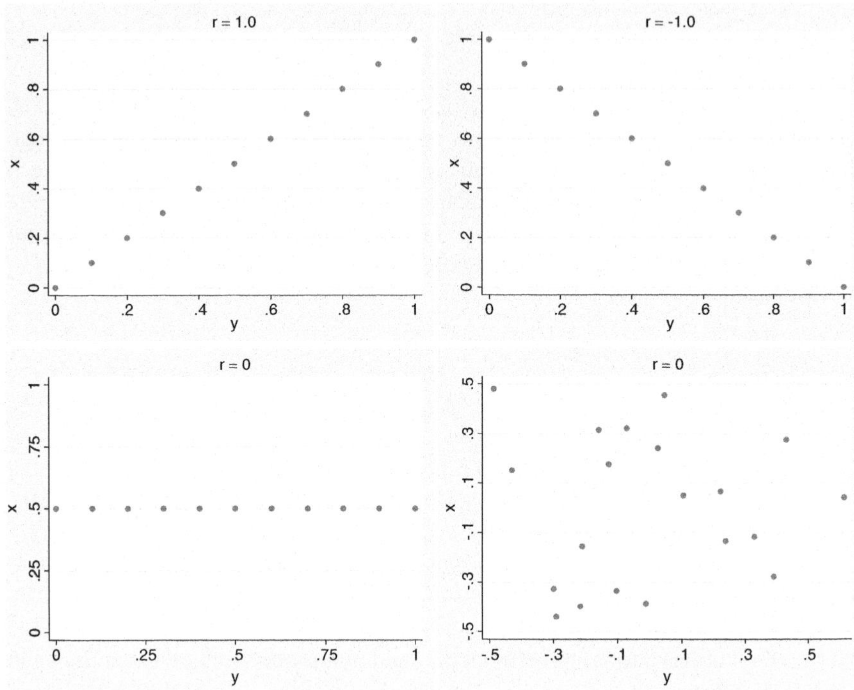

Fig. 5.6 Scatter plots and correlations

scatter plot, the two variables are strongly related.[5] Sometimes, a third variable, corresponding to the color or size (e.g., a *bubble plot*) of the data points, is included, adding another dimension to the plot.

Crosstabs (also referred to as *contingency tables*) are tables in a matrix format that show the frequency distribution of nominal or ordinal variables. They are the equivalent of a scatter plot used to analyze the relationship between two variables. While crosstabs are generally used to show the relationship between two variables, they can also be used for three or more variables, which, however, makes them difficult to grasp. Crosstabs are also part of the χ^2-*test* (pronounced as *chi-square*), which we discuss under nonparametric tests in the ⤓ Web Appendix (→ Downloads).

[5]A similar type of chart is the **line chart**. In a line chart, measurement points are ordered (typically by their x-axis value) and joined with straight line segments.

5.5.4 Bivariate Statistics

Bivariate statistics involve the analysis of two variables to determine the empirical relationship between them. There are two key measures that indicate (linear) associations between two variables; we illustrate their computation in Box 5.2:

- covariance, and
- correlation.

The **covariance** is the degree to which two variables vary together. If the covariance is zero, then two variables do not vary together. The covariance is the sum of the multiplication of the differences between each value of the x_i and y_i variables and their means, divided by the sample size minus 1:

$$Cov(x_i, y_i) = \frac{1}{n-1} \sum_{i=1}^{n} (x_i - \bar{x}) \cdot (y_i - \bar{y})$$

The **correlation** (typically abbreviated as r) is a common measure of how strongly two variables relate to each other. The most common type of correlation, the *Pearson's correlation coefficient*, is calculated as follows:

$$r = \frac{Cov(x_i, y_i)}{s_x \cdot s_y} = \frac{\sum_{i=1}^{n} (x_i - \bar{x}) \cdot (y_i - \bar{y})}{\sqrt{\sum_{i=1}^{n} (x_i - \bar{x})^2} \cdot \sqrt{\sum_{i=1}^{n} (y_i - \bar{y})^2}}$$

The numerator contains the covariance of x_i and y_i ($Cov(x_i, y_i)$), while the denominator contains the product of the standard deviations of x_i and y_i.[6] Thus, the correlation is the covariance divided by the product of the standard deviations. As a result, the correlation is standardized and, unlike the covariance, is no longer dependent on the variables' original measurement. More precisely, the correlation coefficient ranges from -1 to 1, where -1 indicates a perfect negative relationship and 1 indicates the contrary. A correlation coefficient of 0 indicates that there is no relationship, also implying that their covariance is zero.

As a rule of thumb (Cohen 1988), an absolute correlation...

- ... below 0.30 indicates a weak relationship,
- ... between 0.30 and 0.49 indicates a moderate relationship, and
- ... above 0.49 indicates a strong relationship.

[6] Note that the terms $n-1$ in the numerator and denominator cancel each other and are therefore not shown here.

Box 5.2 Sample Calculation of Univariate and Bivariate Statistics

Consider the following list of values for variables x and y, which we treat as ratio-scaled:

x	6	6	7	8	8	8	12	14	14
y	7	6	6	9	8	5	10	9	9

Measures of centrality for x:

$$\text{Median} = 8$$

$$\text{Mean } \bar{x} = \frac{1}{9}(6 + 6 + \ldots + 14 + 14) = \frac{83}{9} \approx 9.22$$

Measures of dispersion for x:

$$\begin{aligned}
&\text{Minimum} &&= 6\\
&\text{Maximum} &&= 14\\
&\text{Range} &&= 14 - 6 = 8\\
&\text{Interquartile range} &&= 6.5
\end{aligned}$$

$$\text{Variance } (s^2) = \frac{\left[(6 - 9.22)^2 + \ldots + (14 - 9.22)^2\right]}{9 - 1} = \frac{83.56}{8} \approx 10.44.$$

$$\text{Standard deviation } (s) = \sqrt{s^2} = \sqrt{10.44} \approx 3.23$$

Measures of association between x and y:

$$\begin{aligned}
\text{Covariance } (\text{cov}(x, y)) &= \frac{1}{9 - 1}[(6 - 9.22) \cdot (7 - 7.67) + \ldots\\
&\quad + (14 - 9.22) \cdot (9 - 7.67)] = \frac{31.67}{8} \approx 3.96\\
\text{Correlation } (r) &= \frac{3.96}{3.23 \cdot 1.73} \approx 0.71
\end{aligned}$$

The scatter plots in Fig. 5.6 illustrate several correlations between two variables x and y. If the observations almost form a straight diagonal line in the scatter plot (upper left and right in Fig. 5.6), the two variables have a high (absolute) correlation. If the observations are uniformly distributed in the scatter plot (lower right in Fig. 5.6), or one variable is a constant (lower left in Fig. 5.6), the correlation is zero.

Pearson's correlation coefficient is the most common coefficient and is generally simply referred to as the correlation (Agresti and Finlay 2014). Pearson's correlation is appropriate for calculating correlations between two variables that are both interval or ratio-scaled. However, it can also be used when one variable is

Table 5.4 Types of descriptive statistics for differently scaled variables

	Nominal	Ordinal	Interval & ratio
Univariate graphs & tables			
Bar chart	X	X	
Histogram			X
Box plot			X
Pie chart	X	X	(X)
Frequency table	X	X	(X)
Univariate statistics: Measures of centrality			
Median		X	X
Mean			X
Univariate statistics: Measures of dispersion			
Range		(X)	X
Interquartile range		(X)	X
Variance			X
Standard deviation			X
Bivariate graphs/tables			
Scatter plot			X
Crosstab	X	X	(X)
Bivariate statistics			
Contingency coefficient	X		
Cramer's V	X		
Phi	X		
Spearman's correlation		X	
Kendall's tau		X	
Pearson's correlation			X

interval or ratio-scale and the other is, for example, binary. There are other correlation coefficients for variables measured on lower scale levels. Some examples are:

- *Spearman's correlation coefficient* and *Kendall's tau* when at least one variable for determining the correlation is measured on an ordinal scale.
- *Contingency coefficient*, *Cramer's V*, and *Phi* for variables measured on a nominal scale. These statistical measures are used with crosstabs; we discuss these in the context of nonparametric tests in the ⭳ Web Appendix (→ Downloads).

In Table 5.4, we indicate which descriptive statistics are useful for differently scaled variables. The brackets X indicate that the use of a graph, table, or statistic is potentially useful while (X) indicates that use is possible, but less likely useful, because this typically requires collapsing data into categories, resulting in a loss of information.

5.6 Transform Data (Optional)

Transforming data is an optional step in the workflow. Researchers transform data as certain analysis techniques require this: it might help interpretation or might help meet the assumptions of techniques that will be discussed in subsequent chapters. We distinguish two types of data transformation:

- variable respecification, and
- scale transformation.

5.6.1 Variable Respecification

Variable respecification involves transforming data to create new variables or to modify existing ones. The purpose of respecification is to create variables that are consistent with the study's objective. *Recoding* a continuous variable into a categorical variable is an example of a simple respecification. For example, if we have a variable that measures a respondent's *number of flights (indicating the number of flights per year)*, we could code those flights below 5 as low (=1), between 5 and 10 flights as medium (=2), and everything above 11 as high (=3). Recoding a variable in such a way always results in a loss of information, since the newly created variable contains less detail than the original. While there are situations that require recoding (e.g., we might be asked to give advice based on different income groups where income is continuous), we should generally avoid recoding.

Another example of respecification is swapping the polarity of a question. If you have a variable measured on a 5-point Likert-scale where: 1 = "strongly agree"; 2 = "agree"; 3 = "undecided"; 4 = "disagree"; 5 = "strongly disagree" and you wish to switch the polarity of the values so that value 1 reverses to 5, value 2 becomes 4, and so on.

Creating a dummy variable is a special way of recoding data. **Dummy variables** (or simply *dummies*) are binary variables that indicate if a certain trait is present or not. For example, we can use a dummy variable to indicate that advertising was used during a period (value of the dummy is 1) or not (value of the dummy is 0). We can also use multiple dummies to capture categorical variables' effects. For example, three levels of *flight intensity* (low, medium, and high) can be represented by two dummy variables: The first takes a value of 1 if the intensity is high (0 else), the second also takes a value of 1 if the intensity is medium (0 else). If both dummies take the value 0, this indicates low flight intensity. We always construct one dummy less than the number of categories. We explain dummies in further detail in the ⭳ Web Appendix (→ Downloads) and more information can be found at http://www.stata.com/support/faqs/data-management/creating-dummy-variables/. Dummies are often used in regression analysis (discussed in Chap. 7).

The creation of *constructs* is a frequently used type of variable respecification. As described in Chap. 3, a construct is a concept that cannot be observed, but can be measured by using multiple items, none of which relate perfectly to the construct. To compute a construct measure, we need to calculate the average (or the sum) of several related items. For example, a traveler's *commitment* to fly with Oddjob Airways can be measured by using the following three items:

- I am very committed to Oddjob Airways.
- My relationship with Oddjob Airways means a lot to me.
- If Oddjob Airways would no longer exist, it would be a true loss for me.

By calculating the average of these three items, we can form a *composite measure* of *commitment*. If one respondent indicated 4, 3, and 4 on the three items' scale, we calculate a **construct score** (also referred to as a composite score) for this person as follows: $(4 + 3 + 4)/3 = 3.67$. Note that we should take the average over the number of nonmissing responses.[7] In Chap. 8 we discuss more advanced methods of doing this by, for example, creating factor scores.

Similar to creating constructs, we can create an **index** of sets of variables. For example, we can create an index of information search activities, which is the sum of the information that customers require from promotional materials, the Internet, and other sources. This measure of information search activities is also referred to as a composite measure, but, unlike a construct, the items in an index define the trait to be measured.

5.6.2 Scale Transformation

Scale transformation involves changing the variable values to ensure comparability with other variables or to make the data suitable for analysis. Different scales are often used to measure different variables. For example, we may use a 5-point Likert scale for one set of variables and a 7-point Likert scale for a different set of variables in our survey. Owing to the differences in scaling, it would not be meaningful to make comparisons across any respondent's measurement scales. These differences can be corrected by **standardizing variables**.

A popular way of standardizing data is by rescaling these to have a mean of 0 and a variance of 1. This type of standardization is called the ***z*-standardization**. Mathematically, standardized scores z_i (also called *z-scores*) can be obtained by

[7]In Stata, this is best done using the `rowmean` command. For example, `egen commitment = rowmean (com1 com2 com3)`. This command automatically calculates the mean over the number of nonmissing responses.

subtracting the mean $\bar{x}$ of every observation x_i and dividing it by the standard deviation s. That is:

$$z_i = \frac{(x_i - \bar{x})}{s}$$

Range standardization (r_i) is another standardization technique which scales the data in a specific range. For example, standardizing a set of values to a range of 0 to 1 requires subtracting the minimum value of every observation x_i and then dividing it by the range (the difference between the maximum and minimum value).

$$r_i = \frac{(x_i - x_{min})}{(x_{max} - x_{min})}$$

The range standardization is particularly useful if the mean, variance, and ranges of different variables vary strongly and are used for some forms of cluster analysis (see Chap. 9).

A **log transformation**—another type of transformation—is commonly used if we have skewed data. **Skewed data** occur if we have a variable that is asymmetrically distributed and can be positive or negative. A *positive skew* (also called *right-skewed data* or data skewed to the right) occurs when many observations are concentrated on the left side of the distribution, producing a long right tail. When data are right-skewed, the mean will be higher than the median. A *negative skew* (also called *left-skewed data* or data skewed to the left) is the opposite, meaning that many observations are concentrated on the right of the distribution, producing a long left tail. When data are negatively skewed, the mean will be lower than the median. A histogram will quickly show whether data are skewed. Skewed data can be undesirable in analyses. Log transformations are commonly used to transform data closer to a normal distribution when the data are right-skewed (i.e., the data are non-negative). Taking a natural logarithm will influence the size of the coefficient related to the transformed variable, but will not influence the value of its outcome.[8]

Finally, **aggregation** is a special type of transformation. Aggregation means that we take variables measured at a lower level to a higher level. For example, if we know the average customer's satisfaction with an airline and the distribution channels from which they buy (i.e., the Internet or a travel agent), we can calculate the average satisfaction at the channel level. Aggregation only works from lower to higher levels and is useful if we want to compare groups at a higher level.

[8]The logarithm is calculated as follows: If $x = y^b$, then $y = log_b(x)$ where x is the original variable, b the logarithm's base, and y the exponent. For example, log 10 of 100 is 2. Logarithms cannot be calculated for negative values (such as household debt) and for the value of zero. In Stata, you can generate a log-transformed variable by typing: `gen loginc = log(income)`, whereby `loginc` refers to the newly created log-transformed variable and `income` refers to the income variable.

While transforming data is often necessary to ensure comparability between variables or to make the data suitable for analysis, there are also drawbacks to this procedure. Most notably, we may lose information during most transformations. For example, recoding the *ticket price* (measured at the ratio scale) as a "low," "medium," and "high" ticket price will result in an ordinal variable. In the transformation process, we have therefore lost information by going from a ratio to an ordinal scale. Another drawback is that transformed data are often more difficult to interpret. For example, the log (*ticket price*) is far more difficult to interpret and less intuitive than simply using the ticket price.

5.7 Create a Codebook

After all the variables have been organized and cleaned, and some initial descriptive statistics have been calculated, we can create a **codebook,** containing essential details of the data collection and data files, to facilitate sharing. Codebooks usually have the following structure:

Introduction: The introduction discusses the goal of the data collection, why the data are useful, who participated, and how the data collection effort was conducted (mail, Internet, etc.).

Questionnaire(s): It is common practice to include copies of all the types of questionnaires used. Thus, if different questionnaires were used for different respondents (e.g., for French and Chinese respondents), a copy of each original questionnaire should be included. Differences in wording may afterwards explain the results of the study, particularly those of cross-national studies, even if a back-translation was used (see Chap. 4). These are not the questionnaires received from the respondents themselves, but blank copies of each type of questionnaire used. Most codebooks include details of each variable as comments close to the actual items used. If a dataset was compiled using secondary measures (or a combination of primary and secondary data), the secondary datasets are often briefly discussed (the version that was used, when it was accessed, etc.).

Description of the variables: This section includes a verbal description of each variable used. It is useful to provide the variable name as used in the data file, a description of what the variable is supposed to measure, and whether the measure has previously been used. You should also describe the measurement level (see Chap. 3).

Summary statistics: This section includes descriptive statistics of each variable. The average (only for interval and ratio-scaled data), minimum, and maximum are often shown. In addition, the number of observations and usable observations (excluding observations with missing values) are included, just like a histogram (if applicable).

Datasets: This last section includes the names of the datasets and sometimes the names of all the revisions of the used datasets. Codebooks sometimes include the file date to ensure that the right files are used.

5.8 The Oddjob Airways Case Study

The most effective way of learning statistical methods is to apply them to a set of data. Before introducing Stata and how to use it, we present the dataset from a fictitous company called *Oddjob Airways* (but with a real website, http://www.oddjobairways.com) that will guide the examples throughout this book. The dataset *oddjob.dta* (⭳ Web Appendix → Downloads) stems from a customer survey of Oddjob Airways. Founded in 1962 by the Korean businessman Toshiyuki Sakata, Oddjob Airways is a small premium airline, mainly operating in Europe, but also offering flights to the US. In an effort to improve its customers' satisfaction, the company's marketing department contacted all the customers who had flown with the airline during the last 12 months and were registered on the company website. A total of 1,065 customers who had received an email with an invitation letter completed the survey online.

The survey resulted in a rich dataset with information about travelers' demographic characteristics, flight behavior, as well as their price/product satisfaction with and expectations in respect of Oddjob Airways. Table 5.5 describes the variables in detail.

5.8.1 Introduction to Stata

Stata is a computer package specializing in quantitative data analysis, and widely used by market researchers. It is powerful, can deal with large datasets, and relatively easy to use. In this book, we use Stata MP4 14.2 (to which we simply refer to as Stata). Prior versions (12 or higher) for Microsoft Windows, Mac or Linux can be used for (almost) all examples throughout the book.

Stata offers a range of versions and packages; your choice of these depends on the size of your dataset and the data processing speed. Stata/SE and Stata/MP are recommended for large datasets. The latter is the fastest and largest version of Stata. Stata/MP can, for example, process large datasets with up to 32,767 variables and 20 billion observations, while Stata/SE can process datasets with the same number of variables, but a maximum of 2.14 billion observations. Stata/IC is recommended

Table 5.5 Variable description and label names of the Oddjob Dataset

Variables	Variable description	Variable name in the dataset
Demographic measures		
Age of the customer	Numerical variable ranging between the ages of 19 and 101.	*Age*
Customer's gender	Dichotomous variable, where 1 = Female; 2 = Male.	*Gender*
Language of customer	Categorical variable, where 1 = German; 2 = English; 3 = French.	*Language*
Home country	Categorical variable, whereby: 1 = Germany (de), 2 = Switzerland (ch); 3 = Austria (at); 4 = France (fr), 5 = the United States (us).	*Country*
Flight behaviour measures		
Flight class	Categorical variable distinguishing between the following categories: 1 = First; 2 = Business; 3 = Economy.	*flight_class*
Latest flight	Categorical variable querying when the customer last flew with Oddjob Airways. Categories are: 1 = Within the last 2 days; 2 = Within the last week; 3 = Within the last month; 4 = Within the last 3 months; 5 = Within the last 6 months; 6 = Within the last 12 months.	*flight_latest*
Flight purpose	Dichotomous variable distinguishing between: 1 = Business; 2 = Leisure.	*flight_purpose*
Flight type	Dichotomous variable, where: 1 = Domestic; 2 = International.	*flight_type*
Number of flights	Numeric variable ranging between 1 and 457 flights per year.	*nflights*
Traveler's status	Categorical variable, where membership status is defined in terms of: 1 = Blue; 2 = Silver; 3 = Gold.	*status*
Perception and satisfaction measures		
Traveler's expectations	23 items reflecting a customer's expectations with the airline: "How high are your expectations that..." All items are measured on a continuous scale ranging from 1 very low to 100 very high.	*e1* to *e23*
Traveler's satisfaction	23 items reflecting a customer's satisfaction with Oddjob Airways regarding the features asked in the expectation items (*e1-e23*) on a continuous scale ranging from 1=very unsatisfied to 100=very satisfied.	*s1* to *s23*
Recommendation	Item on whether a customer is likely to recommend the airline to a friend or colleague. This item is measured on an 11-point Likert-scale ranging from 1 very unlikely to 11 very likely.	*nps*
Reputation	One item stating "Oddjob Airways is a reputable airline." This item is measured on a 7-point Likert-scale ranging from 1 fully disagree to 7 fully agree.	*reputation*
Overall price/ performance satisfaction	One item stating "Overall I am satisfied with the price performance ratio of Oddjob Airways." This item is measured on a 7-point Likert-scale ranging from 1 fully disagree to 7 fully agree.	*overall_sat*

(continued)

Table 5.5 (continued)

Variables	Variable description	Variable name in the dataset
General satisfaction	3 items reflecting a customer's overall satisfaction with the airline. All items are measured on a 7-point Likert scale ranging from 1 fully disagree to 7 fully agree.	*sat1* to *sat3*
Loyalty	5 items reflecting a customer's loyalty to the airline. All items are measured on a 7-point Likert-scale ranging from 1 fully disagree to 7 fully agree.	*loy1* to *loy5*
Commitment	3 items reflecting a customer's commitment to fly with the airline. All items are measured on a 7-point Likert-scale ranging from 1 fully disagree to 7 fully agree.	*com1 to com3*

for moderate-sized datasets. This can process datasets with a maximum of 2,047 variables and up to 2.14 billion observations. Small Stata is only available for students and handles small-sized datasets with a maximum of 99 variables and 1,200 observations. To obtain a copy of Stata, check the stata.com website, or ask the IT department, or your local Stata distributor. Special student and faculty prices are available.

In the next sections, we will use the ► sign to indicate that you should click on something. Options, menu items or drop-down lists that you should look up in dialog boxes are printed in **bold**. Variable names, data files or data formats are printed in *italics* to differentiate them from the rest of the text. Finally, Stata commands are indicated in `Courier`.

5.8.2 Finding Your Way in Stata

5.8.2.1 The Stata Main/Start Up Window and the Toolbar

If you start up Stata for the first time, it presents a screen as shown in Fig. 5.7. This start up screen is the main Stata window in the Mac version.[9]

The main Stata window in Fig. 5.7 consists of five sub-windows. The first sub-window on the left of the start-up screen is called the **Review** window, which displays the history of commands since starting a session. Successful commands are displayed in black, and those containing errors are displayed in red. If you click on one of the past commands displayed in the **Review** window, it will be automatically copied

[9]If you open Stata in the Windows or Linux operating systems, the toolbar looks a bit different, but is structured along the same lines as discussed in this chapter.

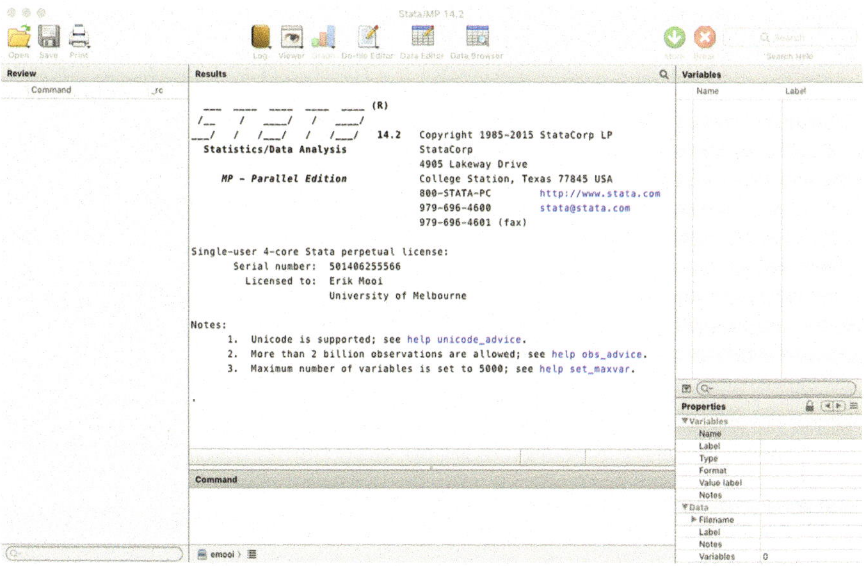

Fig. 5.7 The Stata interface

into the **Command** window, which is located at the bottom of the central screen. The **Review** window stores and displays your commands. It allows you to recall all previous commands, edit, and re-submit them if you wish. The output of your analyses is displayed in the **Results** window, located in the center. The **Variables** window (upper-right) lists all the variables included in the dataset, whereas the **Properties** window (lower-right) displays the features of the selected variable(s).

Stata's toolbar is located underneath its menu bar. This toolbar contains a range of shortcuts to frequently used commands, including opening, saving, printing, viewing, and editing your data. An overview of the toolbar icons is shown in Table 5.6.

5.8.2.2 The Menu Bar

Stata's menu bar includes **File**, **Edit**, **Data**, **Graphics**, **Statistics**, **Users**, **Windows**, and **Help**, which will be discussed briefly in this section. The menu options **Graphics** and **Statistics** will, later in this chapter be discussed in greater detail by means of examples.

> You can open Stata's dialog box by simply typing `db`, followed by the operation (i.e., edit, describe, copy, etc.) or technique (regression) that you wish to carry out. For example, if you wish to open the data editor, you could type: `db edit`. Similarly, if you wish to specify your regression model, you could type `db regress`, which will open the dialog window for the regression model.

Table 5.6 Toolbar icons

Stata's toolbar in detail:

Symbol	Action
	Opens dataset by selecting a dataset from a menu.
	Saves the active dataset.
	Prints contents of a selected window.
	Log begins/closes/suspends/resumes the current log. It also allows for viewing the current log if a log file is open.
	Opens the viewer that provides advice on finding help and how to search through Stata's online resources.
	Brings the graph window to the front.
	Opens a new do-file editor or brings a do-file editor to the front.
	Opens the data editor (edit) or brings the data editor to the front. It allows you to edit variables.
	Opens the data editor (browse) or brings the data editor to the front. It allows you to browse through the variables.
	This tells Stata to show more output.
	This tells Stata to interrupt the current task.
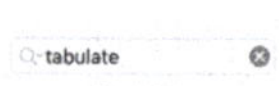	Enables searching for help for Stata commands.

File

Format Types

Stata uses multiple file formats. The *.dta* file format only contains data. Stata can also import other file formats such as Excel (*.xls* and *.xlsx*), SAS, SPSS and can read text files (such as *.txt* and *.dat*). Once these files are open, they can be saved as Stata's *.dta* file format. Under ► File, you find all the commands that deal with the opening, closing, creating, and saving of the different types of files. In the startup screen, Stata shows several options for creating or opening datasets. The options that you should use are either **Open Recent**, under which you can find a list with recently opened data files, or **New Files**.

You can import different types of files into Stata if you go to ► File ► Import, select the file format, and then select the file you wish to open. The **Example dataset** menu item is particularly useful for learning new statistical methods, as it provides access to all the example datasets mentioned under the titles of the various user manuals for a particular statistical method.

> Stata has an extensive number of user-written programs, which, once installed, act as Stata commands. If you know the name of the program, you can directly type `help`, followed by a keyword, which initiates a keyword search. Alternatively, if you do not know the name of the command, but are looking for a specific method, such as cluster analysis, you can type: `help cluster` to initiate a keyword search for this method. This will open a new window containing information about this technique from the help files, the Stata reference manual, the Stata Journal, the Frequently Asked Questions, references to other articles, and other help files.

Stata *.do* Files

In addition to the menu functionalities, we can run Stata by using its command language, called **Stata command**. Think of this as a programming language that can be directly used (if you feel comfortable with its command) or via dialog boxes. Stata's commands can be saved (as a *.do* file) for later use. This is particularly useful if you undertake the same analyses across different datasets. Think, for example, of standardized marketing reports on daily, weekly, or monthly sales. Note that discussing Stata *.do* files in great detail is beyond the scope of this book, but, as we go through the different chapters, we will show all the Stata's commands that we have used in this book.

Edit

This menu option allows you to copy Stata output and analysis, and paste the content into, for example, a Word document. If you want to copy a table from your output, first select the table and then go to ► Edit ► Copy table. Remember that you need to have an output to make use of this option!

View

The **Data Editor** button is listed first in this menu option and brings the data editor to the front. The data editor looks like a spreadsheet (Fig. 5.8) listing variables in columns and the corresponding observations in rows. String variables are displayed in red, value labels and encoded numeric variables with value labels in blue, and numeric variables in black. The data editor can be used to enter new or amend old data across the rows or the columns of the spreadsheet. Entering the new value on the cell and pressing **Enter** will take you to the next row, while, if you press **Tab**, you are able to work across the rows. Blank columns and rows will be marked as missing, so make sure that you do not skip columns and rows when entering new data. Note that there are different types of variables in the data editor and their

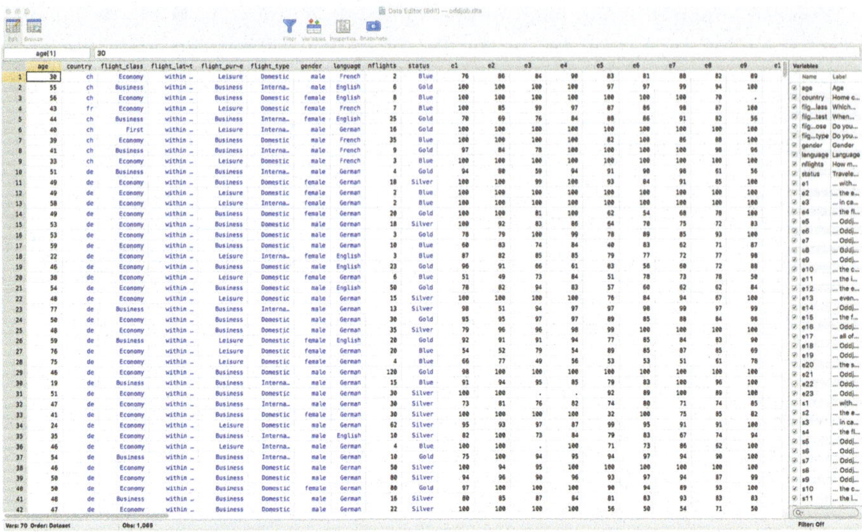

Fig. 5.8 The Stata data editor

properties can be changed in the **Properties** sub-window located at the bottom right of the data editor screen. Another way to bring Data Editor to the front is by typing `edit` in the command window of the Main/Startup Window. Similarly, typing `browse` will open up the data editor (browse) mode, which allows you to navigate through your dataset.

The **View** menu (which is only found on the Mac version) includes other editing options that work in combination with *.do* files (Do-file Editor), graphs (Graph Editor), and structural equation modelling (SEM) (SEM-Builder). These options are beyond the scope of this book and will not be discussed in detail here.

Data

The **Data** menu provides subcommands that allow for summarizing, inspecting, creating, changing or restructuring your dataset.

Under ► Data ► Describe data, you can view or inspect the dataset in a detailed way, including the content, type, and values of the variables included in the dataset. This offers useful information that complements the compact overview that you can obtain from ► Data ► Describe data ► Summary statistics.

Under ► Data ► Data Editor, you can access the different data editor modes (i.e., the edit and browse mode). As discussed above, these options are useful if you want to browse through the specific records in your data (i.e., the browse mode) or wish to change the content of a specific cell (i.e., the edit mode).

By going to ► Data ► Sort, you can sort the data according to the values of one or more specific variable(s). Data are usually sorted according to the respondents' key identifying variable; this a numerical variable that is often called the *id* variable. Depending on the purpose of the analysis, you could, for example, sort

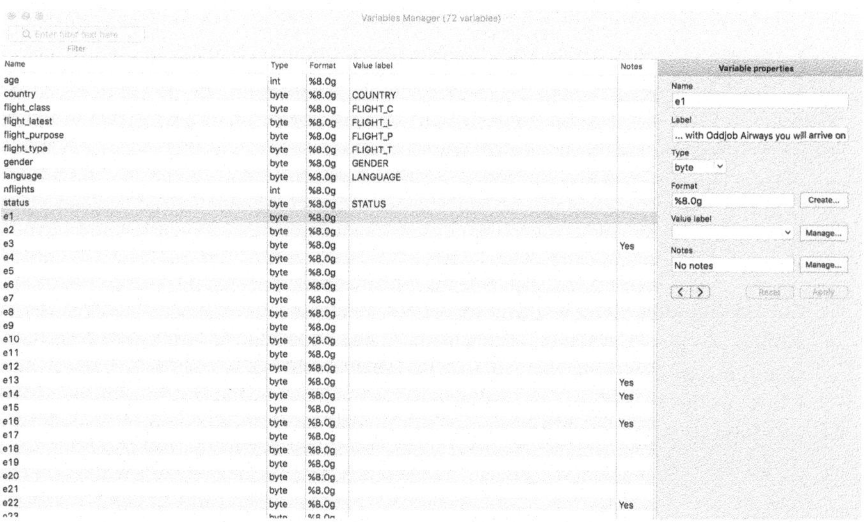

Fig. 5.9 The variables manager

the data according to the *age* of the respondents, which will sort the observations in an ascending order, listing the youngest respondents first, followed by the older respondents in the subsequent rows. Alternatively, **Advanced sort** in the same dialog box allows the observations to be arranged in descending order.

Under ► Data ► Create or change data, you find several options for creating new variables. The options ► Create new variable and ► Create new variable (extended) allow you to create a new variable from one (or more) existing variables. To change or recode the contents of existing variables, you need to select the option ► Other variable-transformation commands ► Recode categorical variable. This command allows you to recode variable values or to combine sets of values into one value. For example, as shown in Fig. 5.12, you could create a new variable called *age_dummy* that splits the sample into young to middle-aged (say, 40 or less) and old passengers (say, more than 40).

An important element of the Data menu is the **Variables Manager**, which is a tool that allows you to organize and structure the variables in your dataset (see Fig. 5.9). Among others it allows you to:

1. Sort variables: One click on the first column of **Variables Manager** will sort the variables in an ascending order, while a second click will sort them in a descending order. If you want to restore the order of the variable list, right-click on the first column and select the option **Restore column defaults.**
2. Select by filtering through the variable list and to change the variable properties: Enter the first letter or name of a variable in the filter box on the upper-left part of the screen to filter through the list of the included variables. This is especially useful if you want to zoom into a series of variables with similar names. Entering *e1* in the filter box, for example, will search everything that contains the value

e1. It will bring all variables to the front that have *e1* in their root, including *e1*, *e10*, *e11* to *e19*. Once you click on a specific variable (e.g., *e1*), the name, label, type, format, value labels, and notes under **Variable properties** will be populated; their content can then be edited if necessary. The properties of each of these boxes is briefly discussed below:

- **Name**: lists the name of the specified variable. Stata is case sensitive, meaning that the variable *e1* differs from *E1*. Variable names must begin with letters (a to z) or an underscore (_). Subsequent characters can include letters (a to z), numbers (0–9) or an underscore (_). Note that neither spaces nor special characters (e.g., %, &, /) are allowed.
- **Label**: allows for a longer description of the specified variable. This can, for example, be the definition of the variable or the original survey question. Click on the tick box to select the option to attach a label to a new variable and type in the preferred label for this variable.
- **Type**: specifies the output format type of the variable. **String** refers to words and is stored as `str#`, indicating the string's maximum length. String variables are useful if you want to include open-ended answers, email addresses or any other type of information that is not a number. Numbers are stored in Stata as `byte`, `int`, `long`, `float`, or `double`.[10]
- **Format**: describes the display format associated with a specified variable. As described in the Stata manual,[11] formats are denoted by a `%` sign, followed by the number of characters (i.e., width) and the number of digits following the decimal points. For example, `%10.2g` means that our display format (indicated by %) should be 10 characters wide (indicated by the number 10) with two digits (indicated by the number 2 following the decimal point). The `g` format indicates that Stata fills the 10 display characters with as much as it can fit. In addition, Stata has a range of formats for dates and time stamps to control how data are displayed.[12]
- **Value label**: presents a description of the values that a specified variable takes. It allows the researcher to create, edit or drop the label of a specified variable. The value labels for *gender*, for example, can be specified as *female* (for values coded as 1) and *male* (for values coded as 0).

3. Keep variables: if you wish to work with a subset of your variables, select all the relevant variables (by dragging the selected variables) and right-click. Then select the option **Keep only selected variables**. Note that this should *only* be done after careful consideration, as dropping relevant variables can ruin the dataset!
4. Drop variables: this is similar to the previous action, but now you only select the variables that you want to drop from the variables list. You can do so by first selecting the variables that you want to drop, then right-click and select the

[10]http://www.stata.com/manuals14/ddatatypes.pdf
[11]http://www.stata.com/manuals14/u.pdf
[12]http://www.stata.com/manuals14/dformat.pdf

option **Drop selected variables**. Here, you also think *very* carefully before dropping variables from the list!

The default missing value in Stata is called *system missing*, which is indicated by a period (.) in the dataset. In addition, there are 26 missing values, also called *extended missing values*, ranging from `.a` to `.z`, depending on how the data are stored. These are necessary to understand the type of the missing data. As discussed in the *Missing Data* section, missing values arise for various reasons. In some occasions, such as in panel studies where the same respondent is approached repeatedly, some respondents may refuse to answer a question at each interview leading to missing observations in some interviews. It could also be that the researcher decides to skip a question from the questionnaire. While both situations lead to missing values, their nature differs. The extended missing values in Stata allow the researcher to distinguish between these different reasons by assigning an `.a` to the first situation and `.b` to the second situation, etc.

Graphics

Under ► Graphics, Stata offers a range of graphical options and tools with which to depict data. These vary from graphical options that support distributional graphs, including two-way graphs, charts, histograms, box and contour plots to graphs that support more advanced statistical methods such as time series, panel, regression, survival techniques, etc. We will discuss the application and interpretation of the different plots as we move through the different chapters and statistical techniques in this book.

Statistics

Under ► Statistics, you find numerous analysis procedures, several of which we will discuss in the remainder of the book. For example, under **Summaries, tables, and tests**, you can request univariate and bivariate statistics. The rest are numerous types of regression techniques, as well as a range of other multivariate analysis techniques. We will discuss descriptive statistics in the next section. In Chap. 6, we will describe models that fall within the group ► Linear models and related, while Chap. 7 will discuss techniques in the ► Multivariate analysis group.

User

Under ► User, you find three empty data, graphs, and statistics subdirectories. Stata programmers can use these to add menu options.

Window

The ► Window option enables you to bring the different types of windows to front. You can also zoom in or minimize the screen.

Help

The ▶ Help function may come in handy if you need further guidance. Under ▶ Help, you find documentations and references that show you how to use most of the commands included in Stata.

5.9 Data Management in Stata

In this section, we will illustrate the application of some of the most commonly used commands for managing data in Stata. These include the following:

- restrict observations,
- create a new variable from existing variable(s), and
- recode variables.

5.9.1 Restrict Observations

You can also use the **Summary Statistics** command to restrict certain observations from the analyses by means of a pre-set condition (e.g., display the summary statistics only for those between 25 and 54 years old). To restrict observations, go to ▶ Data ▶ Describe data ▶ Summary statistics, which opens a screen similar to Fig. 5.10. Under **Variables: (leave empty for all variables)**, enter the condition `if age > 24 & age < 55.` to specify your restriction and then click on **OK**. Stata will now only display the summary statistics for the observations that satisfy this condition. In the Stata command (see below) this restriction appears as an *if* statement.

```
summarize if age>24 & age<55
```

To summarize cases with only valid (non-missing) observations, it is common to add the following rule after the pre-set condition: `& age! = missing ()`. In Stata language, `!=` means "not equal" and `missing()` means "all numerical and string variables included in the dataset." All together, `& age! = missing()` means "if age is not missing from all the included variables in the dataset".

```
summarize if age>24 & age<55 & age !=missing()
```

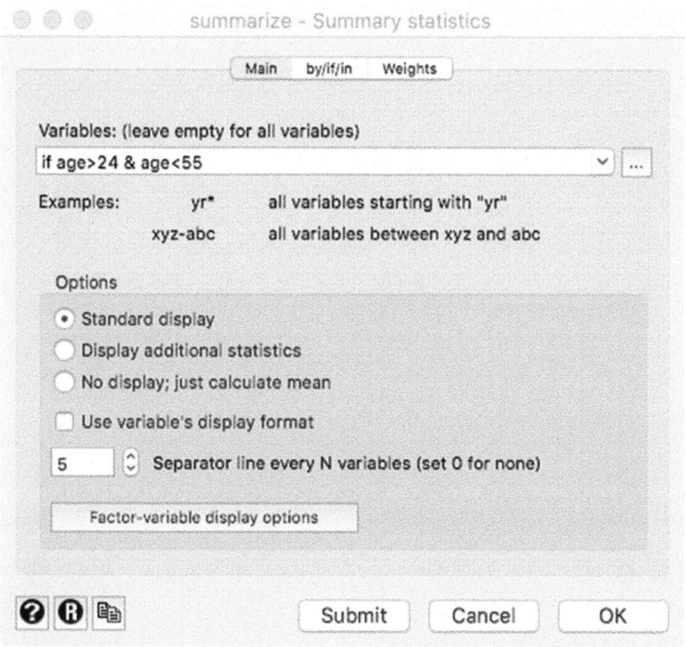

Fig. 5.10 Restrict observations

5.9.2 Create a New Variable from Existing Variable(s)

The **Create new variable (extended)** command enables you to create new variables containing the interquartile range, median, row means, standardized values, and many more different options. We will not discuss all these options in this section, but will demonstrate the power of this tool by creating and index variable from the mean of the following three items related to travelers' satisfaction: *sat1*, *sat2*, and *sat3*. Go to ► Data ► Create or change data ► Create new variable (extended), which will open a dialog box similar to Fig. 5.11.

Next, enter the name of the new variable (e.g., *rating_index*) in the **Generate variable** box on the upper left part of the screen and select **Row Mean** from the **Egen Function** drop-down menu. Under **Generate variable as type**, the variable type **Float** should be automatically selected (i.e., this is Stata's default for numeric variable types). Finally, enter *sat1*, *sat2*, and *sat3* in the **Variables** box and click on **OK**. You have now created a new variable called *rating_index* that appears at the bottom of the variable list. Alternatively, you can type the following command:

```
egen float rating_index = rowmean(sat1 sat2 sat3)
```

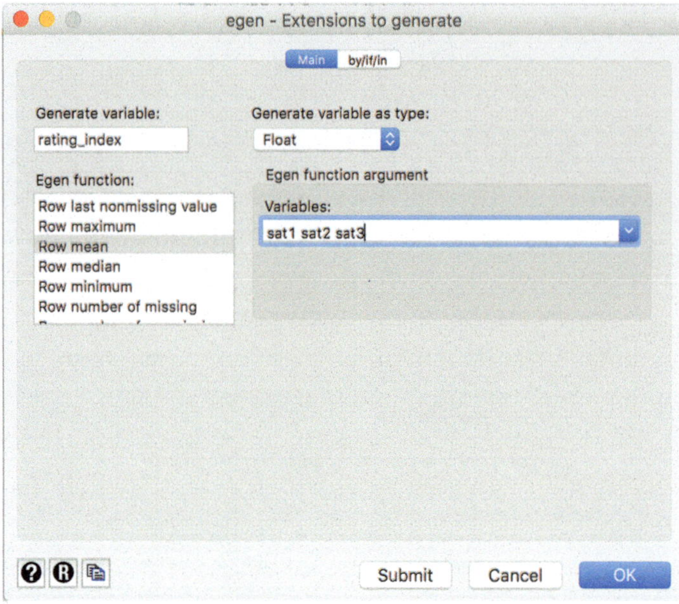

Fig. 5.11 Create new variable(s) extended

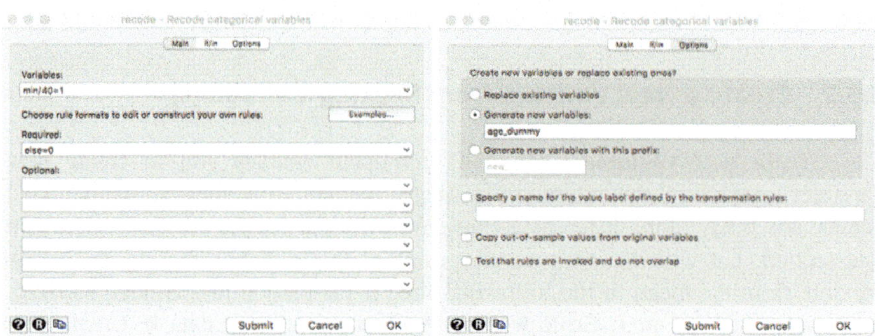

Fig. 5.12 Recode into different variables (Main and Options tabs)

5.9.3 Recode Variables

Recoding (i.e., changing or transforming) the values of an existing variable according to a number of rules is a key data management activity. Numeric variables can be changed by means of the `recode` command. Go to ► Data ► Create or change data ► Other variable-transformation commands ► Recode categorical variables. This will open a dialog box similar to the **Main** Tab left in Fig. 5.12.

Specify the name of the variable that you want to recode (i.e., *age*) under **Variables** in the **Main** tab. Next, specify the values of the new variable under

Required. These are based on the values of the original variable (*age*). The option `(min/40 = 1)` indicates that all values ranging from the smallest age observations to the age of 40 should be coded as 1. Under **Optional** all other age observations are coded as 0 `(else = 0)`.

Next, click on the **Options** tab (right in Fig. 5.12). In the dialog box that follows, you need to indicate whether you want to: (1) **Replace existing variables**, (2) **Generate new variables**, or (3) **Generate new variables with this prefix**. We always recommend using either the second option or the third. If you were to use the first option, any changes you make to the variable will result in the overwriting of the original variable. Consequently, if you thereafter wish to return to the original data, you will either need to revert to a saved previous version, or need to enter all the data again, because Stata cannot undo these actions! Select the second option (i.e., **Generate new Variables**), enter the name of the new variable (i.e., *age_dummy*), and then click on **OK**. Alternatively, the recoding of this variable can be obtained through the following command:

```
recode age(min/40=1)(else=0),generate(age_dummy)
```

You have now created a new dichotomous variable (i.e., *age_dummy*) located at the bottom of the variables list.

Everyone makes mistakes!

If you are worried that commands may not change the data as desired, type `preserve` in the **Command** window. This keeps a snapshot of the data in the computer's memory. Should you wish to revert, simply type `restore` to go back to where you were. You could also save a copy of the dataset under a new name as a milestone and then later delete it manually. This allows you to back up multiple steps and across work sessions.

5.10 Example

We will now examine the dataset *Oddjob.dta* in closer detail by following all the steps in Fig. 5.1. Cleaning the data generally requires checking for interviewer fraud, suspicious response patterns, data entry errors, outliers, and missing data. Several of these steps rely on statistics and graphs, which we discussed in the context of descriptive statistics (e.g., box plots and scatter plots). Note that missing values strategies and the multiple imputation technique will be described and illustrated by means of *Oddjob.dta* in the ⭳ Web Appendix (→ Downloads).

5.10.1 Clean Data

Since the data were cleaned earlier, we need not check for interviewer fraud or suspicious response patterns. Beside double data entries to detect and minimize errors in the process of data entry, exploratory data analysis is required to spot data entry errors that have been overlooked.

A first step in this procedure is to look at the minimum and maximum values of the relevant variables to detect values that are not plausible (i.e., fall outside the expected range of scale categories). To do so, go to ► Data ► Describe Data ► Summary statistics, which opens a dialog box like the one in Fig. 5.13. The **Variables** box should be left empty to obtain the statistics for all the variables in the dataset. Next, select the **Standard display** option that requires the number of observations, the mean, standard deviation, minimum, and maximum values. Proceed by clicking on **OK**. Alternatively, the dialog box for summary statistics can be brought to front by typing `db summarize` in the **Command** window and clicking on enter.

Table 5.7 shows a partial display of the summary statistics, including the number of observations, means, standard deviation, as well as minimum and maximum values. Under **Obs**, we can see that all the listed variables are observed across all 1,065 respondents, meaning that none of the selected variables suffer from missing observations. Among others, it appears that the *age* of the travelers varies between

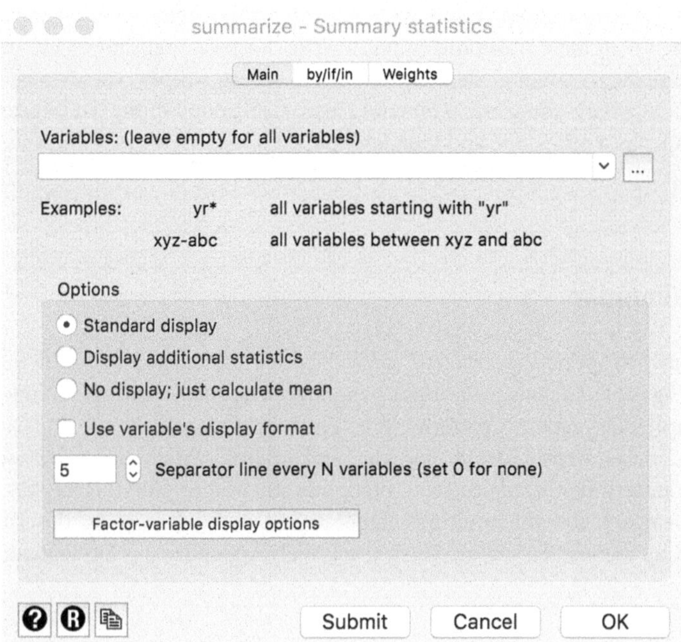

Fig. 5.13 Summary statistics dialog box

Table 5.7 A (partial) output of summary statistics

Variable	Obs	Mean	Std. Dev.	Min	Max
age	1,065	50.41972	12.27464	19	101
country	1,065	2	1.551739	1	5
flight_class	1,065	2.798122	.4352817	1	3
flight_lat~t	1,065	3.788732	1.368779	1	6
flight_pur~e	1,065	1.507042	.5001853	1	2
flight_type	1,065	1.476056	.499661	1	2
gender	1,065	1.737089	.4404212	1	2
language	1,065	1.237559	.4473162	1	3
nflights	1,065	13.41878	20.22647	1	457
status	1,065	1.498592	.7204373	1	3

19 and 101, while the *number of flights* varies between 1 and 457 flights over the past 12 months. Particularly the maximum value in number of flights appears to be implausible. While this observation could represent a flight attendant, it appears more reasonable to consider this observation an outlier, which may need to be eliminated, depending on the type of analysis.

5.10.2 Describe Data

In the next step, we describe the data in more detail, focusing on those statistics and graphs that were not part of the previous step. To do so, we make use of graphs, tables, and descriptive statistics. In Fig. 5.14, we show how you can ask for each previously discussed graph, table, and statistic in Stata.

5.10.2.1 Univariate Graphs and Tables

Bar Charts

To produce a bar chart that plots the *age* of respondents against their *country* of residence, go to ► Graphics ► Bar chart. This will take you to a dialog box where the **Main** tab is displayed by default (left of Fig. 5.15). Under **Type of data** specify the type of bar chart that you want displayed (**Graph of summary statistics**). Next, under **Orientation**, you should select the option **Vertical,** given that it is Stata's default and indicates the direction in which the bar chart is displayed.

Under **Statistics to plot**, select the variable *age* under **Variables** and indicate that you want to display the mean of this variable for all valid observations by selecting the option **mean**. Next, click on the **Categories** tab (displayed to the right of Fig. 5.15) to indicate how you want to categorize the data in the bar chart. Tick the first box, **Group 1** and select the variable *country* from the drop-down menu under **Grouping Variable.**

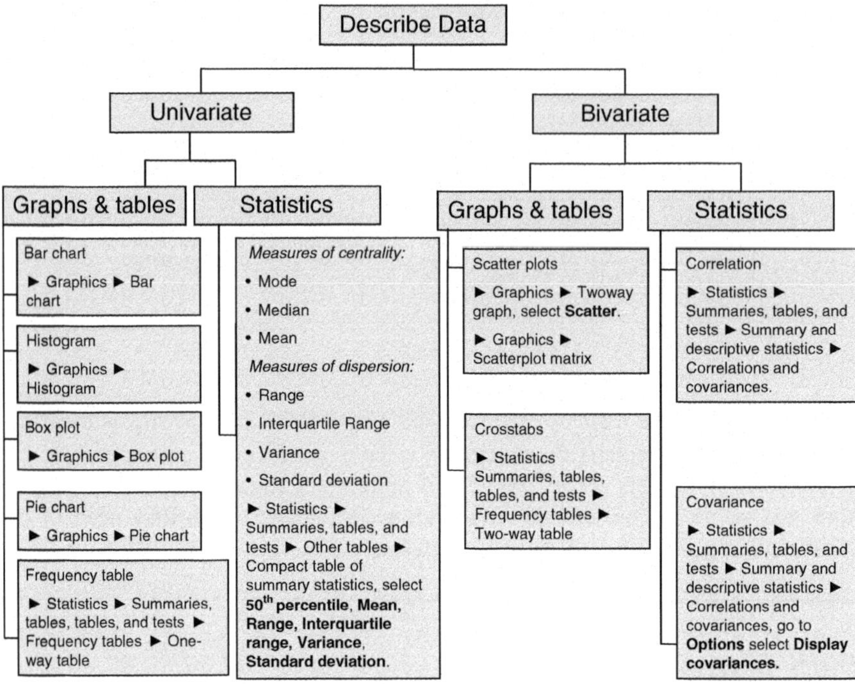

Fig. 5.14 How to ask for graphs, tables, and statistics in Stata

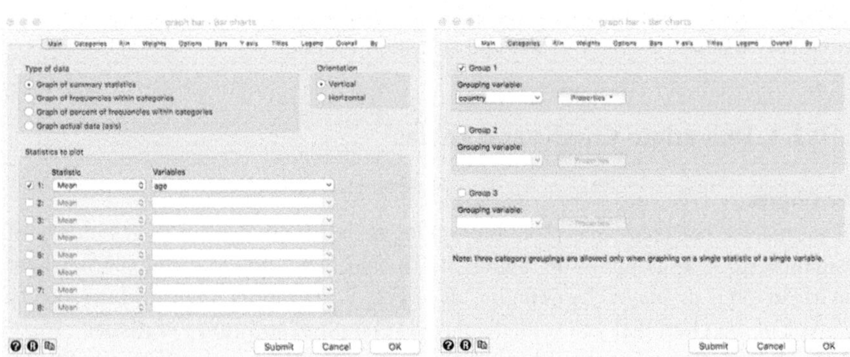

Fig. 5.15 Main dialog box, Bar chart

By default, Stata displays the value labels of the grouping variable horizontally, but you can change this. By clicking on the **Properties** button (see Fig. 5.16), which is next to the **Grouping variable** box, set **Labels** to **Angle: 45°** and then click **Accept**. Stata will then show a graph as in Fig. 5.17.

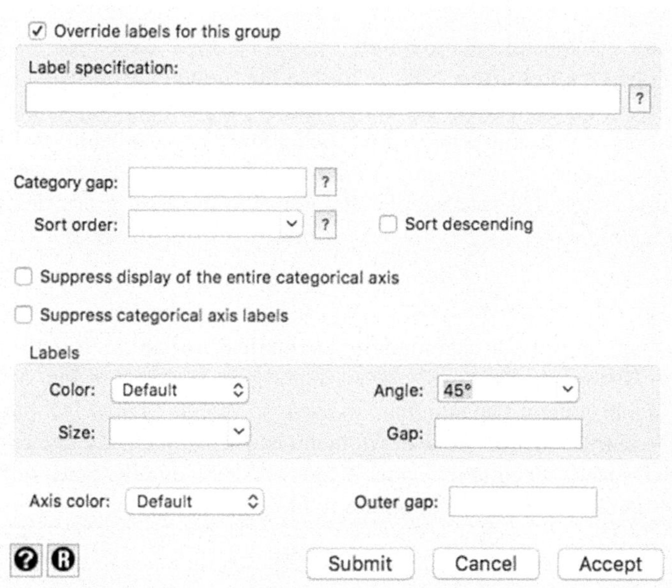

Fig. 5.16 Properties dialog box

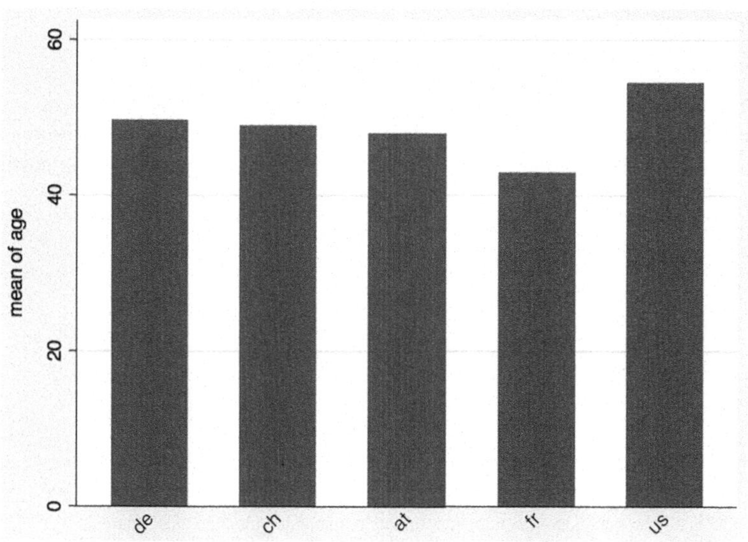

Fig. 5.17 A bar chart

Tip: You can also edit your graph by right-clicking on the graph. Select the option **Start Graph Editor**, which will start the Graph Editing menu. Double click on the x-axis and select the option **Label properties** to adjust the angle of the labels in the same way as described above. You can additionally also adjust the range of the values you want to display on the x-axis by specifying the desired range. The same applies to adjusting the label properties of the y-axis.

Histograms

Histograms are useful for summarizing numerical variables. If you go to ► Graphics ► Histogram, Stata will open a dialog box as shown in Fig. 5.18 on the left. To plot a histogram that summarizes the respondents' *age*, select the relevant variable *age* in the **Variable** drop-down menu and select the **Data are continuous** option. Next, specify **Frequency** under **Y axis** and click on **OK.** Stata will produce a histogram as shown in Fig. 5.18 on the right.

Box Plot

To ask for a box plot, go to ► Graphics ► Box plot, which will open a dialog box similar to the one in Fig. 5.19.

Specify the option **Vertical** under **Orientation** to display the box plot vertically. Next, select the relevant variable *age* in the **Variables** box and then click on **OK**. A box plot like the one in Fig. 5.20 appears.

Pie Charts

Pie charts are useful for displaying categorical or binary variables. Create a pie chart by going to ► Graphics ► Pie chart, which will open a dialog box similar to

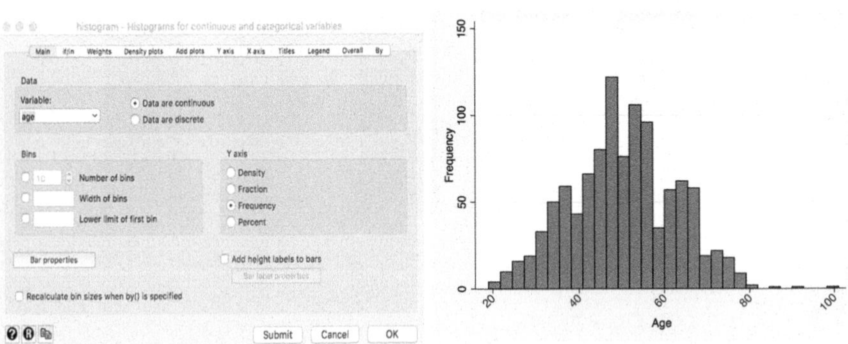

Fig. 5.18 Dialog box, histogram and a histogram

Fig. 5.19 Box plots graph dialog box

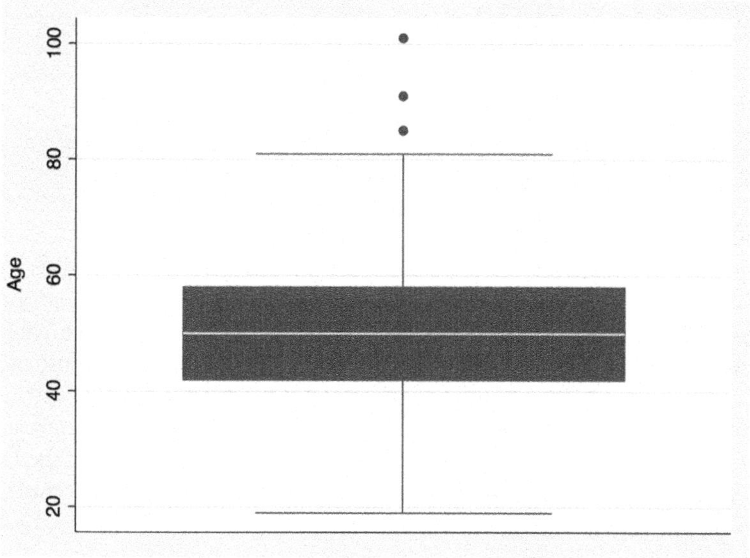

Fig. 5.20 Box plot

that displayed in Fig. 5.21 on the left. In Stata, the default (standard) option is **Graph by Categories**. Plot respondents' membership status (*status*) by selecting *status* under **Category variable** and then clicking on **OK**. Stata will show a pie chart similar to the one on the right in Fig. 5.21.

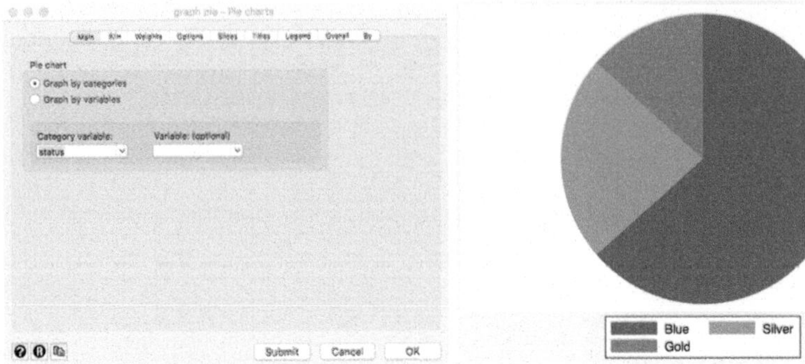

Fig. 5.21 Pie chart

Table 5.8 Example of a frequency table in Stata

```
tabulate country

       Home |
    country |      Freq.     Percent        Cum.
------------+-----------------------------------
         de |        695       65.26       65.26
         ch |         66        6.20       71.46
         at |        108       10.14       81.60
         fr |          1        0.09       81.69
         us |        195       18.31      100.00
------------+-----------------------------------
      Total |      1,065      100.00
```

Frequency Tables

We can produce a frequency table by clicking on ► Statistics ► Summaries, tables, and tests ► Frequency tables ► One-way table. Select the variable *country* under **Categorical variable** and then click on **OK**. This operation will produce Table 5.8, which displays the value of each country with the corresponding absolute number of observations (i.e., **Freq.**), the relative values (i.e., **Percent**), as well as the cumulative relative values (i.e., **Cum.**). It shows that **65.25 percent** of our sample consists of travelers who reside in Germany, followed by travelers from the United States (**18.31 percent**), Austria (**10.14 percent**), Switzerland (**6.20 percent**), and, finally, France (**0.09 percent**).

5.10.2.2 Univariate Statistics

Another useful way of summarizing your data is through the **Tabstat** option, which you can find under ► Statistics ► Summaries, tables, and tests ► Other tables ► Compact table of summary statistics. Selecting this menu option opens a dialog box similar to that in Fig. 5.22. In the **Variables** box, select the relevant variables for which you would like to display summary statistics. These variables range from *age* to *gender*. Next, under **Statistics to display** tick the blank boxes and specify the

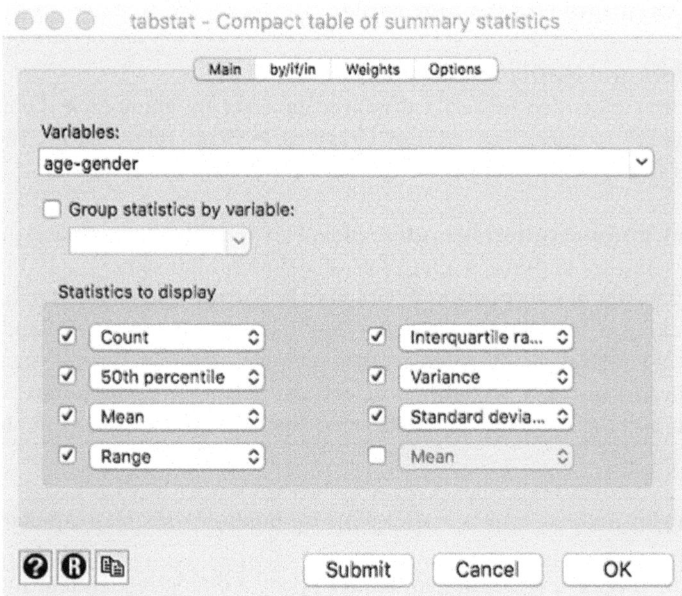

Fig. 5.22 Dialog box for univariate statistics

Table 5.9 Example of a summary table using the tabstat option

```
tabstat age-gender, statistics (count p50 mean range iqr var sd)
```

stats	age	country	flight~s	flight~t	fligh~se	fligh~pe	gender
N	1065	1065	1065	1065	1065	1065	1065
p50	50	1	3	4	2	1	2
mean	50.41972	2	2.798122	3.788732	1.507042	1.476056	1.737089
range	82	4	2	5	1	1	1
iqr	16	2	0	2	1	1	1
variance	150.6667	2.407895	.1894702	1.873557	.2501853	.2496611	.1939708
sd	12.27464	1.551739	.4352817	1.368779	.5001853	.499661	.4404212

descriptive statistics to be displayed from the drop-down menu. In this example, we specify the number of nonmissing observations (**Count** in Stata), the median (**50th percentile** in Stata), the **Mean**, the **Range**, the **Interquartile range**, the **Variance**, and the **Standard deviation**.

Clicking on **OK** will display an output similar to that in Table 5.9, which includes the number of nonmissing observations (**N**), median (**p50**), mean (**mean**), range (**range**), interquartile range (**iqr**), variance (**var**), and standard deviation (**sd**).

Note that `tabstat` arranges the table like a dataset, with the variables as columns. For a more a typical table, go to the **Options** tab (see Fig. 5.22) and change the value of the **Use as columns** from **Variables** to **Statistics**.

5.10.2.3 Bivariate Graphs and Tables

Scatter Plots and Matrix Scatter Plots

Matrix scatter plots can be easily displayed in Stata by going to ▶ Graphics ▶ Twoway graph (scatter, line, etc.) or Graphics ▶ Scatterplot matrix. These two separate graphs differ in the following way:

1. **Twoway graph (scatter, line, etc.)**: plots two variables against each other, but you can display multiple scatter plots at a time on the same graph. Produce a twoway scatter plot by going to ▶ Graphics ▶ Twoway graph (scatter, line, etc.) and clicking on **Create**. In the dialog box that opens (on the left-hand side of Fig. 5.23), select **Scatter**, enter the outcome variable overall satisfaction (*overall_sat*) in the **Y variable** box, and *age* in the **X variable** box, click on **Accept**, and then on **OK**. The right-hand side of Fig. 5.23 shows the resulting scatter plot.

2. **Scatterplot matrix**: creates scatter plots for multiple variables simultaneously.

Going to Graphics ▶ Scatterplot matrix opens a dialog box similar to the one shown on the left of Fig. 5.24. Select *nflights*, *overall_sat*, and *age* under **Variables.** To change the density of the scatter matrix, click on the button **Marker Properties.** This opens the dialog window on the right of Fig. 5.24. Select the option **Point** under **Symbol**, click on **Accept**, and then on **OK.**

This will produce the matrix scatter plot shown in Fig. 5.25. The graph shows distinct bands, because the overall satisfaction variable takes on 7 values (1 to 7).

The first scatter plot in the first row reveals the relationship between the *number of flights* and *overall price satisfaction*. Next to this on the first row, the relationship between the *number of flights* and *age* is displayed. In the second row, the relationship between the *overall price satisfaction* and *number of flights* is shown and so on. In the same row, to the right, the relationship

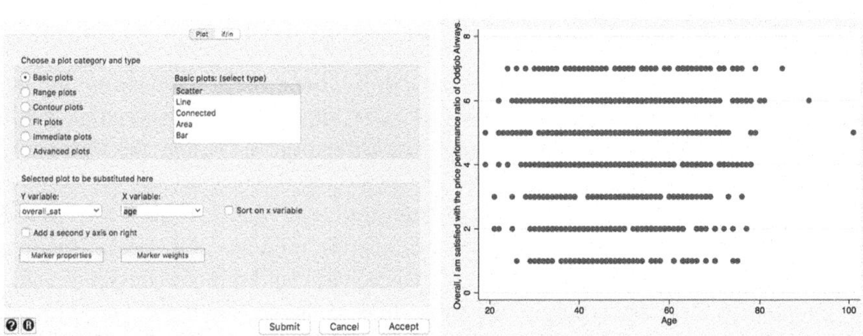

Fig. 5.23 Dialog box and scatter plot

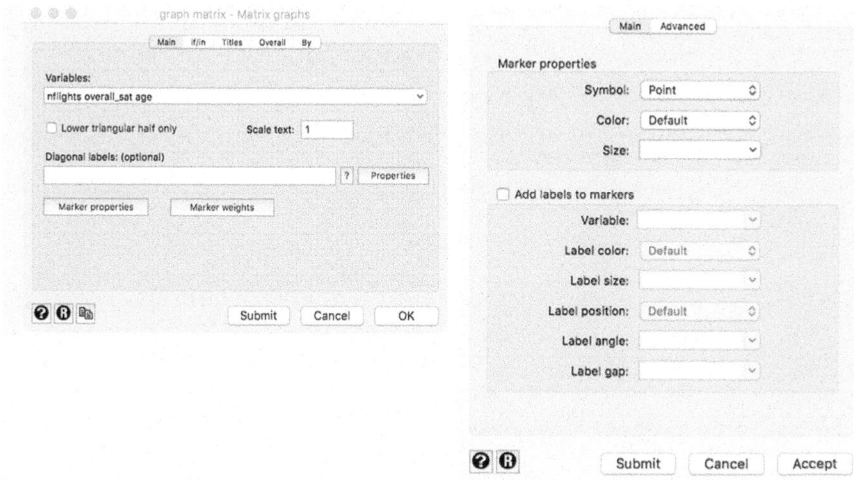

Fig. 5.24 Dialog box scatter plot matrix and the marker properties box

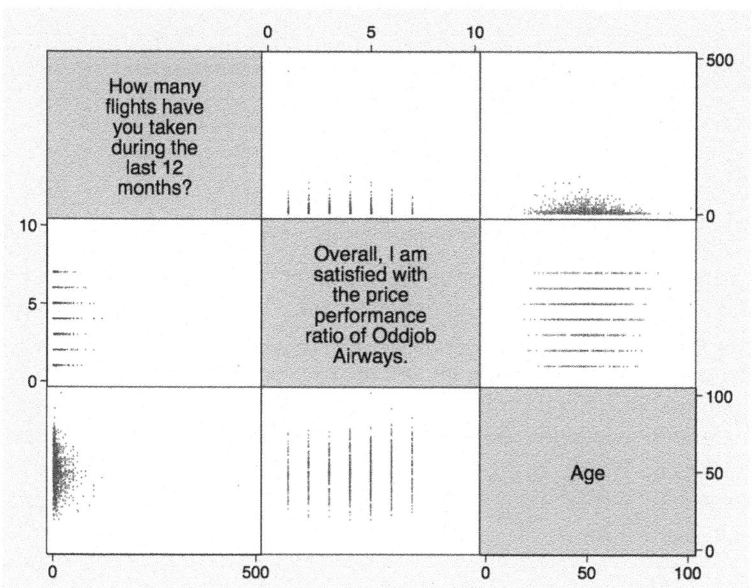

Fig. 5.25 Matrix scatter plot

between the *overall price satisfaction* and *age* is displayed. Finally, the last row displays the relationships between *age* and *number of flights* (first scatter plot), as well as the relationship between *age* and *overall price satisfaction* (second scatter plot).

Table 5.10 Example of a crosstab

```
tabulate country gender, column row

+-------------------+
| Key               |
|-------------------|
|     frequency     |
|  row percentage   |
| column percentage |
+-------------------+

      Home |        Gender
   country |    female       male |     Total
-----------+----------------------+----------
        de |       180        515 |       695
           |     25.90      74.10 |    100.00
           |     64.29      65.61 |     65.26
-----------+----------------------+----------
        ch |        17         49 |        66
           |     25.76      74.24 |    100.00
           |      6.07       6.24 |      6.20
-----------+----------------------+----------
        at |        25         83 |       108
           |     23.15      76.85 |    100.00
           |      8.93      10.57 |     10.14
-----------+----------------------+----------
        fr |         1          0 |         1
           |    100.00       0.00 |    100.00
           |      0.36       0.00 |      0.09
-----------+----------------------+----------
        us |        57        138 |       195
           |     29.23      70.77 |    100.00
           |     20.36      17.58 |     18.31
-----------+----------------------+----------
     Total |       280        785 |     1,065
           |     26.29      73.71 |    100.00
           |    100.00     100.00 |    100.00
```

Cross Tabulation

Cross tabulations are useful for understanding the relationship between two variables scaled on a nominal or ordinal scale. To create a crosstab, go to ► Statistics ► Summaries, tables, and tests ► Frequency tables ► Two-way table with measures of association. It is important that you specify which variable goes in the column and which in the rows. Choose *country* under **Row variable** and *gender* under **Column variable**. Next, select the boxes **Within-column relative frequencies** and **Within-row relative frequencies** under **Cell Contents** to display the column and row percentages. Click on **OK** and Stata produces a table similar to the one in Table 5.10.

5.10.2.4 Bivariate Statistics: Correlation and Covariance

In Stata, we can calculate bivariate correlations by going to ► Statistics ► Summaries, tables, and tests ► Summary and descriptive statistics ► Correlations and covariances. In the dialog box that opens, select the variables to be considered in the analysis. For example, enter *nflights*, *age*, and *overall_sat* in the **Variables** box. When you click on **OK**, Stata will produce a correlation matrix like the one in Table 5.11.

Table 5.11 Correlation matrix produced in Stata

```
correlate nflights age overall_sat
(obs=1,065)

             | nflights      age overal~t
-------------+---------------------------
    nflights |   1.0000
         age |  -0.1158   1.0000
 overall_sat |  -0.1710   0.1207   1.0000
```

Table 5.12 Covariance matrix produced in Stata

```
correlate nflights age overall_sat, covariance
(obs=1,065)

             | nflights      age overal~t
-------------+---------------------------
    nflights |   409.11
         age | -28.7408  150.667
 overall_sat | -5.62173  2.40806  2.64118
```

The correlation matrix in Table 5.11 shows the correlation between each pairwise combination of three variables. For example, the correlation between *nflights* and *age* is −**0.1158**, which is negative and rather weak. Conversely, the relationship between *age* and *overall_sat* is positive (**0.1207**), but still rather weak.

Alternatively, a covariance matrix as in Table 5.12, can be obtained as follows. Go to ► Statistics ► Summaries, tables, and tests ► Summary and descriptive statistics ► Correlations and covariances and ticking the box **Display covariances** in the **Options** tab. This will produce the following covariance matrix.

5.11 Cadbury and the UK Chocolate Market (Case Study)

The UK chocolate market is expected to be £6.46 billion in 2019. Six subcategories of chocolates are used to identify the different chocolate segments: boxed chocolate, molded bars, seasonal chocolate, count lines, straight lines, and "other."

To understand the UK chocolate market for molded chocolate bars, we have a dataset *(chocolate.dta)* that includes a large supermarket's weekly sales of 100g molded chocolate bars from January 2016 onwards. This data file can be downloaded from the book's ⤓ Web Appendix (→ Downloads). This file contains a set of variables. Once you have opened the dataset, you will see the set of variables we discuss in the main Stata window under *Variables* (see Fig. 5.7).

The first variable is *week*, indicating the week of the year and starts with Week 1 of January 2016. The last observation for 2016 ends with observation 52, but the variable continues to count onwards for 16 weeks in 2017.[13] The next variable is

[13]Note an ordinary year has 52 weeks and 1 day, while a leap year has 52 weeks and 2 days. This is because 1 week comprises part of 2016 and part of 2017.

sales, which indicates the weekly sales of 100g Cadbury bars in £. Next, four price variables are included, *price1-price4*, which indicate the price of Cadbury, Nestlé, Guylian, and Milka in £. Next, *advertising1-advertising4* indicate the amount of £ the supermarket spent on advertising each product during that week. A subsequent block of variables, *pop1-pop4*, indicate whether the products were promoted in the supermarket by means of point of purchase advertising. This variable is measured as yes/no. Variables *promo1-promo4* indicate whether the product was put at the end of the supermarket aisle, where it is more noticeable. Lastly, *temperature* indicates the weekly average temperature in degrees Celsius.

You have been tasked with providing descriptive statistics for a client by means of this dataset. To help you with this task, the client has prepared a number of questions:

1. Do Cadbury's chocolate sales vary substantially across different weeks? When are Cadbury's sales at their highest? Please create an appropriate graph to illustrate any patterns.
2. Please tabulate point-of-purchase advertising for Cadbury against point-of-purchase advertising for Nestlé. In addition, create a few more crosstabs. What are the implications of these crosstabs?
3. How do Cadbury's sales relate to the price of Cadbury? What is the strength of the relationship?
4. Which descriptive statistics are appropriate for describing the usage of advertising? Which statistics are appropriate for describing point-of-purchase advertising?

5.12 Review Questions

1. Imagine you are given a dataset on car sales in different regions and are asked to calculate descriptive statistics. How would you set up the analysis procedure?
2. What summary statistics could best be used to describe the change in profits over the last 5 years? What types of descriptive statistics work best to determine the market shares of five different types of insurance providers? Should we use just one or multiple descriptive statistics?
3. What information do we need to determine if a case is an outlier? What are the benefits and drawbacks of deleting outliers?
4. Download the codebook of the Household Income and Labour Dynamics in Australia (HILDA) Survey at: http://melbourneinstitute.unimelb.edu.au/hilda/for-data-users/user-manuals. Is this codebook clear? What do you think of its structure?

5.13 Further Readings

https://www.stata.com/manuals13/mi.pdf

This manual provides a hands-on application of multiple imputation in Stata.

http://www.stata.com/manuals13/u.pdf

There is a detailed description of Stata's properties coupled with hands-on examples in Stata's manual.

http://www.ats.ucla.edu/stat/mult_pkg/whatstat/default.htm

The following link provides a range of general guidelines regarding the type of statistical analyses of and tips about the application of various methods using Stata.

https://www.iriseekhout.com/promotie/thesis/

General strategies on how to deal with missing data.

https://eagereyes.org/pie-charts

This blog provides a good description of the advantages and disadvantages related to pie charts.

This book provides an introduction to multivariate analysis and easy to follow discussions of fundamental statistical concepts.

Hair, J. F., Jr., Black, W. C., Babin, B. J., & Anderson, R. E. (2013). *Multivariate data analysis. A global perspective* (7th ed.). Upper Saddle River: Prentice-Hall.

This book teaches you how to make high-quality graphs in Stata.

Mitchel, M.N. *A Visual Guide to Stata Graphics.* (2008). Stata Press: StataCorp LP.

This book teaches readers how to decipher the meaning of symbols, tables, and figures included in research reports in order to improve their ability to critically assess such reports.

Huck, W.S. (2014). Reading statistics and research (6th ed.). Harlow: Pearson.

SticiGui at http://www.stat.berkeley.edu/~stark/SticiGui/Text/correlation.htm.

These websites interactively demonstrate how strong the correlations between different datasets are.

References

Agarwal, C. C. (2013). *Outlier analysis.* New York: Springer.

Agresti, A., & Finlay, B. (2014). *Statistical methods for the social sciences* (4th ed.). London: Pearson.

Barchard, K. A., & Pace, L. A. (2011). Preventing human error: The impact of data entry methods on data accuracy and statistical results. *Computers in Human Behavior, 27*(5), 1834–1839.

Barchard, K. A., & Verenikina, Y. (2013). Improving data accuracy: Electing the best data checking technique. *Computers in Human Behavior, 29*(50), 1917–1912.

Baumgartner, H., & Steenkamp, J.-B. E. M. (2001). Response styles in marketing research: A cross-national investigation. *Journal of Marketing Research, 38*(2), 143–156.

Carpenter, J., & Kenward, M. (2013). *Multiple imputation and its application.* New York: Wiley.

Cohen, J. (1988). *Statistical power analysis for the behavioral sciences* (2nd ed.). Hillsdale: Lawrence Erlbaum Associates.

Drolet, A. L., & Morrison, D. G. (2001). Do we really need multiple-item measures in service research? *Journal of Service Research, 3*(3), 196–204.

Eekhout, I., de Vet, H. C. W., Twisk, J. W. R., Brand, J. P. L., de Boer, M. R., & Heymans, M. W. (2014). Missing data in a multi-item instrument were best handled by multiple imputation at the item score level. *Journal of Clinical Epidemiology, 67*(3), 335–342.

Gladwell, M. (2008). *Outliers: The story of success*. New York: Little, Brown, and Company.

Graham, J. W. (2012). *Missing data: Analysis and design*. Berlin et al.: Springer.

Hair, J. F., Jr., Black, W. C., Babin, B. J., & Anderson, R. E. (2010). *Multivariate data analysis. A global perspective* (7th ed.). Upper Saddle River: Pearson.

Harzing, A. W. (2005). Response styles in cross-national survey research: A 26-country study. *International Journal of Cross Cultural Management, 6*(2), 243–266.

Johnson, T., Kulesa, P., Lic, I., Cho, Y. I., & Shavitt, S. (2005). The relation between culture and response styles. Evidence from 19 countries. *Journal of Cross-Cultural Psychology, 36*(2), 264–277.

Krippendorff, K. (2012). *Content analysis: An introduction to its methodology*. Thousand Oaks: Sage.

Little, R. J. A. (1998). A test of missing completely at random for multivariate data with missing values. *Journal of the American Statistical Association, 83*(404), 1198–1202.

Paulsen, A., Overgaard, S., & Lauritsen, J. M. (2012). Quality of data entry using single entry, double entry and automated forms processing – An example based on a study of patient-reported outcomes. *PloS One, 7*(4), e35087.

Rubin, D. B. (1987). *Multiple imputation for nonresponse in surveys*. New York: Wiley.

Sarstedt, M., Diamantopoulos, A., Salzberger, T., & Baumgartner, P. (2016). Selecting single items to measure doubly-concrete constructs: A cautionary tale. *Journal of Business Research, 69*(8), 3159–3167.

Schafer, J. L. (1997). *Analysis of incomplete multivariate data*. London: Chapman & Hall.

White, I. R., Royston, P., & Wood, A. M. (2011). Multiple imputation using chained equations: Issues and guidance for practice. *Statistics in Medicine, 30*(4), 377–399.

Hypothesis Testing & ANOVA

6

Keywords
α-Inflation • α error • Analysis of Variance (ANOVA) • β error • Bonferroni correction • Degrees of freedom • eta-squared • F-test • F-test of sample variance • Factor variable • Familywise error rate • Independent samples *t*-test • Interaction effect • Levene's test • Mann-Whitney U test • Main effect • Nonparametric tests • Null and alternative hypothesis • Omega-squared • One-sample *t*-test • One-tailed test • One-way ANOVA • *p*-value • Paired samples *t*-test • Parametric test • Practical significance • Post hoc tests • Power of a test • Shapiro-Wilk test • Significance level • Sampling error • Statistical significance • *t*-test • Test statistic • Tukey's honestly significant difference test • Two-sample *t*-test • Two-tailed test • Two-way ANOVA • Type I and type II error • Welch correction • Wilcoxon signed-rank test • *z*-test

Learning Objectives
After reading this chapter, you should understand:

- The logic of hypothesis testing.
- The steps involved in hypothesis testing.
- What a test statistic is.
- Types of error in hypothesis testing.
- Common types of *t*-tests, one-way, and two-way ANOVA.
- How to interpret Stata output.

6.1 Introduction

Do men or women spend more money on the Internet? Assume that the mean amount that a sample of men spends online is \$200 per year against a women sample's mean of \$250. When we compare mean values such as these, we always

E. Mooi et al., *Market Research*, Springer Texts in Business and Economics,
DOI 10.1007/978-981-10-5218-7_6

expect some difference. But, how can we determine if such differences are statistically significant? Establishing statistical significance requires ascertaining whether such differences are attributable to chance or not. In this chapter, we will introduce hypothesis testing and how this helps determine statistical significance.

6.2 Understanding Hypothesis Testing

A **hypothesis** is a statement about a certain effect or parameter (such as a mean or correlation) that can be tested using a sample drawn from the population. A hypothesis may comprise a claim about the difference between two sample parameters (e.g., there is a difference between males' and females' mean spending). It can also be a test of a judgment (e.g., teenagers spend an average of 4 h per day on the Internet). Data from the sample are used to obtain evidence against, or in favor of, the statement.

Hypothesis testing is performed to infer whether or not a certain effect is statistically significant. **Statistical significance** means that the effect is so large that it is unlikely to have occured by chance. Whether results are statistically significant depends on several factors, including the size of the effect, the variation in the sample data, and the sample size (Agresti and Finlay 2014). When drawing a sample from the population, there is always some probability that we might reach the wrong conclusion due to a sampling error, which is the difference between the sample and the population characteristics. To determine whether the claim is true, we start by setting an acceptable probability (called the **significance level**) that we could incorrectly conclude there is an effect when, in fact, there is none. This significance level is typically set at 0.05, which corresponds to a 5% error probability. Next, subject to the claim made in the hypothesis, we should decide on the correct type of test to perform. This involves making decisions regarding four aspects.

First, we should understand the testing situation. What exactly are we testing: Are we comparing one value against a fixed value, or are we comparing groups, and, if so, how many?

Second, we need to specify the nature of the samples: Is our comparison based on *paired samples* or *independent samples* (the difference is discussed later in this chapter)?

Third, we should check assumptions about the distribution of our data to determine whether parametric or nonparametric tests are appropriate. **Parametric tests** make assumptions about the properties of the population distributions from which the data are drawn, while **nonparametric tests** are not based on any distributional assumptions.

Fourth, we need to decide on the region where we can reject our hypothesis; that is, whether the region of rejection will be on one side or both sides of the sampling distribution.

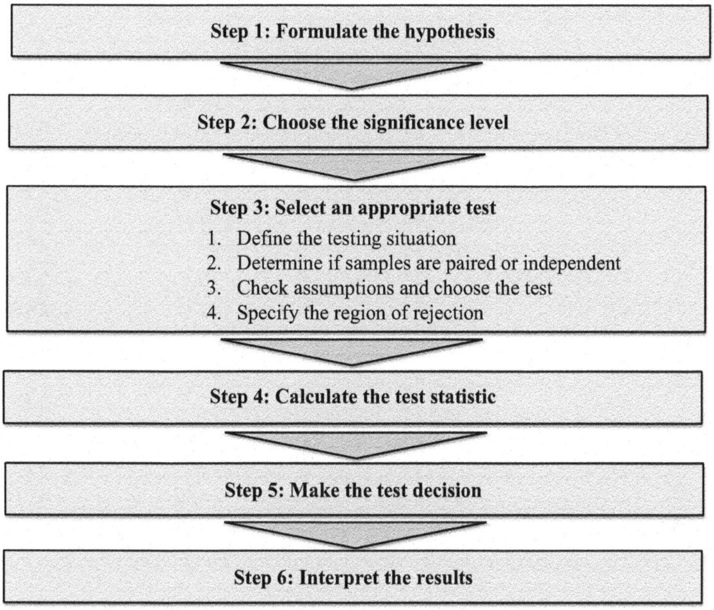

Fig. 6.1 Steps involved in hypothesis testing

Once these four aspects are sorted, we calculate the *test statistic*, which identifies whether the sample supports or rejects the claim stated in the hypothesis. We can then decide to either reject or support the hypothesis. This decision enables us to draw market research conclusions in the final step. Figure 6.1 illustrates the six steps involved in hypothesis testing.

To illustrate the process of hypothesis testing, consider the following example: A department store chain wants to evaluate the effectiveness of three different in-store promotion campaigns that drive the sales of a specific product. These campaigns comprise: (1) a point of sale display, (2) a free tasting stand, and (3) in-store announcements. To help with the evaluation, the management decides to conduct a one-week experiment during which 30 stores are randomly assigned to each campaign type. This random assignment is important to obtain reliable and generalizable results, because randomization should equalize the effect of systematic factors not accounted for in the experimental design (see Chap. 4). Table 6.1 shows the sales of the three different in-store promotion campaigns. The table also contains information on the service type (personal or self-service) in the first column and the *marginal means* representing the means of sales within stores in the last column. The very last row also shows the marginal mean of the type of campaign, while the very last cell shows the grand mean across all the service types and campaigns.

Table 6.1 Sales data

	Sales (units)			
Service type	Point of sale display (stores 1–10)	Free tasting stand (stores 11–20)	In-store announcements (stores 21–30)	Marginal mean
Personal	50	55	45	50.00
Personal	52	55	50	52.33
Personal	43	49	45	45.67
Personal	48	57	46	50.33
Personal	47	55	42	48.00
Self-service	45	49	43	45.67
Self-service	44	48	42	44.67
Self-service	49	54	45	49.33
Self-service	51	54	47	50.67
Self-service	44	44	42	43.33
Marginal mean	47.30	52.00	44.7	48.00 Grand mean

We will use these data to carry out tests to compare the different in-store promotion campaigns' mean sales separately, or in comparison to each other. We first discuss each test theoretically (including the formulas), followed by an empirical illustration. You will realize that the formulas are not as complicated as you might have thought! These formulas contain Greek characters and we have therefore included a table describing each Greek character in the ⭳ Web Appendix (→ Downloads).

6.3 Testing Hypotheses on One Mean

6.3.1 Step 1: Formulate the Hypothesis

Hypothesis testing starts with the formulation of a null and alternative hypothesis. A **null hypothesis** (indicated as H_0) is a statement expecting no difference or effect. Conversely, an **alternative hypothesis** (indicated as H_1) is the hypothesis against which the null hypothesis is tested (Everitt and Skrondal 2010). Examples of potential null and alternative hypotheses on the campaign types are:

1. H_0: The mean sales in stores that installed a point of sale display are equal to or lower than 45 units.
   H_1: The mean sales in stores that installed a point of sale display are higher than 45 units.

2. H_0: There's no difference in the mean sales of stores that installed a point of sale display and those that installed a free tasting stand (statistically, the average sales of the point of sale display = the average sales of the free tasting stand).
H_1: There's a difference in the mean sales of stores that installed a point of sale display and those that installed a free tasting stand (statistically, the average sales of the point of sale display ≠ the average sales of the free tasting stand).

Hypothesis testing can have two outcomes: First, we do not reject the null hypothesis. This suggests there is no difference and that the null hypothesis can be retained. However, it would be incorrect to subsequently conclude that the null hypothesis is true, as it is not possible to "prove" the non-existence of a certain effect or condition. For example, one can examine any number of crows and find that they are all black, yet that would not make the statement "There are no white crows" true. Only sighting one white crow will prove its existence. Second, we could reject the null hypothesis, thus finding support for the alternative hypothesis in which some effect is expected. This outcome is, of course, desirable in most analyses, as we generally want to show that something (such as a promotion campaign) is related to a certain outcome (e.g., sales). Therefore, we frame the effect that we want to investigate as the alternative hypothesis.

Inevitably, each hypothesis test has a certain degree of uncertainty so that even if we reject a null hypothesis, we can never be totally certain that this was the correct decision. Consequently, market researchers should use terms such as "find support for the alternative hypothesis" when they discuss their findings. Terms like "prove" should never be part of hypotheses testing.

Returning to our initial example, the management only considers a campaign effective if the sales it generates are higher than the 45 units normally sold (you can choose any other value, the idea is to test the sample mean against a given standard). One way of formulating the null and alternative hypotheses in respect of this expectation is:

$$H_0\colon \mu \leq 45$$

$$H_1\colon \mu > 45$$

In words, the null hypothesis H_0 states that the population mean, indicated by μ (pronounced as *mu*), is equal to or smaller than 45, whereas the alternative hypothesis H_1 states that the population mean is larger than 45. It is important to note that the hypothesis always refers to a population parameter, in this case, the population mean, represented by μ. It is practice for Greek characters to represent population parameters and for Latin characters to indicate sample statistics (e.g., the Latin $\bar{x}$). In this example, we state a *directional hypothesis* as the alternative hypothesis, which is expressed in a direction (higher) relative to the standard of 45 units. Since we presume that during a campaign, the product sales are higher, we posit a *right-tailed hypothesis* (as opposed to a *left-tailed hypothesis*) for the alternative hypothesis H_1.

Alternatively, presume we are interested in determining whether the mean sales of the point of sale display (μ_1) are equal to the mean sales of the free tasting stand (μ_2). This implies a *non-directional hypothesis*, which can be written as:

$$H_0: \mu_1 = \mu_2$$

$$H_1: \mu_1 \neq \mu_2$$

The difference between the two general types of hypotheses is that a directional hypothesis looks for an increase or a decrease in a parameter (such as a population mean) relative to a specific standard. A non-directional hypothesis tests for *any* difference in the parameter, whether positive or negative.

6.3.2 Step 2: Choose the Significance Level

No type of hypothesis testing can evaluate the validity of a hypothesis with absolute certainty. In any study that involves drawing a sample from the population, there is always some probability that we will erroneously retain or reject the null hypothesis due to **sampling error**, which is a difference between the sample and the population. In statistical testing, two types of errors can occur (Fig. 6.2):

1. a true null hypothesis is incorrectly rejected (**type I** or $\boldsymbol{\alpha}$ **error**), and
2. a false null hypothesis is not rejected (**type II** or $\boldsymbol{\beta}$ **error**).

In our example, a type I error occurs if we conclude that the point of sale displays increased the sales beyond 45 units, when in fact it did not increase the sales, or may have even decreased them. A type II error occurs if we do not reject the null hypothesis, which suggests there was no increase in sales, even though the sales increased significantly.

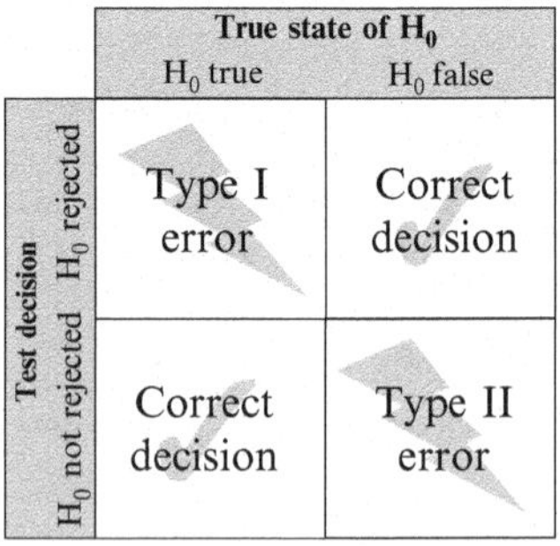

Fig. 6.2 Type I and type II errors

A problem with hypothesis testing is that we don't know the true state of the null hypothesis. Fortunately, we can establish a level of confidence that a true null hypothesis will not be erroneously rejected. This is the maximum probability of a type I error that we want to allow. The Greek character α (pronounced as *alpha*) represents this probability and is called the *significance level*. In market research reports, this is indicated by phrases such as "this test result is significant at a 5% level." This means that the researcher allowed for a maximum chance of 5% of mistakenly rejecting a true null hypothesis.

The selection of an α level depends on the research setting and the costs associated with a type I error. Usually, α is set to 0.05, which corresponds to a 5% error probability. However, when researchers want to be conservative or strict in their testing, such as when conducting experiments, α is set to 0.01 (i.e., 1%). In exploratory studies, an α of 0.10 (i.e., 10%) is commonly used. An α-level of 0.10 means that if you carry out ten tests and reject the null hypothesis every time, your decision in favor of the alternative hypothesis was, on average, wrong once. This might not sound too high a probability, but when much is at stake (e.g., withdrawing a product because of low satisfaction ratings) then 10% may be too high.

Why don't we simply set α to 0.0001% to really minimize the probability of a type I error? Setting α to such a low level would obviously make the erroneous rejection of the null hypothesis very unlikely. Unfortunately, this approach introduces another problem. The probability of a type I error is inversely related to that of a type II error, so that the smaller the risk of a type I error, the higher the risk of a type II error! However, since a type I error is considered more severe than a type II error, we control the former directly by setting α to a desired level (Lehmann 1993).

Sometimes statistical significance can be established even when differences are very small and have little or no managerial implications. Practitioners, usually refer to "significant" as being practically significant rather than statistically significant. **Practical significance** refers to differences or effects that are large enough to influence the decision-making process. An analysis may disclose that a type of packaging increases sales by 10%, which could be practically significant. Whether results are practically significant depends on the management's perception of the difference or effect and whether this warrants action. It is important to separate statistical significance from practical significance. Statistical significance does not imply practical significance.

Another important concept related to this is the **power of a statistical test** (defined by $1 - \beta$, where β is the probability of a type II error), which represents the probability of rejecting a null hypothesis when it is, in fact, false. In other words, the power of a statistical test is the probability of rendering an effect significant when it is indeed significant. Researchers want the power of a test to be as high as

Box 6.1 Statistical Power of a Test

Market researchers encounter the common problem that they, given a predetermined level of α and some fixed parameters in the sample, have to calculate the sample size required to yield an effect of a specific size. Computing the required sample size (called a *power analysis*) can be complicated, depending on the test or procedure used. Fortunately, Stata includes a power and sample size module that allows you to determine sample size under different conditions. In the ⭳ Web Appendix (→ Downloads), we use data from our example to illustrate how to run a power analysis using Stata. An alternative is G * Power 3.0, which is sophisticated and yet easy-to-use. It can be downloaded freely from http://www.psycho.uni-duesseldorf.de/abteilungen/aap/gpower3/.

If these tools are too advanced, Cohen (1992) suggests required sample sizes for different types of tests. For example, detecting the presence of differences between two independent sample means for $\alpha = 0.05$ and a power of $\beta = 0.80$ requires a sample size (n) of $n = 26$ for large differences, $n = 64$ for medium differences, and $n = 393$ for small differences. This demonstrates that sample size requirements increase disproportionally when the effect that needs to be detected becomes smaller.

possible, but when maximizing the power and, therefore, reducing the probability of a type II error, the occurrence of a type I error increases (Everitt and Skrondal 2010). Researchers generally view a statistical power of 0.80 (i.e., 80%) as satisfactory, because this level is assumed to achieve a balance between acceptable type I and II errors. A test's statistical power depends on many factors, such as the significance level, the strength of the effect, and the sample size. In Box 6.1 we discuss the statistical power concept in greater detail.

6.3.3 Step 3: Select an Appropriate Test

Selecting an appropriate statistical test is based on four aspects. First, we need to assess the testing situation: What are we comparing? Second, we need to assess the nature of the samples that are being compared: Do we have one sample with observations from the same object, firm or individual (paired), or do we have two different sets of samples (i.e., independent)? Third, we need to check the assumptions for normality to decide which type of test to use: Parametric (if we meet the test conditions) or non-parametric (if we fail to meet the test conditions)? This step may involve further analysis, such as testing the homogeneity of group variances. Fourth, we should decide on the region of rejection: Do we want to test one side or both sides of the sampling distribution? Table 6.2 summarizes these four aspects with the recommended choice of test indicated in the grey shaded boxes. In the following we will discuss each of these four aspects.

Table 6.2 Selecting an appropriate test

Test #	Testing situation *What do we compare*	Nature of samples *Paired vs. Independent*	Choice of Test[a] *Assumptions*	*Parametric*	*Non-parametric*	Region of rejection *One or two-sided test*
1	One group against a fixed value	Not applicable	Shapiro-Wilk test = normal	One sample *t*-test		One or two-sided
			Shapiro-Wilk test ≠ normal		Wilcoxon signed-rank test	One or two-sided
2	Outcome variable across two groups	Paired sample	Shapiro-Wilk test = normal & Levene's test: $\sigma_1^2 = \sigma_2^2$	Paired *t*-test		One or two-sided
			Shapiro-Wilk test ≠ normal & Levene's test: $\sigma_1^2 = \sigma_2^2$	Paired *t*-test[b]		One or two-sided
			Shapiro-Wilk test = normal & Levene's test: $\sigma_1^2 \neq \sigma_2^2$	Paired *t*-test with Welch's correction		One or two-sided
			Shapiro-Wilk test ≠ normal & Levene's test: $\sigma_1^2 \neq \sigma_2^2$		Wilcoxon matched-pairs signed-rank test	One or two-sided
3	Outcome variable across two groups	Independent samples	Shapiro-Wilk test = normal & Levene's test: $\sigma_1^2 = \sigma_2^2$	Two-sample *t*-test		One or two-sided
			Shapiro-Wilk test ≠ normal & Levene's test: $\sigma_1^2 = \sigma_2^2$	Two-sample *t*-test[b]		One or two-sided
			Shapiro-Wilk test = normal & Levene's test: $\sigma_1^2 \neq \sigma_2^2$	Two-sample *t*-test with Welch's correction		One or two-sided
			Shapiro-Wilk test ≠ normal & Levene's test: $\sigma_1^2 \neq \sigma_2^2$		Wilcoxon rank-sum test	One or two-sided
4	Outcome variable across three or more groups	One factor variable, independent samples	Shapiro-Wilk test = normal & Levene's test: $\sigma_1^2 = \sigma_2^2$	One-way ANOVA: *F*-test		Two-sided*[c]

(continued)

Table 6.2 (continued)

	Testing situation	Nature of samples	Choice of Test[a]			Region of rejection
Test #	*What do we compare*	*Paired vs. Independent*	*Assumptions*	*Parametric*	*Non-parametric*	*One or two-sided test*
			Shapiro-Wilk test $\neq$ normal & Levene's test: $\sigma_1^2 = \sigma_2^2$	One-way ANOVA: *F*-test		Two-sided*
			Shapiro-Wilk test $=$ normal & Levene's test: $\sigma_1^2 \neq \sigma_2^2$	One-way ANOVA: *F*-test with Welch's correction		Two-sided*
			Shapiro-Wilk test $\neq$ normal & Levene's test: $\sigma_1^2 \neq \sigma_2^2$		Kruskal-Wallis rank test	Two-sided*
5	Outcome variable across three or more groups	Two factor variables, independent samples	Shapiro-Wilk test $=$ normal & Levene's test: $\sigma_1^2 = \sigma_2^2$	Two-way ANOVA: *F*-test		Two-sided*
			Shapiro-Wilk test $\neq$ normal & Levene's test: $\sigma_1^2 = \sigma_2^2$	Two-way ANOVA:*F*-test		Two-sided*
			Shapiro-Wilk test $=$ normal & Levene's test: $\sigma_1^2 \neq \sigma_2^2$	Two-way ANOVA: *F*-test with Welch's correction		Two-sided*
			Shapiro-Wilk test $\neq$ normal & Levene's test: $\sigma_1^2 \neq \sigma_2^2$		Kruskal-Wallis rank test	Two-sided*

[a]Selection applies to sample sizes > 30

[b]If the sample size is less than 30, you should transform your outcome variable—through logarithms, square root or power transformations—and re-run the normality test. Alternatively, if normality remains an issue, you should use a non-parametric test that does not rely on the normality assumption

[c]* = Note that although the underlying alternative hypothesis in ANOVA is two-sided, its *F*-statistic is based on the *F*-distribution, which is right-skewed with extreme values only in the right tail of the distribution

6.3.3.1 Define the Testing Situation

When we test hypotheses, we may find ourselves in one of three situations. First, we can test if we want to compare a group to a hypothetical value (test 1). In our example, this can be a pre-determined target of 45 units to establish whether a promotion campaign has been effective or not. Second, we may want to compare the outcome variable (e.g., sales) across two groups (tests 2 or 3). Third, we may wish to compare whether the outcome variable differs between three or more levels of a categorical variable (also called a **factor variable**) with three or more sub-groups (tests 4 or 5). The factor variable is the categorical variable that we use to define the groups (e.g., three types of promotion campaigns). Similarly, we may have situations in which we have two or more factor variables (e.g., three types of promotion campaigns for two types of service). Each of these situations leads to different tests. When assessing the testing situation, we also need to establish the nature of the dependent variable and whether it is measured on an interval or ratio scale. This is important, because parametric tests are based on the assumption that the dependent variable is measured on an interval or ratio scale. Note that we only discuss situations when the test variable is interval or ratio-scaled (see Chap. 3).

6.3.3.2 Determine If Samples Are Paired or Independent

Next, we need to establish whether we compare *paired samples* or *independent samples*. The rule of thumb for determining the samples' nature is to ask if a respondent (or object) was sampled once or multiple times. If a respondent was sampled only once, this means that the values of one sample reveal no information about the values of the other sample. If we sample the same respondent or object twice, it means that the reported values in one period may affect the values of the sample in the next period.[1] Ignoring the "nested" nature of the data increases the probability of type I errors. We therefore need to understand the nature of our samples in order to select a test that takes the dependency between observations (i.e., paired versus independent samples tests) into account. In Table 6.2, test 2 deals with paired samples, whereas tests 3, 4 and, 5 deal with independent samples.

6.3.3.3 Check Assumptions and Choose the Test

Subsequently, we need to check the distributional properties and variation of our data before deciding whether to select a parametric or a non-parametric test.

[1] In experimental studies, if respondents were paired with others (as in a matched case control sample), each person would be sampled once, but it still would be a paired sample.

Normality Test

To test whether the data are normally distributed, we conduct the **Shapiro-Wilk test** (Shapiro and Wilk 1965) that formally tests for normality. Without going into too much detail, the Shapiro-Wilk test compares the correlation between the observed sample scores (which take the covariance between the sample scores into account) with the scores expected under a standard normal distribution. The resulting ratio is called the W-statistic and its scaled version is known as the V-statistic. When the distribution is close to normal, the V statistic will be close to 1 and the associated p-value will be larger than 0.05. Large deviations from a V of 1 will therefore be coupled with p-values that are smaller than 0.05, suggesting that the sample scores are not normally distributed. The Kolmogorov-Smirnov test, which we discuss in Box 6.2, is another popular test to check for normality. An alternative strategy to check for normality is by means of visual inspection, which we discuss in Box 6.3.

Equality of Variances Test

We use **Levene's test** (Levene 1960), also known as the ***F*-test of sample variance**, to test for the equality of the variances between two or more groups of data. The null hypothesis is that population variances across the sub-samples are the same, whereas the alternative hypothesis is that they differ. If the p-value associated with Levene's statistic (referred to as W0 in Stata) is lower than 0.05, we reject the null hypothesis, which implies that the variances are heterogeneous. Conversely, a p-value larger than 0.05 indicates homogeneous variances. In Levene's original paper, the formula for the test statistic is based on the sample mean (Levene

Box 6.2 The Kolmogorov-Smirnov Test

An important (nonparametric) test for normality is the *one-sample Kolmogorov–Smirnov (KS) test*. We can use it to test whether or not a variable is normally distributed. Technically, when assuming a normal distribution, the KS test compares the sample scores with an artificial set of normally distributed scores that has the same mean and standard deviation as the sample data. However, this approach is known to yield biased results, which are modified by means of the Lilliefors correction (1967). The Lilliefors correction takes into consideration that we do not know the true mean and standard deviation of the population. An issue with the KS test with the Lilliefors correction is that it is very sensitive when used on large samples and often rejects the null hypothesis if very small deviations are present. This also holds for Stata's version of the KS test, which only works well for very large sample sizes (i.e., at least 10,000 observations). Consequently, Stata does not recommend the use of a one-sample KS test (for more, read the information in Stata's help file on the KS test: https://www.stata.com/manuals14/rksmirnov.pdf).

Box 6.3 Visual Check for Normality
You can also use plots to visually check for normality. The *normal probability plot* visually contrasts the probability distribution of the test variable's ordered sample values with the cumulative probabilities of a standard normal distribution. The probability plots are structured such that the cumulative frequency distribution of a set of normally distributed data falls in a straight line. This straight line serves as a reference line, meaning that sample values' deviations from this straight line indicate departures from normality (Everitt and Skrondal 2010). The *quantile plot* is another type of probability plot, which differs from the former by comparing the quantiles (Chap. 5) of the sorted sample values with the quantiles of a standard normal distribution. Here, again, the plotted data that do not follow the straight line reveal departures from normality. The normal probability plot assesses the normality of the data in the middle of the distribution well. The quantile plot is better equipped to spot non-normality in the tails. Of the two, the quantile plots are most frequently used. Note that visual checks are fairly subjective and should always be used in combination with more formal checks for normality.

1960), which performs well when the variances are symmetric and have moderate tailed distributions. For skewed data, Brown and Forsythe (1974) proposed a modification of Levene's test whereby the group sample's median (referred to as W50 in Stata) replaces the group sample mean. Alternatively, the group sample's trimmed mean can also replace the group sample mean, whereby 10% of the smallest and largest values of the sample are removed before calculating the mean (referred to as W10 in Stata). If the data are normally distributed, the *p*-values associated with W0, W10, and W50 will all align, thus all be higher than 0.05. If the data are not normally distributed, this might not be the case. In this situation, we should focus on the *p*-values associated with the sample's median (W50), because this measure is robust for samples that are not normally distributed.

Parametric Tests

It is clear-cut that when the normality assumption is met, we should choose a *parametric test.* The most popular parametric test for examining one or two means is the ***t*-test**, which can be used for different purposes. For example, the *t-test* can be used to compare one mean with a given value (e.g., do males spend more than $150 a year online?). The **one-sample *t*-test** is an appropriate test. Alternatively, we can use a *t*-test to test the mean difference between two samples (e.g., do males spend more time online than females?). In this case, a **two-sample *t*-test** is appropriate. **The independent samples *t*-tests** considers two distinct groups, such as males versus females, or users versus non-users. Conversely, **the paired samples *t*-test** relates to the same set of twice observed objects (usually respondents), as in a before-after experimental design discussed in Chap. 4. We are, however, often interested in examining the differences between the means of more than two groups of respondents. Regarding our introductory example, we might be interested in evaluating the differences between the point of sale display, the free tasting stand, and the in-store announcements' mean sales. Instead of making several paired comparisons by means of separate *t*-tests, we should use the **Analysis of Variance (ANOVA)**. The ANOVA is useful when three or more means are compared and, depending on how many variables define the groups to be compared (will be discussed later in this chapter), can come in different forms.

The parametric tests introduced in this chapter are very robust against normality assumption violations, especially when the data are distributed symmetrically. That is, small departures from normality usually translate into marginal differences in the *p*-values, particularly when using sample sizes greater than 30 (Boneau 1960). Thus, even if the Shapiro–Wilk test suggests the data are not normally distributed, we don't have to be concerned that the parametric test results are far off, provided we have sample sizes greater than 30. The same holds for the ANOVA in cases where the sample sizes per group exceed 30.

In sum, with sample sizes greater than 30, we choose a parametric test even when the Shapiro-Wilk test suggests that the data are not normally distributed. When sample sizes are less than 30, we can transform the distribution of the outcome variable—through logarithms, square root, or power transformations (Chap. 5)—so that it approaches normality and re-run a normality test. If violations of normality remain an issue, you should use a non-parametric test that is not based on the normality assumption.

Next, we can also have a situation in which the data are normally distributed, but the variances between two or more groups of data are unequal. This issue is generally unproblematic as long as the group-specific sample sizes are (nearly) equal. If group-specific sample sizes are different, we recommend using parametric tests, such as the two-sample *t*-tests and the ANOVA, in combination with tests that withstand or correct the lack of equal group variances, such as **Welch's correction**. Welch's modified test statistic (Welch 1951) adjusts the underlying parametric tests if the variances are not homogenous in order to control for a type I error. This is particularly valuable when population variances differ and groups comprise very unequal sample sizes.[2] In sum, when samples are normally distributed, but the equality of the variance assumption is violated (i.e., the outcome variable is not distributed equally across three or more groups), we choose a parametric test with Welch's correction. Depending on the testing situation this can be: a paired *t*-test with Welch's correction, a one-way ANOVA *F*-test with Welch's correction, or a two-way ANOVA *F*-test with Welch's correction.

[2]Stata does not directly support Welch's correction for an ANOVA, but a user-written package called `wtest` is readily available and can be installed (see Chap. 5 on how to install user-written packages in Stata). This allows you to perform a test similar to the standard ANOVA test with Welch's correction. For more information see Stata's help file: http://www.ats.ucla.edu/stat/stata/ado/analysis/wtest.hlp

Finally, where both the normality and equality of variance assumptions are violated, non-parametric tests can be chosen directly. In the following, we briefly discuss these non-parametric tests.

Non-parametric Tests

As indicated in Table 6.2, there is a non-parametric equivalent for each parametric test. This would be important if the distributions are not symmetric. For single samples, the **Wilcoxon signed-rank test** is the equivalent of one sample *t*-test, which is used to test the hypothesis that the population median is equal to a fixed value. For two-group comparisons with independent samples, the **Mann-Whitney U test** (also called the *Wilcoxon rank-sum test*, or *Wilcoxon–Mann–Whitney test*) is the equivalent of the independent *t*-test, while, for paired samples, this is the *Wilcoxon matched-pairs signed-rank test*. The Mann-Whitney U test uses the null hypothesis that the distributions of the two independent groups being considered (e.g., randomly assigned high and low performing stores) have the same shape (Mann and Whitney 1947). In contrast to an independent sample *t*-test, the Mann-Whitney U test does not compare the means, but the two groups' median scores. Although we will not delve into the statistics behind the test, it is important to understand its logic. The Mann-Whitney U test is based on ranks and measures the differences in location (Liao 2002). The test works by first combining the separate groups into a single group. Subsequently, each outcome variable score (e.g., sales) is sorted and ranked in respect of each condition based on the values, with the lowest rank assigned to the smallest value. The ranks are then averaged based on the conditions (e.g., high versus low performing stores) and the test statistic *U* calculated. The test statistic represents the difference between the two rank totals. That is, if the distribution of the two groups is identical, then the sum of the ranks in one group will be the same as in the other group. The smaller the *p*-value (which will be discussed later in this chapter), the lower the likelihood that the two distributions' similarities have occurred by chance; the opposite holds if otherwise.

The *Kruskal-Wallis rank test* is the non-parametric equivalent of the ANOVA. The null hypothesis of the *Kruskal-Wallis* rank test is that the distribution of the test variable across group sub-samples is identical (Schuyler 2011). Given that the emphasis is on the distribution rather than on a point estimate, rejecting the null hypothesis implies that such distributions vary in their dispersion, central tendency and/or variability. According to Schuyler (2011) and Liao (2002), the following are the steps when conducting this test: First, single group categories are combined into one group with various categories. Next, objects in this variable (e.g., stores/campaigns) are sorted and ranked based on their associations with the dependent variable (e.g., sales), with the lowest rank assigned to the smallest value. Subsequently, the categorical variable is subdivided to reestablish the original single comparison groups. Finally, each group's sum of its ranks is entered into a formula that yields the calculated test statistic. If this calculated statistic is higher than the critical value, the null hypothesis is rejected. The test statistic of the Kruskal-Wallis rank follows a χ^2 distribution with $k - 1$ degrees of freedom. Use the *Kruskal-Wallis*

H test in situations where the group variances are not equal, as it corrects group variances' heterogeneity.

6.3.3.4 Specify the Region of Rejection

Finally, depending on the formulated hypothesis (i.e., directional versus non-directional), we should decide on whether the region of rejection is on one side (i.e., **one-tailed test**) or on both sides (i.e., **two-tailed test**) of the sampling distribution. In statistical significance testing, a one-tailed test and a two-tailed test are alternative ways of computing the statistical significance of a test statistic, depending on whether the hypothesis is expressed directionally (i.e., $<$ or $>$ in case of a one-tailed test) or not (i.e., $\neq$ in case of a two-tailed test). The word tail is used, because the extremes of distributions are often small, as in the normal distribution or bell curve shown in Fig. 6.3 later in this chapter. Instead of the word tail, the word "sided" is sometimes used.

We need to use two-tailed tests for non-directional hypotheses. Even when directional hypotheses are used, two-tailed tests are used for 75% of directional hypotheses (van Belle 2008). This is because two-tailed tests have strong advantages; they are stricter (and therefore generally considered more appropriate) and can also reject a hypothesis when the effect is in an unexpected direction. The use of two-tailed testing for a directional hypothesis is also valuable, as it identifies significant effects that occur in the opposite direction from the one anticipated. Imagine that you have developed an advertising campaign that you believe is an improvement on an existing campaign. You wish to maximize your ability to detect the improvement and opt for a one-tailed test. In doing so, you do not test for the possibility that the new campaign is significantly less effective than the old campaign. As discussed in various studies (van Belle 2008; Ruxton and Neuhaeuser 2010), one-tailed tests should only be used when the opposite direction is theoretically meaningless or impossible (Kimmel 1957; Field 2013). Such an example would apply to controlled experiments where the intervention (i.e., the drug) can only have a positive and no negative outcome, because such differences are removed beforehand and have no possible meaning (e.g., Lichters et al. 2016). The use of two-tailed tests can seem counter to the idea of hypothesis testing, because two-tailed tests, by their very nature, do not reflect any directionality in a hypothesis. However, in many situations when we have clear expectations (e.g., sales are likely to increase), the opposite is also a possibility.

6.3.4 Step 4: Calculate the Test Statistic

Having formulated the study's main hypothesis, the significance level, and the type of test, we can now proceed with calculating the test statistic by using the sample data at hand. The **test statistic** is a statistic, calculated by using the sample data, to assess the strength of evidence in support of the null hypothesis (Agresti and Finlay 2014). In our example, we want to compare the mean with a given standard of

45 units. Hence, we make use of a *one-sample t-test*, whose test statistic is computed as follows:

$$t = \frac{\bar{x} - \mu}{s_{\bar{x}}}$$

Here $\bar{x}$ is the sample mean, μ is the hypothesized population mean, and $s_{\bar{x}}$ the standard error (i.e., the standard deviation of the sampling distribution). Let's first look at the formula's numerator, which describes the difference between the sample mean $\bar{x}$ and the hypothesized population mean μ. If the point of sale display was highly successful, we would expect $\bar{x}$ to be higher than μ, leading to a positive difference between the two in the formula's numerator. Alternatively, if the point of sale display was not effective, we would expect the opposite to be true. This means that the difference between the hypothesized population mean and the sample mean can go either way, implying a two-sided test. Using the data from the second column of Table 6.1, we can compute the marginal mean of the point of sales display campaign as follows:

$$\bar{x} = \frac{1}{n}\sum_{i=1}^{n} x_i = \frac{1}{10}(50 + 52 + \ldots + 51 + 44) = 47.30$$

When comparing the calculated sample mean (47.30) with the hypothesized value of 45, we obtain a difference of 2.30:

$$\bar{x} - \mu = 47.30 - 45 = 2.30$$

At first sight, it appears as if the campaign was effective as sales during the time of the campaign were higher than those that the store normally experiences. However, as discussed before, we have not yet considered the variation in the sample. This variation is accounted for by the *standard error* of $\bar{x}$ (indicated as $s_{\bar{x}}$), which represents the uncertainty of the sample estimate.

This sounds very abstract, so what does it mean? The sample mean is used as an estimator of the population mean; that is, we assume that the sample mean can be a substitute for the population mean. However, when drawing different samples from the same population, we are likely to obtain different sample means. The standard error tells us how much variance there probably is in the mean across different samples from the same population.

Why do we have to divide the difference $\bar{x} - \mu$ by the standard error $s_{\bar{x}}$? We do so, because when the standard error is very low (there is a low level of variation or uncertainty in the data), the value in the test statistic's denominator is also small, which results in a higher value for the t-test statistic. Higher t-values favor the rejection of the null hypothesis. In other words, the lower the standard error $s_{\bar{x}}$, the greater the probability that the population represented by the sample truly differs from the hypothesized value of 45.

But how do we compute the standard error? We do so by dividing the sample standard deviation (s) by the square root of the number of observations (n), as follows:

$$s_{\bar{x}} = \frac{s}{\sqrt{n}} = \frac{\sqrt{\frac{1}{n-1}\sum_{i=1}^{n}(x_i - \bar{x})^2}}{\sqrt{n}}$$

As we can see, a low standard deviation s decreases the standard error (which means less ambiguity when making inferences from these data). That is, less variation in the data decreases the standard error, thus favoring the rejection of the null hypothesis. Note that the standard error also depends on the sample size n. By increasing the number of observations, we have more information available, thus reducing the standard error.

If you understand this basic principle, you will have no problems understanding most other statistical tests. Let's go back to the example and compute the standard error as follows:

$$s_{\bar{x}} = \frac{\sqrt{\frac{1}{10-1}\left[(50-47.30)^2 + \ldots + (44-47.30)^2\right]}}{\sqrt{10}} = \frac{3.199}{\sqrt{10}} \approx 1.012$$

Thus, the result of the test statistic is:

$$t = \frac{\bar{x} - \mu}{s_{\bar{x}}} = \frac{2.30}{1.012} \approx 2.274$$

This test statistic applies when we compute a sample's standard deviation. In some situations, however, we might know the population's standard deviation, which requires the use of a different test, the **z-test**, (see Box 6.4).

Box 6.4 The *z*-Test

In the previous example, we used sample data to calculate the standard error $s_{\bar{x}}$. If we know the population's standard deviation beforehand, we should use the *z*-test. The *z*-test follows a normal (instead of a *t*-distribution).[3] The *z*-test is also used in situations when the sample size exceeds 30, because the *t*-distribution and normal distribution are similar for $n > 30$. As the *t*-test is slightly more accurate (also when the sample size is greater than 30), Stata uses the *t*-test, which can be accessed by going to ► Statistics ► Summaries, tables, and tests ► Classical tests of hypotheses ► t test (mean-comparison test). We do not show the formulas associated with the *z*-test here, but have included these in the ⭳ Web Appendix (→ Downloads).

[3]The fundamental difference between the *z*- and *t*-distributions is that the *t*-distribution is dependent on sample size n (which the *z*-distribution is not). The distributions become more similar with larger values of n.

6.3.5 Step 5: Make the Test Decision

Once we have calculated the test statistic, we can decide how likely it is that the claim stated in the hypothesis is correct. This is done by comparing the test statistic with the critical value that it must exceed (*Option 1*). Alternatively, we can calculate the actual probability of making a mistake when rejecting the null hypothesis and compare this value with the significance level (*Option 2*). In the following, we discuss both options.

6.3.5.1 Option 1: Compare the Test Statistic with the Critical Value

To make a test decision, we must first determine the critical value, which the test statistic must exceed for the null hypothesis to be rejected. In our case, the critical value comes from a *t*-distribution and depends on three parameters:

1. the significance level,
2. the degrees of freedom, and
3. one-tailed versus two-tailed testing.

We have already discussed the first point, so let's focus on the second. The **degrees of freedom** (usually abbreviated as *df*) represent the amount of information available to estimate the test statistic. In general terms, an estimate's degrees of freedom are equal to the amount of independent information used (i.e., the number of observations) minus the number of parameters estimated. Field (2013) provides a great explanation, which we adapted and present in Box 6.5.

In our example, we count $n - 1$ or $10 - 1 = 9$ degrees of freedom for the *t*-statistic to test a two-sided hypothesis of one mean. Remember that for a two-tailed test,

Box 6.5 Degrees of Freedom

Suppose you have a soccer team and 11 slots on the playing field. When the first player arrives, you have the choice of 11 positions in which you can place him or her. By allocating the player to a position, this occupies one position. When the next player arrives, you can choose from 10 positions. With every additional player who arrives, you have fewer choices where to position him or her. With the very last player, you no longer have the freedom to choose where to put him or her—there is only one spot left. Thus, there are 10 degrees of freedom. You have some degree of choice with 10 players, but for 1 player you don't. The degrees of freedom are the number of players minus 1.

when α is 0.05, the cumulative probability distribution is 1 – α/2 or 1 – 0.05/2 = 0.975. We divide the significance level by two, because half of our alpha tests the statistical significance in the lower tail of the distribution (bottom 2.5%) and half in the upper tail of the distribution (top 2.5%). If the value of the test statistic is greater than the critical value, we can reject the H_0.

We can find critical values for combinations of significance levels and degrees of freedom in the t-distribution table, shown in Table A1 in the ⭳ Web Appendix (→ Downloads). For 9 degrees of freedom and using a significance level of, for example, 5%, the critical value of the t-statistic is 2.262. Remember that we have to look at the $\alpha = 0.05/2 = 0.025$ column, because we use a two-tailed test. This means that for the probability of a type I error (i.e., falsely rejecting the null hypothesis) to be less than or equal to 5%, the value of the test statistic must be 2.262 or greater. In our case, the test statistic (2.274) exceeds the critical value (2.262), which suggests that we should reject the null hypothesis.[4] Even though the difference between the values is very small, bear in mind that hypothesis testing is binary—we either reject or don't reject the null hypothesis. This is also the reason why a statement such as "the result is highly significant" is inappropriate.

Figure 6.3 summarizes this concept graphically. In this figure, you can see that the critical value $t_{critical}$ for an α-level of 5% with 9 degrees of freedoms equals $\pm$ 2.262 on both sides of the distribution. This is indicated by the two dark-shaded rejection regions in the upper 2.5% and bottom 2.5% and the remaining white 95% non-rejection region in the middle. Since the test statistic t_{test} (indicated by the dotted line) falls in the dark-shaded area, we reject the null hypothesis.

Table 6.3 summarizes the decision rules for rejecting the null hypothesis for different types of t-tests, where t_{test} describes the test statistic and $t_{critical}$ the critical value for a specific significance level α. Depending on the test's formulation, test values may well be negative (e.g., −2.262). However, due to the symmetry of the t-distribution, only positive critical values are displayed.

6.3.5.2 Option 2: Compare the *p*-Value with the Significance Level

The above might make you remember your introductory statistics course with horror. The good news is that we do not have to bother with statistical tables when working with Stata. Stata automatically calculates the probability of obtaining a test statistic that is at least as extreme as the actually observed one if the null hypothesis is supported. This probability is also referred to as the ***p*-value** or the probability of observing a more extreme departure from the null hypothesis (Everitt and Skrondal 2010).

---

[4]To obtain the critical value, write `display invt(9,1-0.05/2)` in the command window.

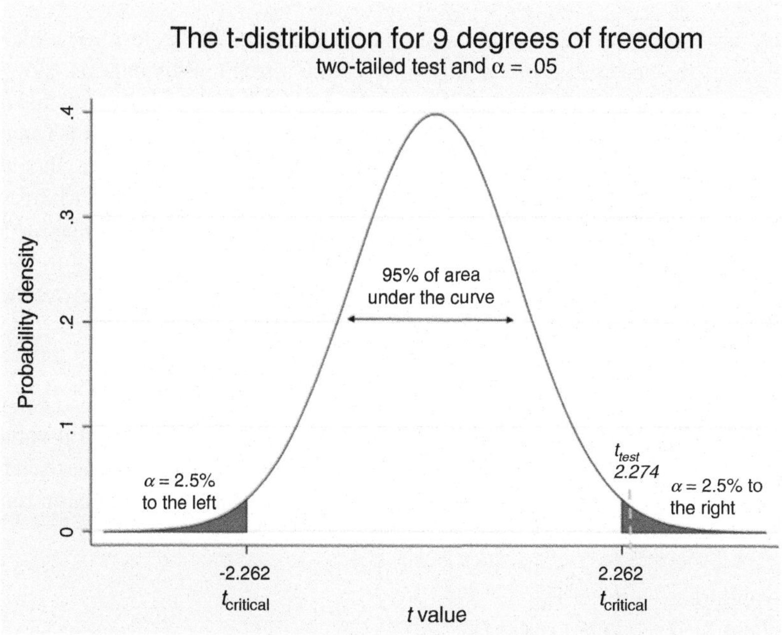

Fig. 6.3 Relationship between test value, critical value, and *p*-value

Table 6.3 Decision rules for testing decisions

Type of test	Null hypothesis (H_0)	Alternative hypothesis (H_1)	Reject H_0 if
Right-tailed test	$\mu \leq$ value	$\mu >$ value	$\|t_{test}\| > t_{critical}\ (\alpha)$
Left-tailed test	$\mu \geq$ value	$\mu <$ value	$\|t_{test}\| > t_{critical}\ (\alpha)$
Two-tailed test	$\mu =$ value	$\mu \neq$ value	$\|t_{test}\| > t_{critical}\ \left(\frac{\alpha}{2}\right)$

In the previous example, the *p*-value is the answer to the following question: If the population mean is not equal to 45 (i.e., therefore, the null hypothesis holds), what is the probability that random sampling could lead to a test statistic value of at least $\pm$ 2.274? This description shows that there is a relationship between the *p*-value and the test statistic. More precisely, these two measures are inversely related; the higher the absolute value of the test statistic, the lower the *p*-value and vice versa (see Fig. 6.3).

The description of the p-value is similar to the significance level α, which describes the acceptable probability of rejecting a true null hypothesis. However, the difference is that the p-value is calculated using the sample, and that α is set by the researcher before the test outcome is observed.[5] The p-value is not the probability of the null hypothesis being supported! Instead, we should interpret it as evidence against the null hypothesis. The α-level is an arbitrary and subjective value that the researcher assigns to the level of risk of making a type I error; the p-value is calculated from the available data. Related to this subjectivity, there has been a revived discussion in the literature on the usefulness of p-values (e.g., Nuzzo 2014; Wasserstein and Lazar 2016).

The comparison of the p-value and the significance level allows the researcher to decide whether or not to reject the null hypothesis. Specifically, if the p-value is smaller than or equal to the significance level, we reject the null hypothesis. Thus, when examining test results, we should make use of the following decision rule—this should become second nature![6]

- p-value $\leq \alpha \rightarrow$ reject H_0
- p-value $> \alpha \rightarrow$ do not reject H_0

Note that this decision rule applies to two-tailed tests. If you apply a one-tailed test, you need to divide the p-value in half before comparing it to α, leading to the following decision rule[7]:

- p-value/2 $\leq \alpha \rightarrow$ reject H_0
- p-value/2 $> \alpha \rightarrow$ do not reject H_0

In our example, the actual two-tailed p-value is 0.049 for a test statistic of $\pm$ 2.274, just at the significance level of 0.05. We can therefore reject the null hypothesis and find support for the alternative hypothesis.[8]

[5]Unfortunately, there is some confusion about the difference between the α and p-value. See Hubbard and Bayarri (2003) for a discussion.

[6]Note that this is convention and most textbooks discuss hypothesis testing in this way. Originally, two testing procedures were developed, one by Neyman and Pearson and another by Fisher (for more details, see Lehmann 1993). Agresti and Finlay (2014) explain the differences between the convention and the two original procedures.

[7]Note that this doesn't apply, for instance, to exact tests for probabilities.

[8]We don't have to conduct manual calculations and tables when working with Stata. However, we can easily compute the p-value ourselves by using the TDIST function in Microsoft Excel. The function has the general form "TDIST(t, df, tails)", where t describes the test value, df the degrees of freedom, and *tails* specifies whether it's a one-tailed test (tails = 1) or two-tailed test (tails = 2). Just open a new spreadsheet for our example and type in "=TDIST(2.274,9,1)". Likewise, there are several webpages with Java-based modules (e.g., http://graphpad.com/quickcalcs/pvalue1.cfm) that calculate p-values and test statistic values.

6.3.6 Step 6: Interpret the Results

The conclusion reached by hypothesis testing must be expressed in terms of the market research problem and the relevant managerial action that should be taken. In our example, we conclude that there is evidence that the point of sale display influenced the number of sales significantly during the week it was installed.

6.4 Two-Samples *t*-Test

6.4.1 Comparing Two Independent Samples

Testing the relationship between two independent samples is very common in market research. Some common research questions are:

- Do heavy and light users' satisfaction with a product differ?
- Do male customers spend more money online than female customers?
- Do US teenagers spend more time on Facebook than Australian teenagers?

Each of these hypotheses aim at evaluating whether two populations (e.g., heavy and light users), represented by samples, are significantly different in terms of certain key variables (e.g., satisfaction ratings).

To understand the principles of the *two independent samples t-test*, let's reconsider the previous example of a promotion campaign in a department store. Specifically, we want to test whether the population mean of the point of sale display's sales (μ_1) differs in any (positive or negative) way from that of the free tasting stand (μ_2). The resulting null and alternative hypotheses are now:

$$H_0: \mu_1 = \mu_2$$

$$H_1: \mu_1 \neq \mu_2$$

The test statistic of the two independent samples *t*-test—which is distributed with $n_1 + n_2 - 2$ degrees of freedom—is similar to the one-sample *t*-test:

$$t = \frac{(\bar{x}_1 - \bar{x}_2) - (\mu_1 - \mu_2)}{s_{\bar{x}_1 - \bar{x}_2}},$$

where $\bar{x}_1$ is the mean of the first sample (with n_1 numbers of observations) and $\bar{x}_2$ is the mean of the second sample (with n_2 numbers of observations). The term $\mu_1 - \mu_2$ describes the hypothesized difference between the population means. In this case, $\mu_1 - \mu_2$ is zero, as we assume that the means are equal, but we could use any other value to hypothesize a specific difference in population means. Lastly, $s_{\bar{x}_1 - \bar{x}_2}$ describes the standard error, which comes in two forms:

– If we assume that the two populations have the same variance (i.e., $\sigma_1^2 = \sigma_2^2$), we compute the standard error based on the so called *pooled* variance estimate:

$$s_{\bar{x}_1 - \bar{x}_2} = \sqrt{\frac{\left[(n_1 - 1) \cdot s_1^2 + (n_2 - 1) \cdot s_2^2\right]}{n_1 + n_2 - 2}} \cdot \sqrt{\frac{1}{\mathrm{n}_1} + \frac{1}{\mathrm{n}_2}}$$

– Alternatively, if we assume that the population variances differ (i.e., $\sigma_1^2 \neq \sigma_2^2$), we compute the standard error as follows:

$$s_{\bar{x}_1 - \bar{x}_2} = \sqrt{\frac{\mathrm{s}_1^2}{\mathrm{n}_1} + \frac{\mathrm{s}_2^2}{\mathrm{n}_2}}$$

How do we determine whether the two populations have the same variance? As discussed previously, this is done using Levene's test, which tests the following hypotheses:

$$\mathrm{H}_0\colon \sigma_1^2 = \sigma_2^2$$

$$\mathrm{H}_1\colon \sigma_1^2 \neq \sigma_2^2$$

The null hypothesis is that the two population variances are the same and the alternative hypothesis is that they differ. In this example, Levene's test provides support for the assumption that the variances in the population are equal, which allows us to proceed with the pooled variance estimate. First, we estimate the variances of the first and second group as follows:

$$s_1^2 = \frac{1}{n_1 - 1} \sum\nolimits_{i=1}^{10} (x_1 - \bar{x}_1)^2 = \frac{1}{10 - 1}\left[(50 - 47.30)^2 + \cdots + (44 - 47.30)^2\right] \approx 10.233$$

$$s_2^2 = \frac{1}{n_2 - 1} \sum\nolimits_{i=1}^{10} (x_2 - \bar{x}_2)^2 = \frac{1}{10 - 1}\left[(55 - 52)^2 + \ldots + (44 - 52)^2\right] \approx 17.556.$$

Using the variances as input, we can compute the standard error:

$$s_{\bar{x}_1 - \bar{x}_2} = \sqrt{\frac{\left[(10 - 1) \cdot 10.233 + (10 - 1) \cdot 17.556\right]}{10 + 10 - 2}} \cdot \sqrt{\frac{1}{10} + \frac{1}{10}} \approx 1.667$$

Inserting the estimated standard error into the test statistic results in:

$$t = \frac{(\bar{x}_1 - \bar{x}_2) - (\mu_1 - \mu_2)}{s_{\bar{x}_1 - \bar{x}_2}} = \frac{(47.30 - 52) - 0}{1.667} \approx -2.819$$

The test statistic follows a *t*-distribution with $n_1 - n_2$ degrees of freedom. In our case, we have 10 + 10 – 2 = 18 degrees of freedom. Looking at the statistical Table A1 in the ⤓ Web Appendix (→ Downloads), we can see that the critical value of a two-sided test with a significance level of 5% is 2.101 (note that we should look at the column labeled $\alpha = 0.025$ and the line labeled $df = 18$). The absolute value of the test statistic (i.e., 2.819) is greater than the critical value of 2.101 and, thus, falls within the bottom 2.5% of the distribution. We can therefore reject the null hypothesis at a significance level of 5% and conclude that the absolute difference between means of the point of sale display's sales (μ_1) and those of the free tasting stand (μ_2) is significantly different from 0.

6.4.2 Comparing Two Paired Samples

In the previous example, we compared the mean sales of two independent samples. If the management wants to compare the difference in the units sold before and after they started the point of sale display campaign. The difference can be either way; that is, it can be higher or lower, indicating a two-sided test. We have sales data for the week before the point of sale display was installed, as well as for the following week when the campaign was in full swing (i.e., the point of sale display had been installed). Table 6.4 shows the sale figures of the 10 stores in respect of both experimental conditions. You can again assume that the data are normally distributed.

At first sight, it appears that the point of sale display generated higher sales: The marginal mean of the sales in the week during which the point of sale display was

Table 6.4 Sales data (extended)

	Sales (units)	
Store	No point of sale display	Point of sale display
1	46	50
2	51	53
3	40	43
4	48	50
5	46	47
6	45	45
7	42	44
8	51	53
9	49	51
10	43	44
Marginal mean	46.10	48

installed (48) is slightly higher than in the week when it was not (46.10). However, the question is whether this difference is statistically significant.

We cannot assume that we are comparing two independent samples, as each set of two samples originates from the same set of stores, but at different points in time and under different conditions. Hence, we should use a *paired samples t-test*. In this example, we want to test whether the sales differ significantly with or without the installation of the point of sale display. We can express this by using the following hypotheses, where μ_d describes the population difference in sales; the null hypothesis assumes that the point of sale display made no difference, while the alternative hypothesis assumes a difference in sales:

$$H_0\colon \mu_d = 0$$

$$H_1\colon \mu_d \neq 0$$

To carry out this test, we define a new variable d_i, which captures the differences in sales between the two conditions (point of sale display – no point of sale display) in each of the stores. Thus:

$$d_1 = 50 - 46 = 4$$

$$d_2 = 53 - 51 = 2$$

$$\ldots$$

$$d_9 = 51 - 49 = 2$$

$$d_{10} = 44 - 43 = 1$$

Based on these results, we calculate the mean difference:

$$\bar{d} = \frac{1}{n}\sum_{i=1}^{10} d_i = \frac{1}{10}(4 + 2 + \ldots + 2 + 1) = 1.9$$

as well as the standard error of this difference:

$$s_{\bar{d}} = \frac{\sqrt{\frac{1}{n-1}\sum_{i=1}^{10}\left(d_i - \bar{d}\right)^2}}{\sqrt{n}}$$

$$= \frac{\sqrt{\frac{1}{9}\left[(4-1.9)^2 + (2-1.9)^2 + \ldots + (2-1.9)^2 + (1-1.9)^2\right]}}{\sqrt{10}} \approx 0.383$$

Next, we compare the mean difference $\bar{d}$ in our sample with the difference expected under the null hypothesis μ_d and divide this difference by the standard error $s_{\bar{d}}$. Thus, the test statistic is:

$$t = \frac{\bar{d} - \mu_d}{s_{\bar{d}}} = \frac{1.9 - 0}{0.383} \approx 4.960.$$

The test statistic follows a t-distribution with $n - 1$ degrees of freedom, where n is the number of pairs that we compare. Recall that for a two-tailed test, when α is 0.05, we need to look at the column labeled $\alpha = 0.025$ and the line labeled $df = 9$. Looking at Table A1 in the ⭳ Web Appendix (→ Downloads), we can see that the critical value of a two-sided test with a significance level of 5% is 2.262 for 9 degrees of freedom. Since the test value (4.960) is larger than the critical value, we can reject the null hypothesis and presume that the point of sale display makes a difference.

6.5 Comparing More Than Two Means: Analysis of Variance (ANOVA)

Researchers are often interested in examining differences in the means between more than two groups. For example:

- Do light, medium, and heavy internet users differ in respect of their monthly disposable income?
- Do customers across four different types of demographic segments differ in their attitude towards a certain brand?
- Is there a significant difference in hours spent on Facebook between US, UK, and Australian teenagers?

Continuing with our previous example on promotion campaigns, we might be interested in whether there are significant sales differences between the stores in which the three different types of campaigns were launched. One way to tackle this research question would be to carry out multiple pairwise comparisons of all the groups under consideration. In this example, doing so would require the following comparisons:

1. the point of sale display versus the free tasting stand,
2. the point of sale display versus the in-store announcements, and
3. the free tasting stand versus the in-store announcements.

While three comparisons seem manageable, you can imagine the difficulties when a greater number of groups are compared. For example, with ten groups, we would have to carry out 45 group comparisons.[9]

[9]The number of pairwise comparisons is calculated as follows: $k \cdot (k - 1)/2$, with k the number of groups to compare.

Making large numbers of comparisons induces the severe problem of **α-inflation**. This inflation refers to the more tests that you conduct at a certain significance level, the more likely you are to claim a significant result when this is not so (i.e., an increase or inflation in the type I error). Using a significance level of $\alpha = 0.05$ and making all possible pairwise comparisons of ten groups (i.e., 45 comparisons), the increase in the overall probability of a type I error (also referred to as **familywise error rate**) is:

$$\alpha^* = 1 - (1 - \alpha)^{45} = 1 - (1 - 0.05)^{45} = 0.901$$

That is, there is a 90.1% probability of erroneously rejecting your null hypothesis in at least some of your 45 *t*-tests—far greater than the 5% for a single comparison! The problem is that you can never tell which of the comparisons' results are wrong and which are correct.

Instead of carrying out many pairwise tests, market researchers use ANOVA, which allows a comparison of three or more groups' averages. In ANOVA, the variable that differentiates the groups is referred to as the *factor variable* (don't confuse this with the factors of factor analysis discussed in Chap. 8!). The values of a factor (i.e., as found in respect of the different groups under consideration) are also referred to as *factor levels*.

In the previous example of promotion campaigns, we considered only one factor variable with three levels, indicating the type of campaign. This is the simplest form of an ANOVA and is called a one-way ANOVA. However, ANOVA allows us to consider more than one factor variable. For example, we might be interested in adding another grouping variable (e.g., the type of service offered), thus increasing the number of treatment conditions in our experiment. In this case, we should use a two-way ANOVA to analyze both factor variables' effect on the sales (in isolation and jointly). ANOVA is even more flexible, because you can also integrate interval or ratio-scaled independent variables and even multiple dependent variables. We first introduce the one-way ANOVA, followed by a brief discussion of the two-way ANOVA.[10] For a more detailed discussion of the latter, you can turn to the ⭳ Web Appendix (→ Downloads).

6.6 Understanding One-Way ANOVA

We now know ANOVA is used to examine the mean differences between more than two groups. In more formal terms, the objective of the **one-way ANOVA** is to test the null hypothesis that the population means of the groups (defined by the factor variable and its levels) are equal. If we compare three groups, as in the promotion campaign example, the null hypothesis is:

[10]Mitchell (2015) provides a detailed introduction to other ANOVA types, such as the analysis of covariance (ANCOVA).

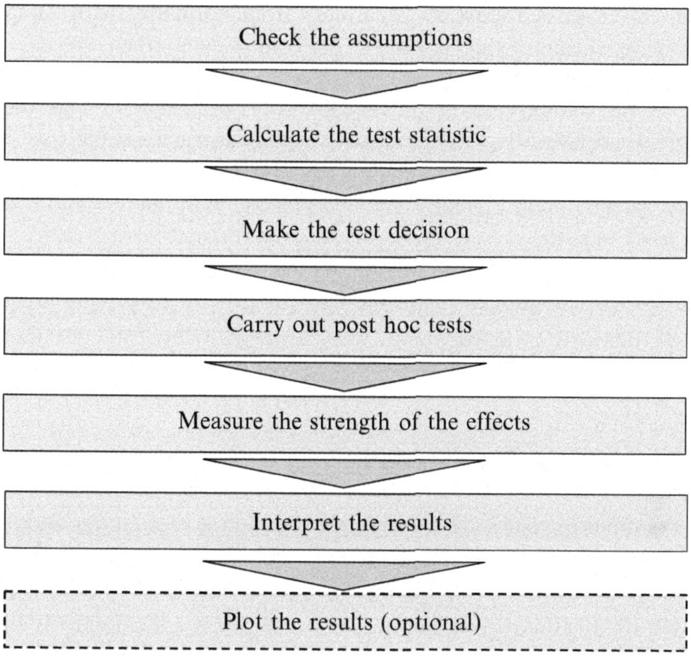

Fig. 6.4 Steps in conducting an ANOVA

$$H_0: \mu_1 = \mu_2 = \mu_3$$

This hypothesis implies that the population means of all three promotion campaigns are identical (which is the same as saying, that the campaigns have the same effect on the mean sales). The alternative hypothesis is:

$$H_1: \text{At least two of } \mu_1, \mu_2, \text{and } \mu_3 \text{ are unequal.}$$

Before we even think of running an ANOVA, we should, of course, produce a problem formulation, which requires us to identify the dependent variable and the factor variable, as well as its levels. Once this task is done, we can dig deeper into ANOVA by following the steps described in Fig. 6.4. We next discuss each step in more detail.

6.6.1 Check the Assumptions

ANOVA is a parametric test that relies on the same distributional assumptions as discussed in Sect. 6.3.3.3. We may use ANOVA in situations when the dependent variable is measured on an ordinal scale and is not normally distributed, but we should then ensure that the group-specific sample sizes are similar. Thus, if possible, it is useful to collect samples of a similar size for each group. As discussed previously, ANOVA is robust to departures from normality with sample sizes greater than 30, meaning that it can be performed even when the data are not normally distributed. Even though ANOVA is rather robust in this respect, violations of the assumption of the equality of variances can bias the results significantly, especially when the groups are of very unequal sample size.[11] Consequently, we should always test for the equality of variances by using Levene's test. We already touched upon Levene's test and you can learn more about it in ⭳ Web Appendix (→ Chap. 6).

Finally, like any data analysis technique, the sample size must be sufficiently high to have sufficient statistical power. There is general agreement that the bare minimum sample size per group is 20. However, 30 or more observations per group are desirable. For more detail, see Box 6.1.

6.6.2 Calculate the Test Statistic

ANOVA examines the dependent variable's variation across groups and, based on this variation, determines whether there is reason to believe that the population means of the groups differ. Returning to our example, each store's sales are likely to deviate from the overall sales mean, as there will always be some variation. The question is therefore whether a specific promotion campaign is likely to cause the difference between each store's sales and the overall sales mean, or whether this is due to a natural variation in sales. To disentangle the effect of the treatment (i.e., the promotion campaign type) and the natural variation, ANOVA separates the total variation in the data (indicated by SS_T) into two parts:

1. the between-group variation (SS_B), and
2. the within-group variation (SS_W).[12]

These three types of variation are estimates of the population variation. Conceptually, the relationship between the three types of variation is expressed as:

$$SS_T = SS_B + SS_W$$

However, before we get into the math, let's see what SS_B and SS_W are all about.

6.6.2.1 The Between-Group Variation (SS_B)

SS_B refers to the variation in the dependent variable as expressed in the variation in the group means. In our example, it describes the variation in the mean values of sales across the three treatment conditions (i.e., point of sale display, free tasting

[11] In fact, these two assumptions are interrelated, since unequal group sample sizes result in a greater probability that we will violate the homogeneity assumption.

[12] SS is an abbreviation of "sum of squares," because the variation is calculated using the squared differences between different types of values.

Box 6.6 Types of Sums of Squares in Stata

Stata allows two options to represent a model's sums of squares. The first, and also the default option, is the *partial sums of squares*. To illustrate what this measure represents, presume we have a model with one independent variable var_1 and we want to know what additional portion of our model's variation is explained if we add independent variable var_2, to the model. The partial sums of squares indicates the portion of the variation that is explained by var_2, given var_1. The second option is the *sequential sums of squares*, which adds variables one at a time to the model in order to assess the model's incremental improvement with each newly added variable. Of the two options, the partial sums of squares is the simpler one and does not rely on the ordering of the variables in the model, but is not suitable for full factorial designs that include interactions between two variables.

stand, and in-store announcements) in relation to the overall mean. What does SS_B tell us? Imagine a situation in which all mean values across the treatment conditions are the same. In other words, regardless of which campaign we choose, the sales are the same, we cannot claim that the promotion campaigns had differing effects. This is what SS_B expresses: it tells us how much variation the differences in observations that truly stem from different groups can explain (for more, see Box 6.6). Since SS_B is the *explained variation* (explained by the grouping of data) and thus reflects different effects, we would want it to be as high as possible. However, there is no given standard of how high SS_B should be, as its magnitude depends on the scale level used (e.g., are we looking at 7-point Likert scales or income in US$?). Consequently, we can only interpret the explained variation expressed by SS_B in relation to the variation that the grouping of data does not explain. This is where SS_W comes into play.

6.6.2.2 The Within-Group Variation (SS_W)

As the name already suggests, SS_W describes the variation in the dependent variable within each of the groups. In our example, SS_W simply represents the variation in sales in each of the three treatment conditions. The smaller the variation within the groups, the greater the probability that the grouping of data can explain all the observed variation. It is obviously the ideal for this variation to be as small as possible. If there is much variation within some or all the groups, then some extraneous factor, not accounted for in the experiment, seems to cause this variation instead of the grouping of data. For this reason, SS_W is also referred to as *unexplained variation.*

Unexplained variation can occur if we fail to account for important factors in our experimental design. For example, in some of the stores, the product might have been sold through self-service, while personal service was available in others. This is a factor that we have not yet considered in our analysis, but which will be used when we look at the two-way ANOVA later in the chapter. Nevertheless, some unexplained variation will always be present, regardless of how sophisticated our experimental design is and how many factors we consider. If the unexplained variation cannot be explained, it is called *random noise* or simply *noise*.

6.6.2.3 Combining SS_B and SS_W into an Overall Picture

The comparison of SS_B and SS_W tells us whether the variation in the data is attributable to the grouping, which is desirable, or due to sources of variation not captured by the grouping, which is not desirable. Figure 6.5 shows this relationship across the stores featuring our three different campaign types:

- point of sale display (•),
- free tasting stand (▪), and
- in-store announcements (▲).

We indicate the group mean of each level by dashed lines. If the group means are all the same, the three dashed lines are horizontally aligned and we then conclude that the campaigns have identical sales. Alternatively, if the dashed lines are very different, we conclude that the campaigns differ in their sales.

At the same time, we would like the variation within each of the groups to be as small as possible; that is, the vertical lines connecting the observations and the dashed lines should be short. In the most extreme case, all observations would lie on

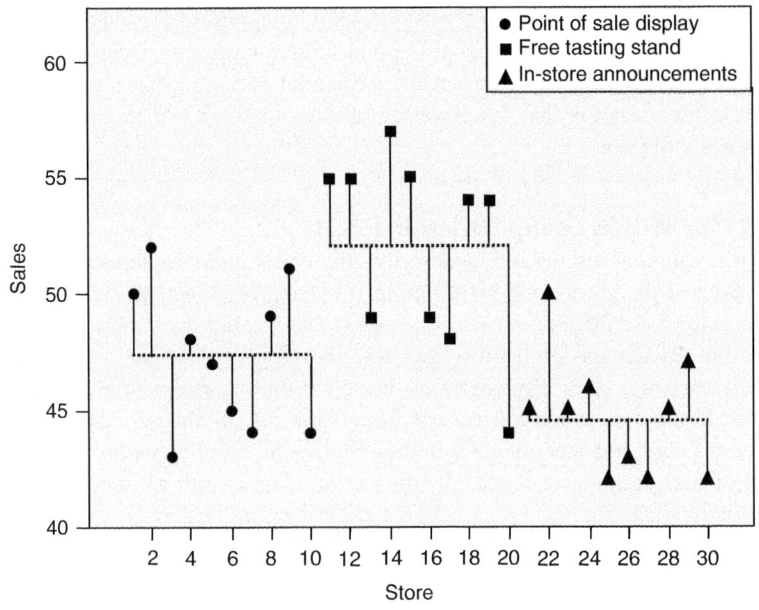

Fig. 6.5 Scatter plot of stores with different campaigns vs. sales

their group-specific dashed lines, implying that the grouping explains the variation in sales perfectly. This, however, hardly ever occurs.

If the vertical bars were all, say, twice as long, it would be difficult to draw any conclusions about the effects of the different campaigns. Too great a variation within the groups then swamps the variation between the groups. Based on the discussion above, we can calculate the three types of variation.

1. The total variation, computed by comparing each store's sales with the overall mean, which is equal to 48 in our example:[13]

$$SS_T = \sum\nolimits_{i=1}^{n} (x_i - \bar{x})^2$$
$$= (50 - 48)^2 + (52 - 48)^2 + \ldots + (47 - 48)^2 + (42 - 48)^2 = 584$$

2. The between-group variation, computed by comparing each group's mean sales with the overall mean, is:

$$SS_B = \sum\nolimits_{j=1}^{k} n_j (\bar{x}_j - \bar{x})^2$$

As you can see, besides index i, as previously discussed, we also have index j to represent the group sales means. Thus, $\bar{x}_j$ describes the mean in the j-th group and n_j the number of observations in that group. The overall number of groups is denoted with k. The term n_j is used as a weighting factor: Groups that have many observations should be accounted for to a higher degree relative to groups with fewer observations. Returning to our example, the between-group variation is then given by:

$$SS_B = 10 \cdot (47.30 - 48)^2 + 10 \cdot (52 - 48)^2 + 10 \cdot (44.70 - 48)^2 = 273.80$$

3. The within-group variation, computed by comparing each store's sales with its group sales mean is:

$$SS_w = \sum_{j=1}^{k} \sum_{i=1}^{n_j} (x_{ij} - \bar{x}_j)^2$$

[13]Note that the group-specific sample size in this example is too small to draw conclusions and is only used to show the calculation of the statistics.

Here, we should use two summation signs, because we want to compute the squared differences between each store's sales and its group sales' mean for all k groups in our set-up. In our example, this yields the following:

$$\begin{aligned} SS_W &= \left[(50-47.30)^2+\ldots+(44-47.30)^2\right] \\ &\quad+\left[(55-52)^2+\ldots+(44-52)^2\right] \\ &\quad+\left[(45-44.70)^2+\ldots+(42-44.70)^2\right] \\ &= 310.20 \end{aligned}$$

In the previous steps, we discussed the comparison of the between-group and within-group variation. The higher the between-group variation is in relation to the within-group variation, the more likely it is that the grouping of the data is responsible for the different levels in the stores' sales instead of the natural variation in all the sales.

A suitable way to describe this relation is by forming an index with SS_B in the numerator and SS_W in the denominator. However, we do not use SS_B and SS_W directly, because they are based on summed values and the scaling of the variables used therefore influence them. Therefore, we divide the values of SS_B and SS_W by their degrees of freedom to obtain the true mean square values MS_B (called *between-group mean squares*) and MS_W (called *within-group mean squares*). The resulting mean square values are:

$$MS_B=\frac{SS_B}{k-1},\text{ and } MS_W=\frac{SS_W}{n-k}$$

We use these mean squares to compute the following test statistic, which we then compare with the critical value:

$$F=\frac{MS_B}{MS_W}$$

Turning back to our example, we calculate the test statistic as follows:

$$F=\frac{MS_B}{MS_W}=\frac{SS_B/_{k-1}}{SS_W/_{n-k}}=\frac{273.80/_{3-1}}{310.20/_{30-3}}\approx 11.916$$

6.6.3 Make the Test Decision

Making the test decision in ANOVA is like the t-tests discussed earlier, with the difference that the test statistic follows an F-distribution (as opposed to a t-distribution). Different from before, however, we don't have to divide α by 2 when looking up the critical value, even though the underlying alternative hypothesis in ANOVA is two-sided! The reason for this is that an F-test statistic is the ratio of the variation explained by systematic variance (i.e., between-group mean squares) to the unsystematic variance (i.e., within-group mean squares), which is always equal to or greater than 0, but never lower than 0. For this reason, and given that no negative values can be taken, it makes no sense to split the significance level in half, although you can always choose a more restrictive alpha (van Belle 2008).

Unlike the t-distribution, the F-distribution depends on two degrees of freedom: One corresponding to the between-group mean squares ($k - 1$) and the other referring to the within-group mean squares ($n - k$). The degrees of freedom of the promotion campaign example are 2 and 27; therefore, on examining Table A2 in the ⭳ Web Appendix (→ Downloads), we see a critical value of 3.354 for $\alpha = 0.05$. In our example, we reject the null hypothesis, because the F-test statistic of 11.916 is greater than the critical value of 3.354. Consequently, we can conclude that at least two of the population sales means of the three types of promotion campaigns differ significantly.

At first sight, it appears that the free tasting stand is most successful, as it exhibits the highest mean sales ($\bar{x}_2 = 52$) compared to the point of sale display ($\bar{x}_1 = 47.30$) and the in-store announcements ($\bar{x}_3 = 44.70$). However, rejecting the null hypothesis does not mean that all the population means differ—it only means that at least two of the population means differ significantly! Market researchers often assume that all means differ significantly when interpreting ANOVA results, but this is wrong. How then do we determine which of the mean values differ significantly from the others? We deal with this problem by using post hoc tests, which is done in the next step of the analysis.

6.6.4 Carry Out Post Hoc Tests

Post hoc tests perform multiple comparison tests on each pair of groups and tests which of the groups differ significantly from each other. The basic idea underlying post hoc tests is to perform tests on each pair of groups and to correct the level of significance of each test. This way, the overall type I error rate across all the comparisons (i.e., the *familywise error rate*) remains constant at a certain level, such as at $\alpha = 0.05$ (i.e., α-inflation is avoided).

There are several post hoc tests, the easiest of which is the **Bonferroni correction.** This correction maintains the familywise error rate by calculating a new pairwise alpha that divides the statistical significance level of α by the number of comparisons made. How does this correction work? In our example, we can compare three groups pairwise: (1) Point of sale display vs. free tasting stand, (2) point of sale display vs. in-store announcements, and (3) free tasting stand vs. in-store announcements. Hence, we would use $0.05/3 \approx 0.017$ as our criterion for significance. Thus, to reject the null hypothesis that the two population means are equal, the p-value would have to be smaller than 0.017 (instead of 0.05!). The Bonferroni adjustment is a very strict way of maintaining the familywise error rate.

However, there is a trade-off between controlling the familywise error rate and increasing the type II error. By reducing the type I error rate, the type II error increases. Hence the statistical power decreases, potentially causing us to miss significant effects in the population.

The good news is that there are alternatives to the Bonferroni method. The bad news is that there are numerous types of post hoc tests—Stata provides no less than nine such methods! All these post hoc tests are based on different assumptions and designed for different purposes, whose details are clearly beyond the scope of this book.[14]

The most widely used post hoc test in market research is **Tukey's honestly significant difference test**, often simply referred to as *Tukey's method*. Tukey's method is a very versatile test controlling for type I error, but is limited in terms of statistical power (Everitt and Skrondal 2010). The test divides the difference between the largest and smallest pairs of means by the data's standard error that combines all possible pairwise differences and produces a value called *Tukey's statistic*. Where Tukey's statistic is larger than the critical value obtained from a normal distribution, the pairwise differences are rendered statistically significant. Tukey's method relies on two important requirements:

1. they require an equal number of observations for each group (differences of only a few observations are not problematic, though), and
2. they assume that the population variances are equal.

Alternative post hoc tests are available if these requirements are not met. When sample sizes clearly differ, we can draw on *Scheffé's method*, which is conservative by nature and thus has low statistical power. Alternatively, we can use *Dunnett's method*, which is useful when multiple pairwise comparisons (i.e., multiple treatment groups) are made with reference to a single control group. This is standard in experiments that distinguish between control and treatment groups, as is often encountered in marketing research.

Post hoc tests thus facilitate pairwise comparisons between groups while maintaining the familywise error rate. However, they do not allow for making statements regarding the strength of a factor variable's effects on the dependent variable. We can only do this after calculating the effect sizes, which we will do next.

6.6.5 Measure the Strength of the Effects

We can compute the η^2 (the **eta-squared**) coefficient to determine the strength of the effect (also referred to as the *effect size*) that the factor variable exerts on the dependent variable. The eta squared is the ratio of the between-group variation

[14]The Stata `help contrast` function provides an overview and references.

(SS_B) to the total variation (SS_T) and therefore indicates the variance accounted for by the sample data. Since η^2 is equal to the *coefficient of determination* (R^2), known from regression analysis (Chap. 7), Stata refers to it as R-squared in the output.

η^2 can take on values between 0 and 1. If all groups have the same mean value, and we can thus assume that the factor has no influence on the dependent variable, η^2 is 0. Conversely, a high value implies that the factor exerts a strong influence on the dependent variable. In our example, η^2 is:

$$\eta^2 = \frac{SS_B}{SS_T} = \frac{273.80}{584} \approx 0.469$$

The outcome indicates that 46.9% of the total variation in sales is explained by the promotion campaigns. The η^2 is often criticized as being too high for small sample sizes of 50 or less. We can compute ω^2 (pronounced **omega-squared**), which corresponds to the *Adjusted* $\boldsymbol{R^2}$ from regression analysis (Chap. 7), to compensate for small sample sizes:

$$\omega^2 = \frac{SS_B - (k-1) \cdot MS_W}{SS_T + MS_W} = \frac{273.80 - (3-1) \cdot 11.916}{584 + 11.916} \approx 0.421$$

This result indicates that 42.1% of the total variation in sales is accounted for by the promotion campaigns. Generally, you should use ω^2 for $n \leq 50$ and η^2 for $n > 50$.

It is difficult to provide firm rules of thumb regarding which values are appropriate for η^2 or ω^2, as this varies from research area to research area. However, since the η^2 resembles Pearson's correlation coefficient (Chap. 7), we follow the suggestions provided in Chap. 7. Thus, we can consider values below 0.30 weak, values from 0.31 to 0.49 moderate, and values of 0.50 and higher strong.

6.6.6 Interpret the Results and Conclude

Just as in any other type of analysis, the final step is to interpret the results. Based on our results, we conclude that not all promotional activities have the same effect on sales. An analysis of the strength of the effects revealed that this association is moderate.

6.6.7 Plotting the Results (Optional)

In a final step, we can plot the estimated means of the dependent variable across the different samples. We could, for example, plot the mean of the sales across the stores with different types of promotion campaigns (i.e., point of sales display, free tasting stand or in-store display). When plotting the estimated group means, it is common to show the *confidence interval*, which is the interval within which the mean estimate falls with a certain probability (e.g., 95%). In this way, we can see whether the mean of the outcome variable across the different groups differ significantly or not without examining the numbers in the table. We will illustrate this optional step in the case study later in this chapter.

6.7 Going Beyond One-Way ANOVA: The Two-Way ANOVA

A logical extension of the one-way ANOVA is to add a second factor variable to the analysis. For example, we could assume that, in addition to the different promotion campaigns, management also varied the type of service provided by offering either self-service or personal service (see column "Service type" in Table 6.1). The two-way ANOVA is similar to the one-way ANOVA, except that the inclusion of a second factor variable creates additional types of variation. Specifically, we need to account for two types of between-group variations:

1. the between-group variation in factor variable 1 (i.e., promotion campaigns), and
2. the between-group variation in factor variable 2 (i.e., service type).

In its simplest form, the two-way ANOVA assumes that these factor variables are unrelated. However, in market research applications, this is rarely so, thereby requiring us to consider *related factors*. When we take two related factor variables into account, we not only have to consider each factor variable's direct effect (also called the **main effect**) on the dependent variable, but also their **interaction effect**. Conceptually, an interaction effect is an additional effect due to combining two (or more) factors. Most importantly, this extra effect cannot be observed when considering each of the factor variables separately and thus reflects a concept known as synergy. There are many examples in everyday life where the whole is more than simply the sum of its parts as we know from cocktail drinks, music, and paintings. For an entertaining example of interaction, see Box 6.7.

In our example, the free tasting stand might be the best promotion campaign when studied separately, but it could well be that when combined with personal service, the point of sale display produces higher sales. A significant interaction effect indicates that the combination of the two factor variables results in higher (or lower) sales than when each factor variable is considered separately. The computation of these effects, as well as a discussion of other technical aspects, lies beyond the scope of this book, but are discussed in the ⭳ Web Appendix (→ Downloads).

Table 6.5 provides an overview of the steps involved when carrying out the following tests in Stata: One-sample *t*-test, paired samples *t*-test, independent samples *t*-test, and the one-way ANOVA. Owing to data limitations to accommodate all types of parametric tests in this chapter by means of Oddjobs Airways, we will use data from the case study in the theory section to illustrate the Stata commands. This data restriction applies only to this chapter.

Box 6.7 A Different Type of Interaction

http://tinyurl.com/interact-coke

Table 6.5 Steps involved in carrying out one, two, or more group comparisons with Stata

Theory	Action
One-sample t-test[a]	
Formulate the hypothesis:	
Formulate the study's hypothesis:	*For example:*
	H_0: $\mu = \#$[b]
	H_1: $\mu \neq \#$
Choose the significance level:	
	Usually, α is set to 0.05, but: if you want to be conservative, α is set to 0.01, and: in exploratory studies, α is set to 0.10. We choose a significance level of 0.05.
Select an appropriate test:	
What is the testing situation?	Determine the fixed value again which that you are comparing.
Is the test variable measured on an interval or ratio scale?	Check Chap. 3 to determine the measurement level of the variables.
Are the observations independent?	Consult Chap. 3 to determine whether the observations are independent.
Is the test variable normally distributed or is $n>30$ and are the group variances the same?	*Check for normality*
	Carry out the Shapiro-Wilk normality test. Go to ► Statistics ► Summaries, tables, and tests ►Distributional plots and tests ►Shapiro-Wilk normality test. Select the test variable *outcome1* under **Variables:** and click on **OK**. A *p*-value below 0.05 indicates non-normality.
	`swilk outcome1`

(continued)

Table 6.5 (continued)

Theory	Action
Specify the type of t-test	Select the one-sample *t*-test.
Is the test one or two-sided?	Determine the region of rejection.
Calculate the test statistic:	
Specify the test variable and the fixed value	Go to ▶ Statistics ▶ Summaries, tables, and tests ▶ Classical tests of hypotheses ▶ t test (mean-comparison test). Select the first option **One-sample** from the dialog box and specify the test variable outcome1 under the **Variable name**:. Next, enter the hypothetical mean under **Hypothesized mean**:, choose the confidence interval **95**, which equates to a statistical significance at < 0.05 and click **OK**.
	`ttest outcome1==#`
Interpret the results:	
Look at the test results	For two-sided tests: compare the *p*-value under **Ha: mean(diff) != 0** with 0.05 and decide whether the null hypothesis is supported. The *p*-value under **Pr(|T| > |t|)** should be lower than 0.05 to reject the null hypothesis.
	For one-sided tests, look at either **Ha: mean (diff) < 0** (left-sided) or **Ha: mean(diff) > 0** (right-sided).
What is your conclusion?	Reject the null hypothesis that the population mean of the outcome variable *outcome1* is equal to the hypothetical known parameter against which you compare (i.e., #) if the *p*-value is lower than 0.05.
Paired samples t-test	
Formulate the hypothesis:	
Formulate the study's hypothesis:	*For example:*
	$H_0: \mu_1 = \mu_2$
	$H_1: \mu_1 \neq \mu_2$
Choose the significance level:	
	Usually, α is set to 0.05, but: if you want to be conservative, α is set to 0.01, and: in exploratory studies, α is set to 0.10. We choose a significance level of 0.05.
Select an appropriate test:	
What is the testing situation?	Determine the number of groups you are comparing.
Are the test variables measured on an interval or ratio scale?	Check Chap. 3 to determine the measurement level of the variables.
Are the observations dependent?	Next, consult Chap. 3 to determine whether the observations are independent.
	Check for normality

(continued)

Table 6.5 (continued)

Theory	Action
Are the test variables normally distributed or is $n > 30$ in each of the groups and are the group variances the same?	Run the Shapiro-Wilk normality test. Go to ► Statistics ► Summaries, tables, and tests ►Distributional plots and tests ►Shapiro-Wilk normality test. Select the test variable *outcome1* under **Variables.** Next, under the tab **by/if/in** tick the box **Repeat command by groups**, specify the grouping variable *groupvar* under **Variables that define groups:** and click on **OK**. A *p*-value below 0.05 indicates non-normality.
	`by groupvar,`[c] `sort: swilk outcome1`
	Check for equality of variances assumptions
	To perform Levene's test, go to ► Statistics ► Classical tests of hypotheses ► Robust equal-variance test. Specify the dependent variable under **Variable** *outcome1*, and the grouping variable *groupvar* under **Variable defining comparison groups** and click **OK**. To validate the equality of variances, the assumption *p*-values should lie above 0.05 for **W0**, **W50**, and **W10**.
	`robvar outcome1, by(groupvar)`
Specify the type of t-test	The data appear to be normally distributed with equal group variances. We can now proceed with the paired sample *t*-test.
Is the test one or two-sided?	Determine the region of rejection.
Calculate the test statistic:	
Select the paired test variables	Go to ► Statistics ► Summaries, tables, and tests ► Classical tests of hypotheses ► t test (mean-comparison test). Select the fourth option **Paired** from the dialog box and specify the test variable "*variable1*" under the First Variable. Next, enter the second comparison group "*variable2*" under **Second variable**, choose the confidence interval **95**, which equates to a statistical significance of $\alpha = 0.05$ and click **OK.**
	`ttest variable1==variables2`[d]
Interpret the results:	
Look at the test results	For two-sided tests: compare the *p*-value under **Ha: mean(diff) != 0** with 0.05 and decide whether the null hypothesis is supported. The *p*-value under **Pr(|T| > |t|)** should be lower than 0.05 to reject the null hypothesis.
	For one-sided tests, look at either **Ha: mean (diff) < 0** (left-sided) or **Ha: mean(diff) > 0** (right-sided).
What is your conclusion?	Reject the null hypothesis if the *p*-value is lower than 0.05.

(continued)

Table 6.5 (continued)

Theory	Action
Independent sample t-test	
Formulate the hypothesis:	
Formulate the study's hypothesis:	*For example:*
	H_0: $\mu_1 = \mu_2$
	H_1: $\mu_1 \neq \mu_2$
Choose the significance level:	
	Usually, α is set to 0.05, but: if you want to be conservative, α is set to 0.01, and: in exploratory studies, α is set to 0.10. We choose a significance level of 0.05.
Select an appropriate test:	
What is the testing situation?	Determine the number of groups you are comparing.
Are the test variables measured on an interval or ratio scale?	Check Chap. 3 to determine the measurement level of the variables.
Are the observations dependent?	Next, consult Chap. 3 to determine whether the observations are independent.
Are the test variables normally distributed or is $n > 30$ in each of the groups and are the group variances the same?	*Check for normality*
	Run the Shapiro-Wilk normality test. Go to ► Statistics ► Summaries, tables, and tests ► Distributional plots and tests ► Shapiro-Wilk normality test. Select the test variable *overall_sat* under **Variables.** Next, under the tab **by/if/in** tick the box **Repeat command by groups**, specify the grouping variable *gender* under **Variables that define groups** and click on **OK**. A *p*-value below 0.05 indicates non-normality.
	`by gender, sort: swilk overall_sat`
	Check for equality of variances assumptions
	Next, to perform Levene's test, go to ► Statistics ► Classical tests of hypotheses ►Robust equal-variance test. Specify the dependent variable *overall_sat* under **Variable:** and the variable *gender* under **Variable defining comparison groups:** aand click **OK**. The *p*-values of **W0**, **W50** and **W10** should be above 0.05 to validate the equality of the variances assumption.
	`robvar overall_sat, by(gender)`
Specify the type of *t*-test	The data appear to be normally distributed with equal group variances. We can now proceed with the two-sample *t*-test.
Is the test one or two-sided?	Determine the region of rejection.

(continued)

Table 6.5 (continued)

Theory	Action
Calculate the test statistic:	
Select the test variable and the grouping variable	Go to ► Statistics ► Summaries, tables, and tests ► Classical tests of hypotheses ► t test (mean-comparison test). Select the second option **Two-sample using groups** from the dialog box and specify the test variable *overall_sat* under the **Variable name**. Next, enter the variable *gender* under **Group variable name**. Select the confidence interval **95**, which equates to a statistical significance of $\alpha = 0.05$ and then click **OK**.
	`ttest overall_sat, by(gender)`
Interpret the results:	
Look at the test results	For two-sided tests: Compare the *p*-value under **Ha: mean(diff)! = 0** with 0.05 and decide whether the null hypothesis is supported. The *p*-value under **Pr(|T| > |t|)** should be lower than 0.05 to reject the null hypothesis.
	For one-sided tests, look either at **Ha: mean (diff) < 0** (left-sided) or **Ha: mean(diff) > 0** (right-sided).
What is your conclusion?	Reject the null hypothesis that the population mean of the overall satisfaction score among female travelers is equal to the population overall mean satisfaction score of male traverlers if the *p*-value is lower than *0.05*.
One-way ANOVA	
Formulate the hypothesis:	
Formulate the study's hypothesis:	*For example:*
	H_0: $\mu_1 = \mu_2 = \mu_3$
	H_1: At least two of the population means are different.
Choose the significance level:	
	Usually, α is set to 0.05, but: if you want to be conservative, α is set to 0.01, and: in exploratory studies, α is set to 0.10. We choose a significance level of 0.05.
Select an appropriate test:	
What is the testing situation?	Determine the number of groups you are comparing.
Are there at least 20 observations per group?	Check Chap. 5 to determine the sample size in each group.
Is the dependent variable measured on an interval or ratio scale?	Determine the type of test that you need to use for your analyses by checking the underlying assumptions first. Check Chap. 3 to determine the measurement level of the variables.

(continued)

Table 6.5 (continued)

Theory	Action
Are the observations independent?	Next, consult Chap. 3 to determine whether the observations are independent.
Is the test variable normally distributed or is *n* larger than 30 per group and are the group variances the same?	*Check for normality*
	Carry out the Shapiro-Wilk normality test. Go to ► Statistics ► Summaries, tables, and tests ►Distributional plots and tests ►Shapiro-Wilk normality test. Select the test variable *overall_sat* under **Variables.** Next, under the tab **by/if/in** tick the box **Repeat command by groups**, specify the grouping variable *status* under **Variables that define groups** and click on **OK.** Values (**V**) larger than **1** with *p*-values below 0.05 indicate non-normality.
	`by status, sort: swilk overall_sat`
	Check for Equality of Variances Assumption
	Perform Levene's test. Go to ► Statistics ► Classical tests of hypotheses ►Robust equal-variance test. Specify the dependent variable *overall_sat* under **Variable:** aand the grouping variable *status* under **Variable defining comparison groups:** aand then click on **OK**. The *p*-values of **W0**, **W50** and **W10** should be above 0.05 to validate the equality of the variances assumption.
	`robvar overall_sat, by (status)`
Select the type of the test	Now that the assumption of normality and equality of the variance are met, proceed with the one-way ANOVA analysis.
Calculate the test statistic:	
Specify the dependent variable and the factor (grouping variable)	Go to ►Statistics ►Linear models and related ANOVA/MANOVA ►Analysis of variance and covariance. Specify the dependent variable *overall_sat* under the **Dependent variable:** aand the variable *status* under the **Model:**. Next, select the **Partial Sums of squares** and then click on **OK.**
	`anova overall_sat status`
Interpret the results:	
Look at the test results	Compare the *p*-value under **Model** with the significance level. The *p*-value should be lower than 0.05 to reject the null hypothesis.
Carry out pairwise comparisons	You can only carry out post hoc tests *after* you have carried out the ANOVA analysis. After the ANOVA analysis, go to ► Statistics ► Postestimation. In the next window that follows, go to ►Tests, contrasts, and comparisons of parameter estimates ► Pairwise comparisons and click on **Launch.**

(continued)

Table 6.5 (continued)

Theory	Action
	Select the variable *status* under **Factor terms to compute pairwise comparisons for:** aand select **Tukey's method** option in the **Multiple comparisons** box. Finally, go to the **Reporting** tab, tick the box **Show effects table with confidence intervals and p-values** and the box **Sort the margins/differences in each term** and click on **OK**. Now check whether the pairwise mean comparisons differ significantly if the *p*-values (under **P>|t|**) are lower than 0.05.
	`pwcompare status, effects sort mcompare (tukey)`
	If unequal variances are assumed, use Scheffé's method.
Look at the strength of the effects	Check for the strengths of the effects under **R-squared** and **Adjusted R-squared** in the output.
What is your conclusion?	Based on pairwise comparisons: Check which pairs differ significantly from each other. If the *p*-values tied to the pairwise mean comparisons are < 0.05, reject the null hypothesis that the mean comparisons between the two groups are equal.
	Based on the output from the one-way ANOVA table, reject the null hypothesis that at least two population means are equal if the *p*-value is lower than 0.05.
Plotting the ANOVA results (optional)	To plot the results from the one-way ANOVA, go to ► Statistics ► Postestimation. In the next window that follows, go to ► Marginal analysis ► Marginal means and marginal effects, fundamental analyses and click on **Launch**. Enter the variable *status* under **Covariate**, tick the box **Draw profile plots of results** and then click on **OK**.

[a]Note that the Oddjob Airways dataset is not well suited to perform (1) the one-sample *t*-test and (2) the paired samples *t*-test. We therefore use hypothetical variables to illustrate the Stata commands

[b]# = refers to a hypothetical constant (number) against which you want to compare

[c]**Outcome1** refers to the dependent variable, with **groupvar** representing the two groups with and without treatment

[d]**Variable1** and **Variable2** represent the outcome variable with and without treatment

6.8 Example

Let's now turn to the Oddjob Airways case study and apply the materials discussed in this chapter. Our aim is to identify the factors that influence customers' overall price/performance satisfaction with the airline and explore the relevant target groups for future advertising campaigns. Based on discussions with the Oddjob Airways management, answering the following three research questions will help achieve this aim:

1. Does the overall price/performance satisfaction differ by gender?
2. Does the overall price/performance satisfaction differ according to the traveler's status?
3. Does the impact of the traveler's status on the overall price/performance satisfaction depend on the different levels of the variable gender?

The following variables (variable names in parentheses) from the Oddjob Airways dataset (⤓ Web Appendix → Downloads) are central to this example:

- overall price/performance satisfaction (*overall_sat*),
- respondent's gender (*gender*), and
- traveler's status (*status*).

6.8.1 Independent Samples *t*-Test

6.8.1.1 Formulate Hypothesis

We start by formulating a non-directional hypothesis. The null hypothesis of the first research question is that the overall price/performance satisfaction means of male and female travelers are the same (H_0), while the alternative hypothesis (H_1) expects the opposite.

6.8.1.2 Choose the Significant Level

Next, we decide to use a significance level (α) of 0.05, which means that we allow a maximum chance of 5% of mistakenly rejecting a true null hypothesis.

6.8.1.3 Select an Appropriate Test

We move to the next step to determine the type of test, which involves assessing the testing situation, the nature of the measurements, checking the assumptions, and selecting the region of rejection. We start by defining the testing situation of our analysis, which concerns comparing the mean overall price/performance satisfaction scores (measured on a ratio scale) of male and female travelers. In our example, we know that the sample is a random subset of the population and we also know that other respondents' responses do not influence those of the respondents (i.e., they are independent). Next, we need to check if the dependent variable *overall_sat* is normally distributed between male and female travelers (i.e., normality assumption) and whether male travelers show the same variance in their overall price satisfaction as female travelers (i.e., equality of variance assumption). We use the Shapiro-Wilk

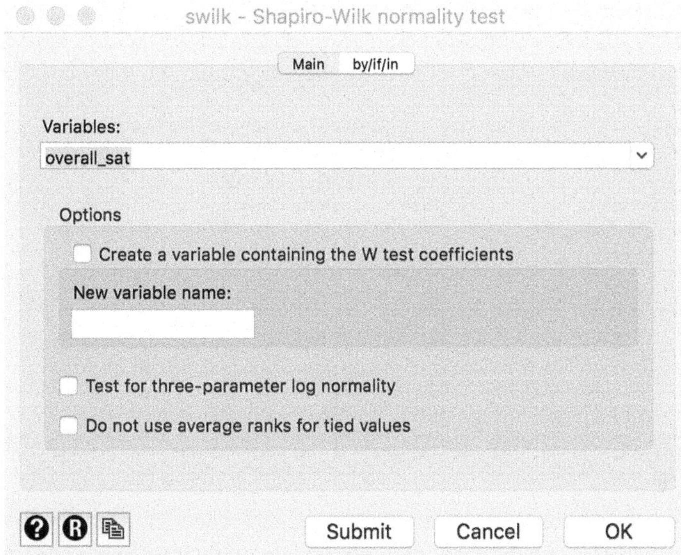

Fig. 6.6 Shapiro-Wilk normality test dialog box

Table 6.6 Shapiro-Wilk normality test output

```
by gender, sort: swilk overall_sat

-------------------------------------------------------------------------------------------
---
-> gender = female

                   Shapiro-Wilk W test for normal data

    Variable |        Obs       W           V         z       Prob>z
-------------+------------------------------------------------------
 overall_sat |        280    0.98357      3.293     2.788    0.00265

-------------------------------------------------------------------------------------------
---
-> gender = male

                   Shapiro-Wilk W test for normal data

    Variable |        Obs       W           V         z       Prob>z
-------------+------------------------------------------------------
 overall_sat |        785    0.99050      4.805     3.848    0.00006
```

test for the normality test. Go to ► Statistics ► Summaries, tables, and tests ► Distributional plots and tests ► Shapiro-Wilk normality test. In the **Main** dialog box that follows (Fig. 6.6), select the variable *overall_sat* under **Variables**. Next, in the **by/if/in** tab, tick the box **Repeat command by groups** and enter the variable *gender* under **Variables that define groups** and then click on **OK**.

Table 6.6 displays the Stata output that follows. Stata reports the Shapiro-Wilk test statistic in its original version (**W**) and scaled version (**V**) with their corresponding *z*-values (*z*) and *p*-values under (**Prob** > **z**). Table 6.6 shows that

Table 6.7 Levene's test output

```
robvar overall_sat, by(gender)

            | Summary of Overall, I am satisfied
            | with the price performance ratio of
            |          Oddjob Airways.
     Gender |        Mean   Std. Dev.        Freq.
------------+------------------------------------
     female |         4.5   1.6461098          280
       male |   4.2369427   1.6130561          785
------------+------------------------------------
      Total |   4.3061033   1.6251693        1,065

W0  =  0.41763753   df(1, 1063)     Pr > F = 0.51825772
W50 =  0.23527779   df(1, 1063)     Pr > F = 0.62773769
W10 =  0.02419285   df(1, 1063)     Pr > F = 0.87642474
```

the *p*-values under (**Prob** > **z**) of both female (**0.00265**) and male (**0.00006**) samples are smaller than 0.05, indicating that the normality assumption *is* violated.

Next, we need to check for the equality of the variances assumption. Go to ▶ Statistics ▶ Summaries, tables, and tests ▶ Classical tests of hypotheses ▶ Robust equal-variance test. Enter the dependent variable *overall_sat* into the **Variable** box and *gender* in the **Variable defining the comparison groups** box and click on **OK**.

As you can see in Table 6.7, Stata calculates Levene's test for the mean (**W0**), the median (**W50**), and for the 10% trimmed mean replacement (**W10**), with their corresponding *p*-values (**Pr** > **F**) at the bottom of the table. We can see that the *p*-values of **W0**, **W50**, and **W10** are higher than 0.05 and, thus, not significant. This means that there is no reason to think that the variances for male and female travelers are different. Overall, we conclude that the data are not normally distributed, but that the variances across the male and the female groups are equal, allowing us to utilize a parametric test for group differences, because these are robust against violations of normality when sample sizes are larger than 30. Having checked the underlying assumptions, we can now decide on the region of rejection of our study's main research questions. This relates does, of course, relate to our study's main hypothesis, which was formulated as non-directional, implying a two-sided test.

6.8.1.4 Calculate the Test Statistic and Make the Test Decision

In the next step, and given that the equal variances assumption was tenable, we decide to use an independent samples *t*-test. To run this test, go to ▶ Statistics ▶ Summaries, tables, and tests ▶ Classical tests of hypotheses ▶ t test (mean-comparison test). In the **Main** dialog box that follows (Fig. 6.7), select the second option **Two-sample using groups** from the dialog box and specify the outcome variable (*overall_sat*) under **Variable name**. Next, enter the grouping variable *gender* under **Group variable name**. Select the confidence interval **95**, which equates to a significance level of 5% and then click **OK**.

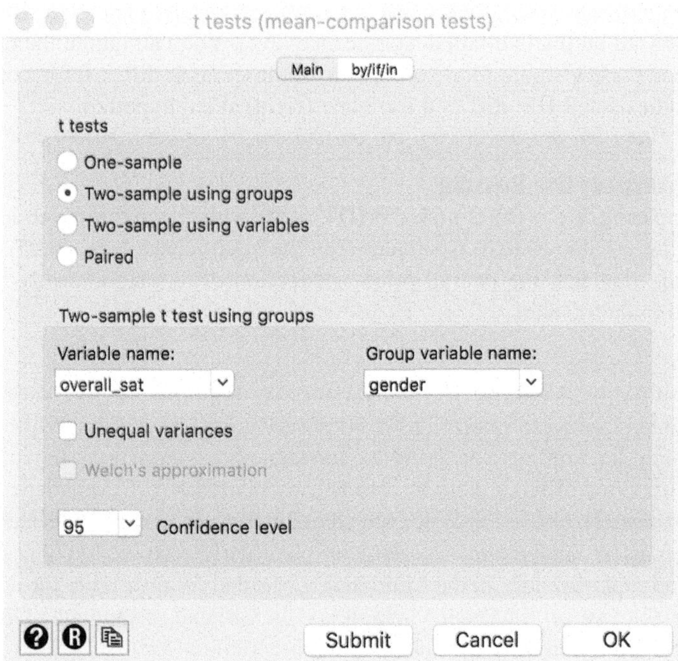

Fig. 6.7 Dialog box, independent sample *t*-test

Table 6.8 Output of the independent sample *t*-test in Stata

```
ttest overall_sat, by(gender)

Two-sample t test with equal variances
------------------------------------------------------------------------------
   Group |     Obs        Mean    Std. Err.   Std. Dev.   [95% Conf. Interval]
---------+--------------------------------------------------------------------
  female |     280         4.5    .0983739     1.64611    4.306351    4.693649
    male |     785    4.236943    .0575724    1.613056    4.123928    4.349957
---------+--------------------------------------------------------------------
combined |   1,065    4.306103    .0497994    1.625169    4.208387    4.403819
---------+--------------------------------------------------------------------
    diff |            .2630573    .1128905                .0415438    .4845708
------------------------------------------------------------------------------
    diff = mean(female) - mean(male)                              t =   2.3302
Ho: diff = 0                                     degrees of freedom =     1063

    Ha: diff < 0                 Ha: diff != 0                 Ha: diff > 0
 Pr(T < t) = 0.9900         Pr(|T| > |t|) = 0.0200          Pr(T > t) = 0.0100
```

The output that follows (Table 6.8) provides diverse information on the male and female travelers, including their mean overall price/performance satisfaction, the standard error, standard deviation, and the 95% confidence intervals (see Chap. 5). At the bottom of the table, Stata shows the results of the *t*-test for both one-sided and two-sided tests, and leaves it to the researcher to decide which results to interpret. In our case, our main hypothesis was formulated non-directionally (i.e., two-sided) and we therefore focus on the test results under **Ha: diff! = 0**, which are based on the two-tailed significance level. You can ignore the other test results, since these tests are for directional hypotheses (**Ha: diff < 0** for a one-(left) sided hypothesis and **Ha: diff > 0** for a one-(right) sided hypothesis).

6.8.1.5 Interpret the Results

When comparing the *p*-value under **Pr(|T| > |t|)** with the significance level, we learn that the *p*-value (**0.020**) is smaller than the significance level (0.05). Hence, we conclude that that the overall price satisfaction differs significantly between female and male travelers.

If the normality and equality of the variance assumptions were violated, and the sample sizes were small (i.e., < 30 observations), the Mann-Whitney U test, which Stata refers to as the Wilcoxon signed rank-sum test, should have been performed. This is the non-parametric counterpart of the independent sample *t*-test. To obtain the Wilcoxon signed rank-sum test, go to ►Statistics ► Summaries, tables, and tests ► Non-parametric tests of hypotheses ► Wilcoxon rank-sum test. In the **Main** dialog box that follows, enter the dependent variable *overall_sat* under **Variable** and the variable *gender* under **Grouping variable**. Stata lists the number of observations for female and male travelers separately, followed by their corresponding observed and expected rank sums. The test statistic can be found at the bottom of the table under the **H0**. The *p*-value under **Prob > |z|** is **0.0125** and, thus, smaller than 0.05. This result indicates that the medians of male and female travelers differ significantly.

6.8.2 One-way ANOVA

In the second research question, we examine whether customers' membership status influences their overall price/performance satisfaction (i.e., *overall_sat)* with Oddjob Airways. The membership can have three forms: *Blue*, *Silver*, and *Gold*. Again, we start by formulating a null hypothesis that is again non-directional in nature, expecting that the mean of the overall price/performance satisfaction is the same between the status groups, while the alternative hypothesis states that at least two status groups differ. Next, we decide to use a significance level (α) of 0.05. We have already established that a comparison of three or more groups involves a one-way ANOVA and we therefore follow the steps as indicated in Fig. 6.4.

6.8.2.1 Check the Assumptions

In checking the assumptions, we already know that the sample is a random subset of the population and we also know that other respondents' responses do not influence those of the respondents (i.e., they are independent). Next, we check the normality and equality of the variance assumptions by focusing directly on Stata's output tables (see previous research question for the menu options). We start with the results of the Shapiro-Wilk test displayed in Table 6.9.

Table 6.9 Stata ouput of the Shapiro-Wilk test

```
by status, sort: swilk overall_sat
-----------------------------------------------------------------------------
-> status = Blue
                   Shapiro-Wilk W test for normal data

     Variable |       Obs       W           V         z       Prob>z
 -------------+--------------------------------------------------------
  overall_sat |       677    0.98403     7.067     4.764    0.00000

-----------------------------------------------------------------------------
-> status = Silver
                   Shapiro-Wilk W test for normal data

     Variable |       Obs       W           V         z       Prob>z
 -------------+--------------------------------------------------------
  overall_sat |       245    0.99353     1.153     0.331    0.37021

-----------------------------------------------------------------------------
-> status = Gold
                   Shapiro-Wilk W test for normal data

     Variable |       Obs       W           V         z       Prob>z
 -------------+--------------------------------------------------------
  overall_sat |       143    0.98998     1.119     0.255    0.39927
```

We can see that the Shapiro-Wilk test produces significant effects for *Blue* members (**Prob** > **z** = **0.00000**), but not for *Silver* members (**Prob** > **z** = **0.37021**) and *Gold* members (**Prob** > **z** = **0.39927**). This means that in the three different samples, the overall price satisfaction is normally distributed in the *Silver* and *Gold* groups, but *not* in the *Blue* group. As we mentioned previously, the ANOVA is robust to violations from normality when samples are greater than 30, meaning that we can move to the next step to test the equality of variance assumptions, even though one of our samples violates the normality assumption. The output of Levene's test is shown in Table 6.9.

As we can see in Table 6.10, the *p*-values of **W0**, **W50**, and **W10** under **PR** > **F** are higher than 0.05 (thus not significant), which means that the variances between travelers with different statuses are the same. Overall, we conclude that the equal variance assumption is tenable, we can therefore move ahead and test our study's second research question.

6.8.2.2 Calculate the Test Statistic

To run an ANOVA, go to ► Statistics ► Linear models and related ► ANOVA/MANOVA ► Analysis of variance and covariance. In the dialog box that follows (Fig. 6.8), select *overall_sat* from the drop-down menu under **Dependent variable** and *status* (**status**) from the drop-down menu under **Model**. Next, then click on **OK**.

Stata will produce the output as shown in Table 6.11. The top part of the table reports several measures of model fit, which we discuss in Box 6.6. Under **Partial SS**, Stata lists different types of variation. **Model** represents the between-group variation (SS_B), whereas **Residual** indicates the within-group variation (SS_W). Next

Table 6.10 Stata output of Levene's test

```
robvar overall_sat, by(status)

            | Summary of Overall, I am satisfied
            | with the price performance ratio of
   Traveler |          Oddjob Airways.
     status |        Mean   Std. Dev.       Freq.
------------+------------------------------------
       Blue |   4.4726736   1.6411609         677
     Silver |   4.0326531   1.5599217         245
       Gold |    3.986014   1.5563863         143
------------+------------------------------------
      Total |   4.3061033   1.6251693       1,065

W0  =  0.90696260   df(2, 1062)     Pr > F = 0.4040612
W50 =  0.06775398   df(2, 1062)     Pr > F = 0.93449438
W10 =  0.88470222   df(2, 1062)     Pr > F = 0.41314113
```

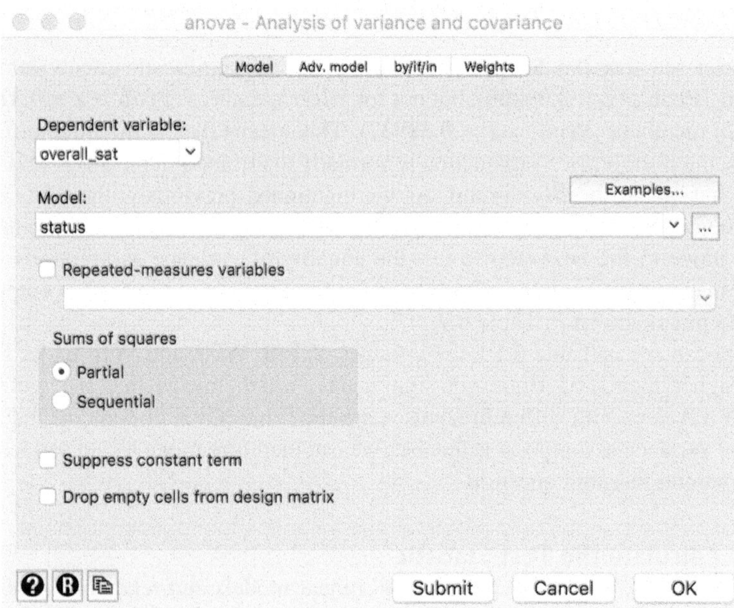

Fig. 6.8 ANOVA dialog box

to **status**, Stata lists this variable's partial contribution to the total variance. Given that the model has only one variable at this stage, the partial sums of squares explained by *status* (**51.755064**) is exactly the same as the partial sums of squares of the model (**51.755064**).

Table 6.11 One-way ANOVA

```
anova overall_sat status

                        Number of obs =      1,065    R-squared     =  0.0184
                        Root MSE      =    1.61165    Adj R-squared =  0.0166

                 Source | Partial SS         df         MS          F    Prob>F
             -----------+----------------------------------------------------
                  Model |  51.755064          2   25.877532      9.96  0.0001
                 status |  51.755064          2   25.877532      9.96  0.0001
               Residual |  2758.4553      1,062   2.5974155
             -----------+----------------------------------------------------
                  Total |  2810.2103      1,064   2.6411751
```

6.8.2.3 Make the Test Decision

Let's now focus on the F-test result with respect to the overall model. The model has an F-value of **9.96**, which yields a p-value of **0.0001** (see **Prob** > **F**), suggesting a statistically significant model.

6.8.2.4 Carry Out Post Hoc Tests

Next, we carry out pairwise group comparisons using Tukey's method. In Stata, this is a post estimation command, which means that comparisons can only be carried out after estimating the ANOVA. To run Tukey's method, go to ► Statistics ► Postestimation. In the window that follows, go to ► Tests, contrasts, and comparisons of parameter estimates ► Pairwise comparisons and click on **Launch**. In the dialog box that opens, select the variable *status* under **Factor terms to compute pairwise comparisons for** and select the **Tukey's method** option from the **Multiple comparisons** drop-down menu. Next, go to the **Reporting** tab and first tick **Specify additional tables (default is effects table with confidence intervals)** and then tick **Show effects table with confidence intervals and p-values**. Finally, in the same window, tick the box **Sort the margins/differences in each term**, and then click on **OK**. This produces the following output as in Table 6.12.

To check whether the means differ significantly, we need to inspect the p-values under **Tukey P** > **|t|.** The results in Table 6.12 indicate that the overall price/performance satisfaction differs significantly between *Gold* and *Blue* members, as well as between *Silver* and *Blue* members. The p-values of these pairwise comparisons are respectively **0.003** and **0.001** and, thus, smaller than 0.05. In contrast, the pairwise mean difference between *Gold* and *Silver* members does not differ significantly, as the p-value of **0.959** is above 0.05, indicating that members from these two status groups share similar views about the overall price/performance satisfaction with Oddjob Airways.

6.8.2.5 Measure the Strength of the Effects

The upper part of Table 6.11 reports the model fit. Besides the **R-squared** Stata reports the effect size η^2 **(0.0184)**. This means that differences in the travelers' status explain **1.841%** of the total variation in the overall price satisfaction. The ω^2 displayed under **Adj R-squared** is **0.0166**.

6.8.2.6 Interpret the Results

Overall, based on the outputs of the ANOVA in Tables 6.11 and 6.12, we conclude that:

1. *Gold* and *Blue* members, as well as *Silver* and *Blue* members, differ significantly in their mean overall price satisfaction.
2. Membership status explains only a minimal share of the customers' price satisfaction. Hence, other factors—presently not included in the model—explain the remaining variation in the outcome variable.

6.8.2.7 Plot the Results

Next, we plot the results, but we first should save the estimated parameters. Note that these estimated parameters capture the instantaneous change in the overall satisfaction level in respect of every unit change in the membership status, also termed the *marginal effect* (Greene 1997; Bartus 2005). Stata allows a fast and efficient way of saving the estimated parameters from the ANOVA. To do so, go to ► Statistics ► Postestimation. In the dialog box that follows, go to ► Marginal analysis ► Marginal means and marginal effects, fundamental analyses and click on **Launch**. Enter the variable *status* under **Covariate**, tick the box **Draw profile plots of results** and then click on **OK**. Stata will simultaneously produce Table 6.13 and the plot shown in Fig. 6.9.

The plot depicts the predicted group means (listed in Table 6.13) surrounded by their confidence intervals, which the vertical lines through the dots in Fig. 6.9 indicate. These intervals are very useful, because they immediately reveal whether the predicted group means differ significantly. More precisely, if the vertical bars do *not* overlap vertically with each other, we can say there is a significant difference. Overall, Fig. 6.9 indicates that there is a significant mean difference in the overall price satisfaction between *Blue* and *Silver* members and between *Blue* and *Gold* members, but no significant difference between *Silver* and *Gold* members. This is exactly what Tukey's pairwise comparison indicated in Table 6.12.

Table 6.12 Output of Tukey's method

```
pwcompare status, effects sort mcompare(tukey)

Pairwise comparisons of marginal linear predictions
Margins      : asbalanced
---------------------------
             |    Number of
             |  Comparisons
-------------+-------------
      status |            3
---------------------------
--------------------------------------------------------------------------------
                |                            Tukey                Tukey
                |   Contrast   Std. Err.      t    P>|t|     [95% Conf. Interval]
----------------+---------------------------------------------------------------
         status |
   Gold vs Blue |  -.4866596   .1483253    -3.28   0.003    -.8347787   -.1385404
 Silver vs Blue |  -.4400205   .1201597    -3.66   0.001     -.722035    -.158006
 Gold vs Silver |  -.0466391   .1696038    -0.27   0.959    -.4446987    .3514205
--------------------------------------------------------------------------------
```

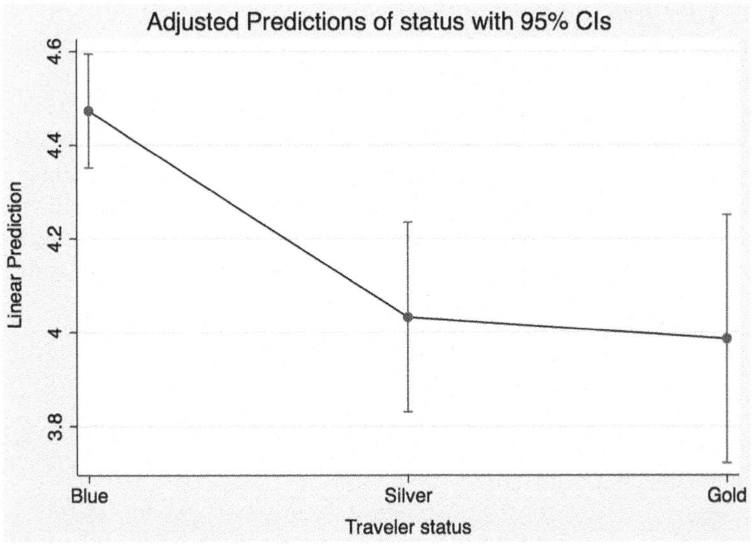

Fig. 6.9 Mean predicted level of overall satisfaction by flight frequency

Table 6.13 Average marginal effects by flight frequency

```
margins status

Adjusted predictions                              Number of obs     =      1,065

Expression    : Linear prediction, predict()

------------------------------------------------------------------------------
             |            Delta-method
             |     Margin   Std. Err.      t    P>|t|     [95% Conf. Interval]
-------------+----------------------------------------------------------------
      status |
       Blue  |   4.472674   .0619407    72.21   0.000     4.351133    4.594214
     Silver  |   4.032653   .1029645    39.17   0.000     3.830616     4.23469
       Gold  |   3.986014   .1347729    29.58   0.000     3.721562    4.250465
------------------------------------------------------------------------------
```

6.8.3 Two-way ANOVA

The final research question asks whether the impact of *status* on *overall price satisfaction* depends on the different levels of the variable *gender* (i.e., male versus female). The two-way ANOVA allows answering this research question. The null hypothesis for this combined effect of status and gender (i.e., their interaction effect) is that the difference in the overall price satisfaction between *Blue*, *Silver*, and *Gold* members is the same regardless of the travelers' gender. We decide to use a significance level (α) of 0.05 and directly calculate the test statistic, given that we already checked the ANOVA assumptions in the second research question.

6.8.3.1 Calculate the Test Statistic

In Stata, interaction effects between two variables are indicated by a hashtag (#) between the variables that we interact. Now, let us test for interaction effects. Go to ► Statistics ► Linear models and related ► ANOVA/MANOVA ► Analysis of variance and covariance. The dialog box that opens is similar to that shown in Fig. 6.10, only this time we enter **status##gender** (instead of only **status**) in the **Model** box and click on **OK**. This produces the output shown in Table 6.14. Note that it is essential to have two hashtags (i.e., ##), as this tells Stata to include the two main variables **status** and **gender**, plus the interaction variable **status#gender**. The

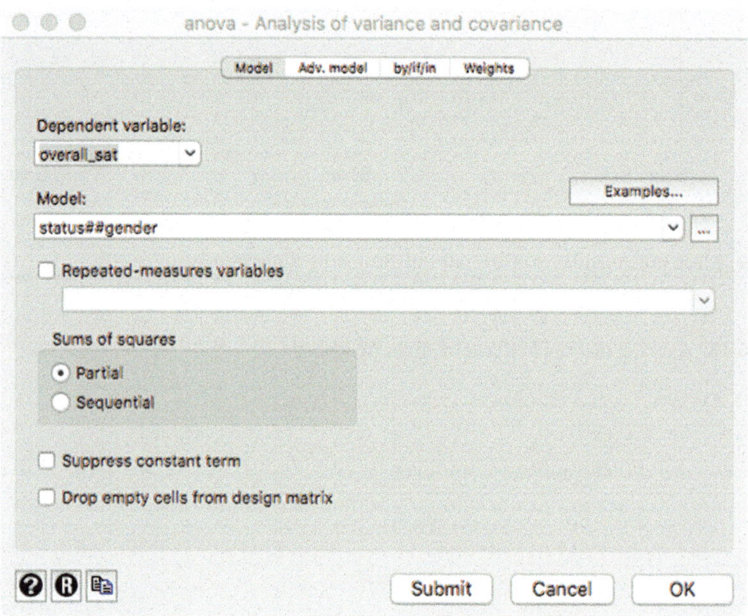

Fig. 6.10 ANOVA dialog box

Table 6.14 Output ANOVA

```
anova overall_sat status##gender
```

Number of obs =	1,065	R-squared =	0.0255
Root MSE =	1.6081	Adj R-squared =	0.0209

Source	Partial SS	df	MS	F	Prob>F
Model	71.656085	5	14.331217	5.54	0.0000
status	53.277593	2	26.638796	10.30	0.0000
gender	.2061234	1	.2061234	0.08	0.7777
status#gender	14.512817	2	7.2564083	2.81	0.0609
Residual	2738.5542	1,059	2.5859813		
Total	2810.2103	1,064	2.6411751		

reason for this is that *status* and *gender* are conditional main effects, whereby the effect of the one main variable is conditional on the effect of the other main variable with a value of 0. Thus, in this example, the conditional main effect of *status* represents the effect of *status* when *gender* is equal to 0, while the conditional effect of gender represents the effect of *gender* when *status* is equal to 0. When the included variables have no 0 category, such conditional effects have no interpretation. It is therefore important to set a meaningful 0 category.

6.8.3.2 Make the Test Decision

The output in Table 6.14 shows an F-value of **2.81** with a corresponding p-value of **0.0609** for the interaction term (**status#gender**). This p-value is higher than 0.05 and thereby not statistically significant. Note that had we decided to use a significance level (α) of 0.10, we would have found a significant interaction effect! However, as this is not the case in our example, we conclude that the overall level of price satisfaction by membership status does not depend on gender.

6.8.3.3 Carry Out Post Hoc Tests

To run Tukey's method, go to ► Statistics ► Postestimation. In the window that follows, go to ► Tests, contrasts, and comparisons of parameter estimates ► Pairwise comparisons and click on **Launch**. In the dialog box that opens, select the variable *status#gender* under **Factor terms to compute pairwise comparisons for** and select the **Tukey's method** option from the **Multiple comparisons** drop-down menu. Finally, go to the **Reporting** tab and first tick **Specify additional tables (default is effects table with confidence intervals)** and then tick **Show effects table with confidence intervals and p-values**. Finally, in the same window, tick the box **Sort the margins/differences in each term**, and then click on **OK**. This produces the output as shown in Table 6.15.

In this special case, post hoc tests are carried out across 15 distinct combinations within and between gender and membership status groups. We can check whether the pairwise mean comparisons differ significantly if the p-values (under **Tukey P > |t|**) are lower than 0.05. The results in Table 6.15 indicate that the overall price/performance satisfaction is significantly lower between: (1) female travelers with a *Silver* and *Blue* membership status, (2) female travelers with a *Gold* and *Blue* membership status, and (3) male and female travelers with a *Silver* membership status. The p-values of these pairwise comparisons are respectively **0.006**, **0.002**, and **0.002**, thus smaller than 0.05. All the other effects have a p-value higher than 0.05 and are not significant.

6.8.3.4 Measure the Strength of the Effects

In the next step, we focus on the model's effect strength. In Table 6.14 this is shown under **R-squared** and is **0.0255**, indicating that our model, which includes the interaction term, explains 2.5% of the total variance in the overall price satisfaction.

Table 6.15 Output of Tukey's method

```
pwcompare status#gender, mcompare(tukey) effects sort

Pairwise comparisons of marginal linear predictions

Margins      : asbalanced

---------------------------
             |    Number of
             |  Comparisons
-------------+-------------
status#gender |           15
---------------------------

---------------------------------------------------------------------------------------------------
                                  |                              Tukey                 Tukey
                                  |   Contrast   Std. Err.      t    P>|t|     [95% Conf. Interval]
----------------------------------+----------------------------------------------------------------
                    status#gender |
(Silver#female) vs (Blue#female)  |  -.9876923   .2789273    -3.54   0.006    -1.784013   -.1913714
  (Gold#female) vs (Blue#female)  |     -.7425   .4160734    -1.78   0.476    -1.930365    .4453648
    (Gold#male) vs (Blue#female)  |   -.687874   .1784806    -3.85   0.002    -1.197425   -.1783227
  (Silver#female) vs (Blue#male)  |  -.6771613   .2683811    -2.52   0.118    -1.443373    .0890507
  (Silver#male) vs (Blue#female)  |  -.5829126   .1550695    -3.76   0.002    -1.025627   -.1401984
    (Gold#female) vs (Blue#male)  |   -.431969   .4090783    -1.06   0.899    -1.599863     .735925
      (Gold#male) vs (Blue#male)  |   -.377343   .1615031    -2.34   0.180    -.8384248    .0837387
    (Blue#male) vs (Blue#female)  |   -.310531   .1312038    -2.37   0.169    -.6851101    .0640482
    (Silver#male) vs (Blue#male)  |  -.2723816   .1351832    -2.01   0.334    -.6583217    .1135584
  (Gold#female) vs (Silver#male)  |  -.1595874   .4173454    -0.38   0.999    -1.351083    1.031909
    (Gold#male) vs (Silver#male)  |  -.1049614   .1814259    -0.58   0.992    -.6229216    .4129988
    (Gold#male) vs (Gold#female)  |    .054626    .426598     0.13   1.000    -1.163286    1.272538
(Gold#female) vs (Silver#female)  |   .2451923   .4774212     0.51   0.996    -1.117817    1.608201
  (Gold#male) vs (Silver#female)  |   .2998183   .2943965     1.02   0.912     -.540666    1.140303
(Silver#male) vs (Silver#female)  |   .4047797   .2808212     1.44   0.702    -.3969479    1.206507
---------------------------------------------------------------------------------------------------
```

6.8.3.5 Interpret the Results

The mere increase of 0.7% in the explained variance (from 1.8% in Table 6.11 to 2.5% in Table 6.14) indicates that the interaction term does not add much to the model's fit. This is not surprising, as the interaction term reiterates the main effects in a slightly different form.

6.8.3.6 Plot the Results

Finally, we move to plotting the results, which is optional. Like research question 2, we go to ► Statistics ► Postestimation. In the dialog box that follows, go to ► Marginal analysis ► Marginal means and interaction analysis ► At sample means and click on **Launch**. Next, enter the first variable *status* under **Covariate** and tick the box **Interaction analysis with another covariate** where you enter the second variable *gender*. Then tick the box **Draw profile plots of results** and click **OK**. Stata will produce the output in Table 6.16 and the plot displayed in Fig. 6.11.

Under **status#gender** in Table 6.16, we see the *average marginal effects* of overall price satisfaction in respect of all status groups as they vary by gender. Here we find that the mean overall price satisfaction of female *Blue* status members (indicated by **Blue#female**) is **4.68**, while the mean of male *Blue* status members equals **4.369469**, and so on. Figure 6.11 depicts exactly these margins and their corresponding confidence intervals. Overall, we conclude that the relationship between the overall price satisfaction and travelers' membership status does not vary by gender.

Table 6.16 Predicted average marginal effects of a combination between flight frequency and gender

```
margins status#gender, plot

  Adjusted predictions                              Number of obs     =      1,065

  Expression   : Linear prediction, predict()
  at           : 1.status        =     .6356808 (mean)
                 2.status        =     .2300469 (mean)
                 3.status        =     .1342723 (mean)
                 1.gender        =     .2629108 (mean)
                 2.gender        =     .7370892 (mean)
------------------------------------------------------------------------------
               |            Delta-method
               |     Margin   Std. Err.      t    P>|t|     [95% Conf. Interval]
---------------+--------------------------------------------------------------
 status#gender |
   Blue#female |       4.68   .1072066    43.65   0.000     4.469639    4.890361
     Blue#male |   4.369469   .0756386    57.77   0.000      4.22105    4.517888
 Silver#female |   3.692308   .2575019    14.34   0.000     3.187036     4.19758
   Silver#male |   4.097087   .1120415    36.57   0.000     3.877239    4.316936
   Gold#female |     3.9375   .4020247     9.79   0.000     3.148645    4.726355
     Gold#male |   3.992126   .1426957    27.98   0.000     3.712128    4.272124
------------------------------------------------------------------------------
```

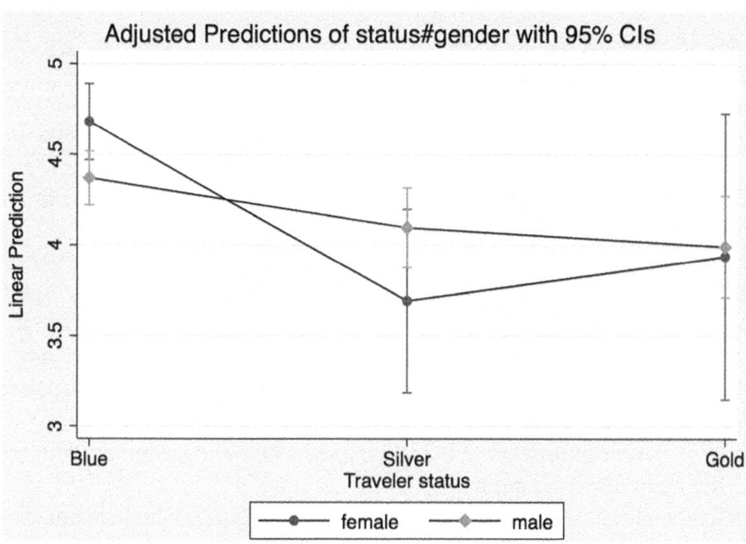

Fig. 6.11 Mean predicted level of overall satisfaction by membership status and gender

6.9 Customer Analysis at Crédit Samouel (Case Study)

CS
Crédit Samouel

In 2017, Crédit Samouel, a globally operating bank underwent a massive re-branding campaign. In the course of this campaign, the bank's product range was also restructured and its service and customer orientation improved. In addition, a comprehensive marketing campaign was launched, aimed at increasing the bank's customer base by one million new customers by 2025.

In an effort to control the campaign's success and to align the marketing actions, the management decided to conduct an analysis of newly acquired customers. Specifically, the management is interested in evaluating the segment customers aged 30 and below. To do so, the marketing department surveyed the following characteristics of 251 randomly drawn new customers (variable names in parentheses):

- Gender (*gender*: male/female).
- Bank deposit in Euro (*deposit*: ranging from 0 to 1,000,000).
- Does the customer currently attend school/university? (*training*: yes/no).
- Customer's age specified in three categories (*age_cat*: 16–20, 21–25, and 26–30).

Use the data provided in *bank.dta* (⭳ Web Appendix → Downloads) to answer the following research questions:

1. Which test do we have to apply to find out whether there is a significant difference in bank deposits between male and female customers? Do we meet the assumptions required to conduct this test? Also use an appropriate normality test and interpret the result. Does the result give rise to any cause for concern? Carry out an appropriate test to answer the initial research question.
2. Is there a significant difference in bank deposits between customers who are still studying and those who are not?
3. Which type of test or procedure would you use to evaluate whether bank deposits differ significantly between the three age categories? Carry out this procedure and interpret the results.
4. Reconsider the previous question and, using post hoc tests, evaluate whether there are significant differences between the three age groups.
5. Is there a significant interaction effect between the variables *training* and *age_cat* in terms of the customers' deposit?

6. Estimate and plot the average marginal effects of bank deposits over the different combinations of training groups and the customers' different age categories.
7. Based on your analysis results, please provide recommendations for the management team on how to align their future marketing actions.

6.10 Review Questions

1. Describe the steps involved in hypothesis testing in your own words.
2. Explain the concept of the *p*-value and explain how it relates to the significance level α.
3. What level of α would you choose for the following types of market research studies? Give reasons for your answers.
   (a) An initial study on preferences for mobile phone colors.
   (b) The production quality of Rolex watches.
   (c) A repeat study on differences in preference for either Coca Cola or Pepsi.
4. Write two hypotheses for each of the example studies in question 3, including the null hypothesis and alternative hypothesis.
5. Describe the difference between independent and paired samples *t*-tests in your own words and provide two examples of each type.
6. What is the difference between an independent samples *t*-test and an ANOVA?
7. What are post hoc test and why is their application useful in ANOVA?

6.11 Further Readings

Hubbard, R., & Bayarri, M. J. (2003). Confusion over measure of evidence (p's) versus errors (α's) in classical statistical testing. *The American Statistician, 57*(3), 171–178.

The authors discuss the distinction between p-value and α and argue that there is general confusion about these measures' nature among researchers and practitioners. A very interesting read!

Kanji, G. K. (2006). *100 statistical tests* (3rd ed.). London: Sage.

If you are interested in learning more about different tests, we recommend this best-selling book in which the author introduces various tests with information on how to calculate and interpret their results using simple datasets.

Mooi, E., & Ghosh, M. (2010). Contract specificity and its performance implications. *Journal of Marketing, 74*(2), 105–120.

This is an interesting article that demonstrates how directional hypotheses are formulated based on theory-driven arguments about contract specificity and performance implications.

Stata.com (http://www.stata.com/manuals14/rmargins.pdf).

Stata offers a very thorough explanation of marginal effects and the corresponding Stata syntaxes for the estimation of marginal means, predictive margins, and marginal effects.

References

Agresti, A., & Finlay, B. (2014). *Statistical methods for the social sciences* (4th ed.). London: Pearson.

Bartus, T. (2005). Estimation of marginal effects using marge off. *The Stata Journal, 5*(3), 309–329.

Boneau, C. A. (1960). The effects of violations of assumptions underlying the t test. *Psychological Bulletin, 57*(1), 49–64.

Brown, M. B., & Forsythe, A. B. (1974). Robust tests for equality of variances. *Journal of the American Statistical Association, 69*(346), 364–367.

Cohen, J. (1992). A power primer. *Psychological Bulletin, 112*(1), 155–159.

Everitt, B. S., & Skrondal, A. (2010). *The Cambridge dictionary of statistics* (4th ed.). Cambridge: Cambridge University Press.

Field, A. (2013). *Discovering statistics using SPSS* (4th ed.). London: Sage.

Greene, W. H. (1997). *Econometric analysis* (3rd ed.). Upper Saddle River: Prentice Hall.

Hubbard, R., & Bayarri, M. J. (2003). Confusion over measure of evidence (p's) versus errors (α's) in classical statistical testing. *The American Statistician, 57*(3), 171–178.

Kimmel, H. D. (1957). Three criteria for the use of one-tailed tests. *Psychological Bulletin, 54*(4), 351–353.

Lehmann, E. L. (1993). The Fischer, Neyman-Pearson theories of testing hypotheses: One theory or two? *Journal of the American Statistical Association, 88*(424), 1242–1249.

Levene, H. (1960). Robust tests for equality of variances. In I. Olkin (Ed.), *Contributions to probability and statistics* (pp. 278–292). Palo Alto: Stanford University Press.

Liao, T. F. (2002). *Statistical group comparison.* New York: Wiley-InterScience.

Lichters, M., Brunnlieb. C., Nave, G., Sarstedt, M., & Vogt, B. (2016). The influence of serotonin defficiency on choice deferral and the compromise effect. *Journal of Marketing Research,* 53(2), 183–198.

Lilliefors, H. W. (1967). On the Kolmogorov–Smirnov test for normality with mean and variance unknown. *Journal of the American Statistical Association, 62*(318), 399–402.

Mann, H. B., & Whitney, D. R. (1947). On a test of whether one of two random variables is stochastically larger than the other. *The Annals of Mathematical Statistics, 18*(1), 50–60.

Mitchell, M. N. (2015). *Stata for the behavioral sciences.* College Station: Stata Press.

Nuzzo, R. (2014). Scientific method: Statistical errors. *Nature, 506*(7487), 150–152.

Ruxton, G. D., & Neuhaeuser, M. (2010). When should we use one-tailed hypothesis testing? *Methods in Ecology and Evolution, 1*(2), 114–117.

Schuyler, W. H. (2011). *Readings statistics and research* (6th ed.). London: Pearson.

Shapiro, S. S., & Wilk, M. B. (1965). An analysis of variance test for normality (complete samples). *Biometrika, 52*(3/4), 591–611.

Van Belle, G. (2008). *Statistical rules of thumb* (2nd ed.). Hoboken: Wiley.

Wasserstein, R. L., & Lazar, N. A. (2016). The ASA's statement on p-values: Context, process, and purpose. *The American Statistician, 70*(2), 129–133.

Welch, B. L. (1951). On the comparison of several mean values: An alternative approach. *Biometrika, 38*(3/4), 330–336.

Regression Analysis

7

Keywords

Adjusted R^2 • Akaike information criterion (AIC) • Autocorrelation • Bayes information criterion (BIC) • Binary logistic regression • Breusch-Pagan test • Coefficient of determination • Constant • Collinearity • Cross validation • Disturbance term • Dummy variable • Durbin-Watson test • Error • Estimation sample • η^2 (eta-squared) • F-test • Heteroskedasticity • Interaction effects • Intercept • Moderation analysis • (Multi)collinearity • Multinomial logistic regression • Multiple regression • Nested models • Ordinary least squares • Outlier • Ramsey's RESET test • Residual • R^2 • Robust regression • Simple regression • Split-sample validation • Standard error • Standardized effects • Unstandardized effects • Validation sample • Variance inflation factor (VIF) • White's test

Learning Objectives

After reading this chapter, you should understand:

- The basic concept of regression analysis.
- How regression analysis works.
- The requirements and assumptions of regression analysis.
- How to specify a regression analysis model.
- How to interpret regression analysis results.
- How to predict and validate regression analysis results.
- How to conduct regression analysis with Stata.
- How to interpret regression analysis output produced by Stata.

E. Mooi et al., *Market Research*, Springer Texts in Business and Economics,
DOI 10.1007/978-981-10-5218-7_7

7.1 Introduction

Regression analysis is one of the most frequently used analysis techniques in market research. It allows market researchers to analyze the relationships between dependent variables and independent variables (Chap. 3). In marketing applications, the dependent variable is the outcome we care about (e.g., sales), while we use the independent variables to achieve those outcomes (e.g., pricing or advertising). The key benefits of using regression analysis are it allows us to:

1. Calculate if one independent variable or a set of independent variables has a significant relationship with a dependent variable.
2. Estimate the relative strength of different independent variables' effects on a dependent variable.
3. Make predictions.

Knowing whether independent variables have a significant effect on dependent variables, helps market researchers in many different ways. For example, this knowledge can help guide spending if we know promotional activities relate strongly to sales.

Knowing effects' relative strength is useful for marketers, because it may help answer questions such as: Do sales depend more on the product price or on product promotions? Regression analysis also allows us to compare the effects of variables measured on different scales, such as the effect of price changes (e.g., measured in dollars) and the effect of a specific number of promotional activities.

Regression analysis can also help us make predictions. For example, if we have estimated a regression model by using data on the weekly supermarket sales of a brand of milk in dollars, the milk price (which changes with the season and supply), as well as an index of promotional activities (comprising product placement, advertising, and coupons), the results of the regression analysis could answer the question: what would happen to the sales if the prices were to increase by 5% and the promotional activities by 10%? Such answers help (marketing) managers make sound decisions. Furthermore, by calculating various scenarios, such as price increases of 5%, 10%, and 15%, managers can evaluate marketing plans and create marketing strategies.

7.2 Understanding Regression Analysis

In the previous paragraph, we briefly discussed what regression can do and why it is a useful market research tool. We now provide a more detailed discussion. Look at Fig. 7.1, which plots a dependent (y) variable (the weekly sales of a brand of milk in dollars) against an independent (x_1) variable (an index of promotional activities). Regression analysis is a way of fitting a "best" line through a series of observations. With a "best" line we mean one that is fitted in such a way that it minimizes the sum of the squared differences between the observations and the line itself. It is

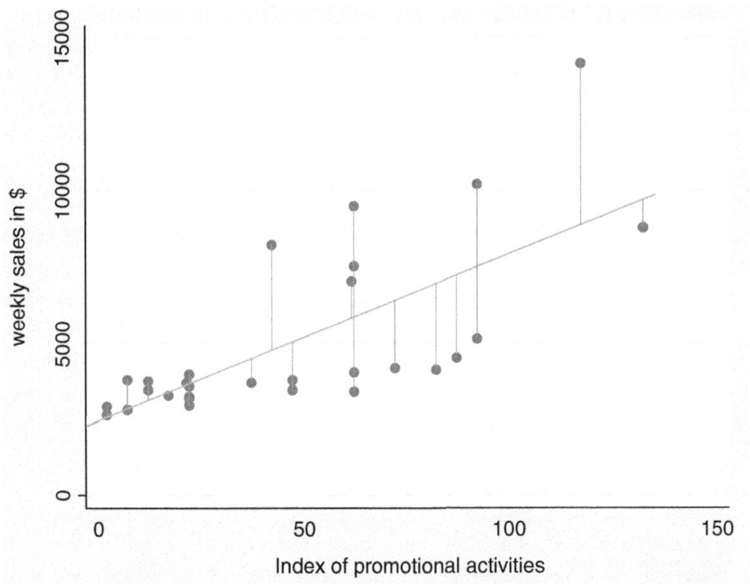

Fig. 7.1 A visual explanation of regression analysis

important to know that the best line fitted by means of regression analysis is not necessarily the true line (i.e., the line that represents the population). Specifically, if we have data issues, or fail to meet the regression assumptions (discussed later), the estimated line may be biased.

Before we discuss regression analysis further, we should discuss regression notation. Regression models are generally denoted as follows:

$$y = \alpha + \beta_1 x_1 + e$$

What does this mean? The y represents the dependent variable, which is the outcome you are trying to explain. In Fig. 7.1, we plot the dependent variable on the vertical axis. The α represents the **constant** (or **intercept**) of the regression model, and indicates what your dependent variable would be if the independent variable were zero. In Fig. 7.1, you can see the constant is the value where the fitted straight (sloping) line crosses the y-axis. Thus, if the index of promotional activities is zero, we expect the weekly supermarket sales of a specific milk brand to be $2,500. It may not always be realistic to assume that independent variables are zero (prices are, after all, rarely zero), but the constant should always be included to ensure the regression model's best possible fit with the data.

The independent variable is indicated by x_1, while the β_1 (pronounced beta) indicates its (regression) coefficient. This coefficient represents the slope of the line, or the slope of the diagonal grey line in Fig. 7.1. A positive β_1 coefficient indicates an upward sloping regression line, while a negative β_1 coefficient indicates a downward sloping line. In our example, the line slopes upward. This makes sense, since sales tend to increase with an increase in promotional activities. In our example, we estimate the β_1 as 54.59, meaning that if we increase the promotional activities by one unit, the weekly supermarket sales of a brand of milk will go up by an average of $54.59. This β_1 value has a degree of associated uncertainty called the **standard error**. This standard error is assumed to be normally distributed. Using a *t*-test (see Chap. 6), we can test if the β_1 is indeed significantly different from zero.

The last element of the notation, the e, denotes the equation **error** (also called the **residual** or **disturbance term**). The error is the distance between each observation and the best fitting line. To clarify what a regression error is, examine Fig. 7.1 again. The error is the difference between the regression line (which represents our regression prediction) and the actual observation (indicated by each dot). The predictions made by the "best" regression line are indicated by $\widehat{y}$ (pronounced *y*-hat). Thus, the error of each observation is:[1]

$$e = y - \widehat{y}$$

In the example above, we have only one independent variable. We call this **simple regression**. If we include multiple independent variables, we call this **multiple regression**. The notation for multiple regression is similar to that of simple regression. If we were to have two independent variables, say the price (x_1), and an index of promotional activities (x_2), our notation would be:

$$y = \alpha + \beta_1 x_1 + \beta_2 x_2 + e$$

We need one regression coefficient for each independent variable (i.e., β_1 and β_2). Technically the βs indicate how a change in an independent variable influences the dependent variable if all other independent variables are held constant.[2]

The Explained Visually webpage offers an excellent visualization of how regression analysis works, see http://setosa.io/ev/ordinary-least-squares-regression/

Now that we have introduced a few regression analysis basics, it is time to discuss how to execute a regression analysis. We outline the key steps in Fig. 7.2. We first introduce the regression analysis data requirements, which will determine if regression analysis can be used. After this first step, we specify and estimate the regression model. Next, we discuss the basics, such as which independent variables to select. Thereafter, we discuss the assumptions of regression analysis, followed by

[1]Strictly speaking, the difference between the predicted and the observed *y*-values is $\widehat{e}$.
[2]This only applies to the standardized βs.

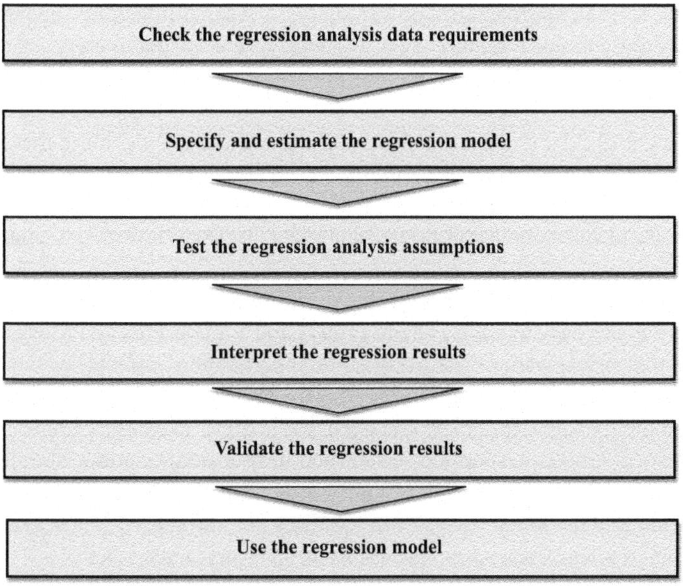

Fig. 7.2 Steps to conduct a regression analysis

how to interpret and validate the regression results. The last step is to use the regression model to, for example, make predictions.

7.3 Conducting a Regression Analysis

7.3.1 Check the Regression Analysis Data Requirements

Various data requirements must be taken into consideration before we undertake a regression analysis. These include the:

- sample size,
- variables need to vary,
- scale type of the dependent variable, and
- collinearity.

We discuss each requirement in turn.

7.3.1.1 Sample Size

The first data requirement is that we need an "acceptable" sample size. "Acceptable" relates to a sample size that gives you a good chance of finding significant results if they are possible (i.e., the analysis achieves a high degree of statistical power; see Chap. 6). There are two ways to calculate "acceptable" sample sizes.

- The first, formal, approach is a power analysis. As mentioned in Chap. 6 (Box 6. 1), these calculations require you to specify several parameters, such as the expected effect size and the maximum type I error you want to allow for. Generally, you also have to set the power—0.80 is an acceptable level. A power level of 0.80 means there is an 80% probability of deciding that an effect will be significant, if it is indeed significant. Kelley and Maxwell (2003) discuss sample size requirements in far more detail, while: https://stats.idre.ucla.edu/stata/dae/multiple-regression-power-analysis/ discusses how to calculate sample sizes precisely.
- The second approach is by using rules of thumb. These rules are not specific or precise, but are easy to apply. Green (1991) and VanVoorhis and Morgan (2007) suggest that if you want to test for individual parameters' effect (i.e., whether one coefficient is significant or not), you need a sample size of 104 + k. Thus, if you have ten independent variables, you need 104 + 10 = 114 observations. Note that this rule of thumb is best applied when you have a small number of independent variables, say less than 10 and certainly less than 15. VanVoorhis and Morgan (2007) add that having at least 30 observations per variable (i.e., $30 \times k$) allows for detecting smaller effects (an expected R^2 of 0.10 or smaller) better.

7.3.1.2 Variables Need to Vary

A regression model cannot be estimated if the variables have no variation. If there is no variation in the dependent variable (i.e., it is constant), we also do not need regression, as we already know what the dependent variable's value is! Likewise, if an independent variable has no variation, it cannot explain any variation in the dependent variable.

> No variation can lead to epic failures! Consider the admission tests set by the University of Liberia: Not a single student passed the entry exams. In such situations, a regression analysis will clearly make no difference! http://www.independent.co.uk/student/news/epic-fail-all-25000-students-fail-university-entrance-exam-in-liberia-8785707.html

7.3.1.3 Scale Type of the Dependent Variable

The third data requirement is that the dependent variable needs to be interval or ratio scaled (Chap. 3 discusses scaling). If the data are not interval or ratio scaled, alternative types of regression should be used. You should use **binary logistic regression** if the dependent variable is binary and only takes two values (zero and one). If the dependent variable is a nominal variable with more than two levels, you should use **multinomial logistic regression**. This should, for example, be used if you want to explain why people prefer product A over B or C. We do not discuss these different methods in this chapter, but they are related to regression. For a discussion of regression methods with dependent variables measured on a nominal or ordinal scale, see Cameron and Trivedi (2010).

7.3.1.4 Collinearity

The last data requirement is that no or little collinearity should be present.[3] **Collinearity** is a data issue that arises if two independent variables are highly correlated. Perfect collinearity occurs if we enter two or more independent variables containing exactly the same information, therefore yielding a correlation of 1 or −1 (i.e., they are perfectly correlated). Perfect collinearity may occur if you enter the same independent variable twice, or if one variable is a linear combination of another (e.g., one variable is a multiple of another variable, such as sales in units and sales in thousands of units). If this occurs, regression analysis cannot estimate one of the two coefficients; this means one coefficient will not be estimated. In practice, however, weaker forms of collinearity are common. For example, if we study what drives supermarket sales, variables such as price reductions and promotions are often used together. If this occurs very often, the variables price and promotion may be collinear, which means there is little uniqueness or new information in each of the variables. The problem with having collinearity is that it tends to regard significant parameters as insignificant. Substantial collinearity can even lead to sign changes in the regression coefficients' estimates. When three or more variables are strongly related to each other, we call this **multicollinearity**.

Fortunately, collinearity is relatively easy to detect by calculating the **variance inflation factor (VIF)**. The VIF indicates the effect on the standard error of the regression coefficient for each independent variable. Specifically, the square root of the VIF indicates you how much larger the standard error is, compared to if that variable were uncorrelated with all other independent variables in the regression model. Generally, a VIF of 10 or above indicates that (multi) collinearity is a problem (Hair et al. 2013).[4] Some research now suggests that VIFs far above 10—such as 20 or 40—can be acceptable if the sample size is large and the R^2 (discussed later) is high (say 0.90 or more) (O'brien 2007). Conversely, if the sample sizes are below 200 and the R^2 is low (0.25 or less), collinearity is more problematic (Mason and Perreault 1991). Consequently, in such situations, lower VIF values—such as 5—should be the maximum.

You can remedy collinearity in several ways. If perfect collinearity occurs, drop one of the perfectly overlapping variables. If weaker forms of collinearity occur, you can utilize two approaches to reduce collinearity (O'brien 2007):

- The first option is to use principal component or factor analysis on the collinear variables (see Chap. 8). By using principal component or factor analysis, you create a small number of factors that comprise most of the original variables'

[3]This is only a requirement if you are interested in the regression coefficients, which is the dominant use of regression. If you are only interested in prediction, collinearity is not important.

[4]The VIF is calculated using a completely separate regression analysis. In this regression analysis, the variable for which the VIF is calculated is regarded as a dependent variable and all other independent variables are regarded as independents. The R^2 that this model provides is deducted from 1 and the reciprocal value of this sum (i.e., $1/(1 - R^2)$) is the VIF. The VIF is therefore an indication of how much the regression model explains one independent variable. If the other variables explain much of the variance (the VIF is larger than 10), collinearity is likely a problem.

information, but are uncorrelated. If you use factors, collinearity between the previously collinear variables is no longer an issue.

- The second option is to re-specify the regression model by removing highly correlated variables. Which variables should you remove? If you create a correlation matrix of all the independent variables entered in the regression model, you should first focus on the variables that are most strongly correlated (see Chap. 5 for how to create a correlation matrix). First try removing one of the two most strongly correlated variables. The one you should remove depends on your research problem—pick the most relevant variable of the two.
- The third option is not to do anything. In many cases removing collinear variables does not reduce the VIF values significantly. Even if we do, we run the risk of mis-specifying the regression model (see Box 7.1 for details). Given the trouble researchers go through to collect data and specify a regression model, it is often better to accept collinearity in all but the most extreme cases.

7.3.2 Specify and Estimate the Regression Model

We need to select the variables we want to include and decide how to estimate the model to conduct a regression analysis. In the following, we will discuss each step in detail.

7.3.2.1 Model Specification

The model specification step involves choosing the variables to use. The regression model should be simple yet complete. To quote Albert Einstein: "Everything should be made as simple as possible, but not simpler!" How do we achieve this? By focusing on our ideas of what relates to the dependent variable of interest, the availability of data, client requirements, and prior regression models. For example, typical independent variables explaining the sales of a particular product include the price and promotions. When available, in-store advertising, competitors' prices, and promotions are usually also included. Market researchers may, of course, choose different independent variables for other applications. Omitting important variables (see Box 7.1) has substantial implications for the regression model, so it is best to be inclusive. A few practical suggestions:

- If you have many variables available in the data that overlap in terms of how they are defined—such as satisfaction with the waiter/waitress and with the speed of service—try to pick the variable that is most distinct or relevant for the client. Alternatively, you could conduct a principal component or factor analysis (see Chap. 8) first and use the factors as the regression analysis's independent variables.
- If you expect to need a regression model for different circumstances, you should make sure that the independent variables are the same, which will allow you to compare the models. For example, temperature can drive the sales of some supermarket products (e.g., ice cream). In some countries, such as Singapore, the temperature is relatively constant, so including this variable is not important.

Box 7.1 Omitting Relevant Variables

Omitting key variables from a regression model can lead to biased results. Imagine that we want to explain weekly sales by only referring to promotions. From the introduction, we know the β of the regression model only containing promotions is estimated as 54.59. If we add the variable price (arguably a key variable), the estimated β of promotions drops to 42.27. As can be seen, the difference between the estimated βs in the two models (i.e., with and without price) is 12.32, suggesting that the "true" relationship between promotions and sales is weaker than in a model with only one independent variable. This example shows that omitting important independent variables leads to biases in the value of the estimated βs. That is, if we omit a relevant variable x_2 from a regression model that only includes x_1, we cause a bias in the β_1 estimate. More precisely, the β_1 is likely to be inflated, which means that the estimated value is higher than it should be. Thus, the β_1 itself is biased because we omit x_2!

In other countries, such as Germany, the temperature can fluctuate far more. If you are intent on comparing the ice cream sales in different countries, it is best to include variables that may be relevant to all the countries you want to compare (e.g., by including temperature, even if it is not very important in Singapore).

- Consider the type of advice you want to provide. If you want to make concrete recommendations regarding how to use point-of-sales promotions and free product giveaways to boost supermarket sales, both variables need to be included.
- Take the sample size rules of thumb into account. If practical issues limit the sample size to below the threshold that the rules of thumb recommend, use as few independent variables as possible. Larger sample sizes allow you more freedom to add independent variables, although they still need to be relevant.

7.3.2.2 Model Estimation

Model estimation refers to how we estimate a regression model. The most common method of estimating regression models is **ordinary least squares** (**OLS**). OLS fits a regression line to the data that minimizes the sum of the squared distances to it. These distances are squared to stop negative distances (i.e., below the regression line) from cancelling out positive distances (i.e., above the regression line), because squared values are always positive. Moreover, by using the square, we emphasize observations that are far from the regression much more, while observations close to the regression line carry very little weight. The rule to use squared distances is an effective (but also arbitrary) way of calculating the best fit between a set of observations and a regression line (Hill et al. 2008). If we return to Fig. 7.1., we see the vertical "spikes" from each observation to the regression line. OLS estimation is aimed at minimizing the squares of these spikes.

Table 7.1 Regression data

Week	Sales	Price	Promotion
1	3,454	1.10	12.04
2	3,966	1.08	22.04
3	2,952	1.08	22.04
4	3,576	1.08	22.04
5	3,692	1.08	21.42
6	3,226	1.08	22.04
7	3,776	1.09	47.04
8	14,134	1.05	117.04
9	5,114	1.10	92.04
10	4,022	1.08	62.04
11	4,492	1.12	87.04
12	10,186	1.02	92.04
13	7,010	1.08	61.42
14	4,162	1.06	72.04
15	3,446	1.13	47.04
16	3,690	1.05	37.04
17	3,742	1.10	12.04
18	7,512	1.08	62.04
19	9,476	1.08	62.04
20	3,178	1.08	22.04
21	2,920	1.12	2.04
22	8,212	1.04	42.04
23	3,272	1.09	17.04
24	2,808	1.11	7.04
25	2,648	1.12	2.04
26	3,786	1.11	7.04
27	2,908	1.12	2.04
28	3,395	1.08	62.04
29	4,106	1.04	82.04
30	8,754	1.02	132.04

We use the data behind Fig. 7.1—as shown in Table 7.1—to illustrate the method with which OLS regressions are calculated. This data has 30 observations, with information on the supermarket's sales of a brand of milk (*sales*), the price (*price*), and an index of promotional activities (*promotion*) for weeks 1–30. This dataset is small and only used to illustrate how OLS estimates are calculated. The data *regression.dta* can be downloaded, but are also included in Table 7.1. (see ⭳ Web Appendix → Downloads).

To estimate an OLS regression of the effect of *price* and *promotion* on *sales*, we need to calculate the βs, of which the estimate is noted as $\widehat{\beta}$ (pronounced as beta-hat). The $\widehat{\beta}$ indicates the estimated association between each independent variable (*price* and *promotion*) and the dependent variable *sales*. We can estimate $\widehat{\beta}$ as follows:

$$\widehat{\beta} = \left(x^T x\right)^{-1} . \, x^T y$$

In this equation to solve the $\widehat{\beta}$, we first multiply the transposed matrix indicated as x^T. This matrix has three elements, a vector of 1s, which are added to estimate the intercept and two vectors of the independent variables *price* and *promotion*. Together, these form a 30 by 3 matrix. Next, we multiply this matrix with the untransposed matrix, indicated as x, consisting of the same elements (as a 3 by 30 matrix). This multiplication results in a 3×3 matrix of which we calculate the inverse, indicated by the power of −1 in the equation. This also results in a 3 by 3 matrix $(x^T x)^{-1}$. Next, we calculate $x^T y$, which consists of the 30 by 3 matrix and the vector with the dependent variables' observations (a 1 by 30 matrix). In applied form:[5]

$$x = \begin{bmatrix} 1 & 1.10 & 12.04 \\ 1 & 1.08 & 22.04 \\ \vdots & \vdots & \vdots \\ 1 & 1.02 & 132.04 \end{bmatrix},$$

$$x^T = \begin{bmatrix} 1 & 1 & \ldots & 1 \\ 1.10 & 1.08 & \ldots & 1.02 \\ 12.04 & 22.04 & \ldots & 132.04 \end{bmatrix},$$

$$\left(x^T x\right)^{-1} = \begin{bmatrix} 77.97 & -70.52 & -0.04 \\ -70.52 & 63.86 & 0.03 \\ -0.04 & 0.03 & 0.00 \end{bmatrix},$$

$$x^T y = \begin{bmatrix} 147615.00 \\ 158382.64 \\ 8669899.36 \end{bmatrix},$$

$$\text{Hence,} \quad \left(x^T x\right)^{-1} \cdot x^T y = \begin{bmatrix} 30304.05 \\ -25209.86 \\ 42.27 \end{bmatrix}.$$

This last matrix indicates the estimated βs with 30304.05 representing the intercept, −25209.86 representing the effect of a one-unit increase in the price on sales, and 42.27 the effect of a one-unit increase in promotions on sales. This shows how the OLS estimator is calculated.

[5]This term can be calculated manually, but also by using the function *mmult* in Microsoft Excel where $x^T x$ is calculated. Once this matrix has been calculated, you can use the *minverse* function to arrive at $(x^T x)^{-1}$.

As discussed before, each $\widehat{\beta}$ has a standard error, which expresses the uncertainty associated with the estimate. This standard error can be expressed in standard deviations and, as discussed in Chap. 6, with more than 100 degrees of freedom and $\alpha = 0.05$, t-values outside the critical value of ± 1.96 indicate that the estimated effect is *significant* and that the null hypothesis can be rejected. If this the t-value falls within the range of $\pm$ 1.96, the $\widehat{\beta}$ is said to be *insignificant*.

While OLS is an effective estimator, there are alternatives that work better in specific situations. These situations occur if we violate one of the regression assumptions. For example, if the regression errors are heteroskedastic (discussed in Sect. 7.3.3.3), we need to account for this by, for example, using **robust regression** (White 1980).[6] Random-effects estimators allow for estimating a model with correlated errors. There are many more estimators, but these are beyond the scope of this book. Greene (2011) discusses these and other estimation procedures in detail. Cameron and Trivedi (2010) discuss their implementation in Stata.

7.3.3 Test the Regression Analysis Assumptions

We have already discussed several issues that determine whether running a regression analysis is useful. We now discuss regression analysis assumptions. If a regression analysis fails to meet its assumptions, it can provide invalid results. Four regression analysis assumptions are required to provide valid results:

1. the regression model can be expressed linearly,
2. the regression model's expected mean error is zero,
3. the errors' variance is constant (homoscedasticity), and
4. the errors are independent (no autocorrelation).

There is a fifth assumption, which is, however optional. If we meet this assumption, we have information on how the regression parameters are distributed, which allows straightforward conclusions regarding their significance. If the regression analysis fails to meet this assumption, the regression model will still be accurate, but it becomes we cannot rely on the standard errors (and t-values) to determine the regression parameters' significance.

5. The errors need to be approximately normally distributed.

We next discuss these assumptions and how we can test each of them.

[6]In Stata this can be done by using the `, robust` option.

7.3.3.1 First Assumption: Linearity

The first assumption means that we can write the regression model as $y = \alpha + \beta_1 x_1 + e$. Thus, non-linear relationships, such as $\beta_1^2 x_1$, are not permissible. However, logarithmic expressions, such as $\log(x_1)$, are possible as the regression model is still specified linearly. If you can write a model whose regression parameters (the βs) are linear, you satisfy this assumption.

A separate issue is whether the relationship between the independent variable x and the dependent variable y is linear. You can check the linearity between x and y variables by plotting the independent variables against the dependent variable. Using a scatter plot, we can then assess whether there is some type of non-linear pattern. Fig. 7.3 shows such a plot. The straight, sloping line indicates a linear relationship between *sales* and *promotions*. For illustration purposes, we have also added a curved upward sloping line. This line corresponds to a x_1^2 transformation. It visually seems that a linear line fits the data best. If we fail to identify non-linear relationships as such, our regression line does not fit the data well, as evidenced in a low model fit (e.g., the R^2, which we will discuss later) and nonsignificant effects. After transforming x_1 by squaring it (or using any other transformation), you still satisfy the assumption of specifying the regression model linearly, despite the non-linear relationship between x and y.

Ramsey's RESET test is a specific linearity test (Ramsey 1969; Cook and Weisberg 1983). This test includes the squared values of the independent variables

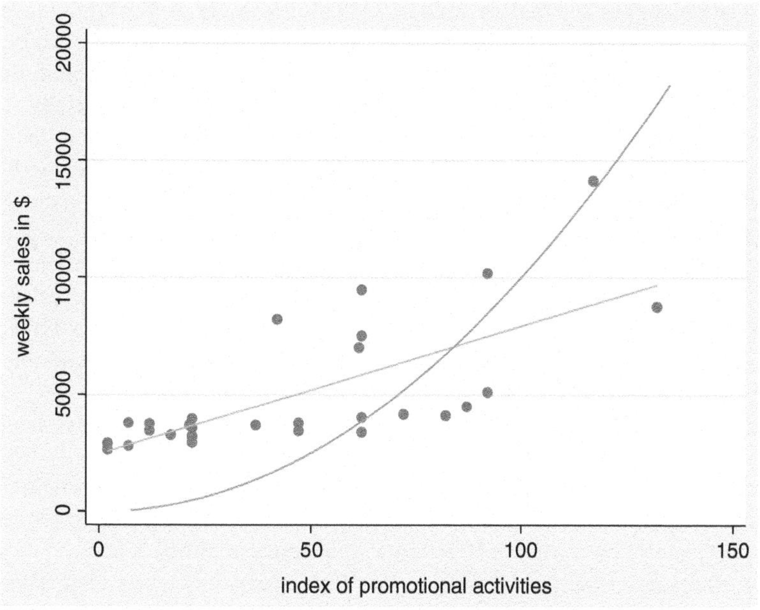

Fig. 7.3 Different relationships between promotional activities and weekly sales

(i.e., x_1^2 and third powers (i.e., x_1^3), and tests if these are significant (Baum 2006).[7] While this test can detect these specific types of non-linearities, it does not indicate which variable(s) has(ve) a non-linear relationship with the dependent variable. Sometimes this test is (falsely) called a test for omitted variables, but it actually tests for non-linearities.

7.3.3.2 Second Assumption: Expected Mean Error is Zero

The second assumption is that the expected (not the estimated!) mean error is zero. If we do not expect the sum of the errors to be zero, we obtain a biased line. That is, we have a line that consistently overestimates or underestimates the true relationship. This assumption is not testable by means of statistics, as OLS always renders a best line with a calculated mean error of exactly zero. This assumption is important, because if the error's expected value is not zero, there is additional information in the data that has not been used in the regression model. For example, omitting important variables, as discussed in Box 7.1, or autocorrelation may cause the expected error to no longer be zero (see Sect. 7.3.3.4).

7.3.3.3 Third Assumption: Homoscedasticity

The third assumption is that the errors' variance is constant, a situation we call homoscedasticity. Imagine that we want to explain various supermarkets' weekly sales in dollars. Large stores obviously have a far larger sales spread than small supermarkets. For example, if you have average weekly sales of $50,000, you might see a sudden jump to $60,000, or a fall to $40,000. However, a very large supermarket could see sales move from an average of $5 million to $7 million. This causes the weekly sales' error variance of large supermarkets to be much larger than that of small supermarkets. We call this non-constant variance **heteroskedasticity**. If we estimate regression models on data in which the variance is not constant, they will still result in correct βs. However, the associated standard errors are likely to be too large and may cause some βs to not be significant, although they actually are.

Figure 7.4 provides a visualization of heteroskedasticity. As the dependent variable increases, the error variance also increases. If heteroskedasticity is an issue, the points are typically funnel shaped, displaying more (or less) variance as the independent variable increases (decreases). This funnel shape is typical of heteroskedasticity and indicates that, as a function of the dependent variable, the error variance changes.

We can always try to visualize heteroskedasticity, whose presence is calculated by means of the errors, but it is often very difficult to determine visually whether heteroskedasticity is present. For example, when datasets are large, it is hard to see a funnel shape in the scatterplot. We can formally test for the presence of heteroskedasticity by using the **Breusch-Pagan test** and **White's test**.

The Breusch-Pagan test (1980) is the most frequently used test. This test determines whether the errors' variance depends on the variables in the model by

[7]The test also includes the predicted values squared and to the power of three.

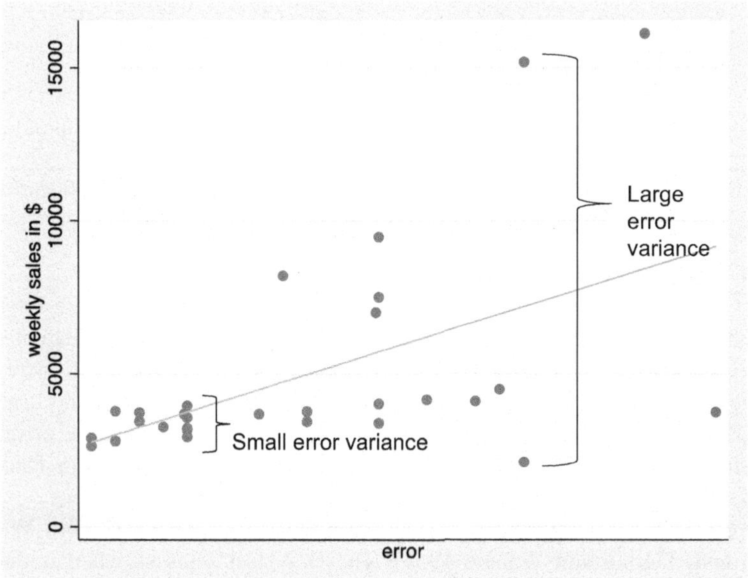

Fig. 7.4 An example of heteroskedasticity

using a separate regression analysis (Greene 2011). This tests the null hypothesis that the errors' variance does not depend on the variables in the regression model.[8] Rejecting this null hypothesis suggests that heteroskedasticity is present. If the Breusch-Pagan test indicates that heteroskedasticity is an issue, *robust regression* remedy for this (see Sect. 7.3.2.2). Note that to illustrate heteroskedasticity, we have used slightly different data than in the other examples.

White's test, as extended by Cameron and Trivedi (1990), is a different test for heteroskedasticity. This test does not test if the error goes up or down, but adopts a more flexible approach whereby errors can first go down, then up (i.e., an hourglass shape), or first go up, then down (diabolo-shaped). Compared to the Breusch-Pagan test, White's test considers more shapes that can indicate heteroskedasticity. This has benefits in that more forms of heteroskedasticity can be detected, but in small samples; however, White's test may not detect heteroskedasticity, even if it is present. It is therefore best to use both tests, as they have slightly different strengths. Generally, the two tests are comparable, but if they are not, it is best to rely on White's test.

7.3.3.4 Fourth Assumption: No Autocorrelation

The fourth assumption is that the regression model errors are independent; that is, the error terms are uncorrelated for any two observations. Imagine that you want to explain the supermarket sales of a brand of milk by using the previous week's sales

[8]Specifically, in the mentioned regression model $y = \alpha + \beta_1 x_1 + \beta_2 x_2 + \beta_3 x_3 + e$, the Breusch-Pagan test determines whether $\widehat{e}^2 = \alpha + \beta_{BP1} x_1 + \beta_{BP2} x_2 + \beta_{BP3} x_3 + e_{BP}$.

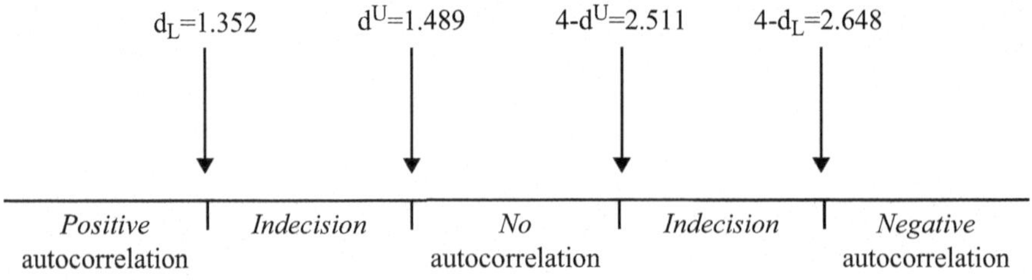

Fig. 7.5 Durbin-Watson test values ($n = 30$, $k = 1$)

of that milk. It is very likely that if sales increased last week, they will also increase this week. This may be due to, for example, the growing economy, an increasing appetite for milk, or other reasons that underlie the growth in supermarket sales of milk. This issue is called **autocorrelation** and means that regression errors are correlated positively (or negatively) over time. For example, the data in Table 7.1 are taken from weeks 1 to 30, which means they have a time component.

We can identify the presence of autocorrelation by using the **Durbin-Watson (D-W) test** (Durbin and Watson 1951). The D-W test assesses whether there is autocorrelation by testing the null hypothesis of no autocorrelation, which is tested for negative autocorrelation against a lower and upper bound and for positive autocorrelation against a lower and upper bound. If we reject the null hypothesis of no autocorrelation, we find support for an alternative hypothesis that there is some degree of positive or negative autocorrelation. Essentially, there are four situations, which we indicate in Fig. 7.5.

First, the errors may be positively related (called positive autocorrelation). This means that if we have observations over time, we observe that positive errors are generally followed by positive errors and negative errors by negative errors. For example, supermarket sales usually increase over certain time periods (e.g., before Christmas) and decrease during other periods (e.g., the summer holidays).

Second, if positive errors are commonly followed by negative errors and negative errors by positive errors, we have negative autocorrelation. Negative autocorrelation is less common than positive autocorrelation, but also occurs. If we study, for example, how much time salespeople spend on shoppers, we may see that if they spend much time on one shopper, they spend less time on the next, allowing the salesperson to stick to his/her schedule, or to simply go home on time.

Third, if no systematic pattern of errors occurs, we have no autocorrelation. This absence of autocorrelation is required to estimate standard (OLS) regression models.

Fourth, the D-W values may fall between the lower and upper critical value. If this occur, the test is inconclusive.

The situation that occurs depends on the interplay between the D-W test statistic (d) and the lower (d_L) and upper (d^U) critical value.

1. If the test statistic is lower than the lower critical value ($d < d_L$), we have positive autocorrelation.
2. If the test statistic is higher than 4 minus the lower critical value ($d > 4 - d_L$), we have negative autocorrelation.
3. If the test statistic falls between the upper critical value and 4 minus the upper critical value ($d^U < d < 4 - d^U$), we have no autocorrelation.
4. If the test statistic falls between the lower and upper critical value ($d_L < d < d^U$), or it falls between 4 minus the upper critical value and 4 minus the lower critical value ($4 - d^U < d < 4 - d^L$), the test does not inform on the presence of autocorrelation and is undecided.

The critical values d_L and d^U can be found on the website accompanying this book (⤓ Web Appendix → Downloads). From this table, you can see that the lower critical value d_L of a model with one independent variable and 30 observations is 1.352 and the upper critical value d^U is 1.489. Figure 7.5 shows the resulting intervals. Should the D-W test indicate autocorrelation, you should use models that account for this problem, such as panel and time series models. We do not discuss these methods in this book, but Cameron and Trivedi (2010) is a useful source of further information.

7.3.3.5 Fifth (Optional) Assumption: Error Distribution

The fifth, optional, assumption is that the regression model errors are approximately normally distributed. If this is not the case, the *t*-values may be incorrect. However, even if the regression model errors are not normally distributed, the regression model still provides good estimates of the coefficients. Consequently, we consider this assumption an optional one. Potential reasons for regression errors being non-normally distributed include **outliers** (discussed in Chap. 5) and a non-linear relationship between the independent and (a) dependent variable(s) as discussed in Sect. 7.3.3.1.

There are two main ways of checking for normally distributed errors: you can use plots or carry out a formal test. Formal tests of normality include the Shapiro-Wilk test (see Chap. 6), which needs to be run on the saved errors. A formal test may indicate non-normality and provide absolute standards. However, formal test results reveal little about the source of non-normality. A histogram with a normality plot may can help assess why errors are non-normally distributed (see Chap. 5 for details). Such plots are easily explained and interpreted and may suggest the source of non-normality (if present).

7.3.4 Interpret the Regression Results

In the previous sections, we discussed how to specify a basic regression model and how to test regression assumptions. We now discuss the regression model fit, followed by the interpretation of individual variables' effects.

7.3.4.1 Overall Model Fit

The model significance is the first aspect that should be determined. While model significance is not an indicator of (close) fit, it makes little sense to discuss model fit if the model itself is not significant. The **F-test** determines the model significance. The test statistic's F-value is the result of a one-way ANOVA (see Chap. 6) that tests the null hypothesis that all the regression coefficients equal zero. Thus, the following null hypothesis is tested:[9]

$$H_0 = \beta_1 = \beta_2 = \beta_3 = \ldots = 0$$

If the regression coefficients are all equal to zero, then all the independent variables' effect on the dependent variable is zero. In other words, there is no (zero) relationship between the dependent variable and the independent variables. If we do not reject the null hypothesis, we need to change the regression model, or, if this is not possible, report that the regression model is non-significant. A p-value of the F-test below 0.05 (i.e., the model is significant) does not, however, imply that all the regression coefficients are significant, or even that one of them is significant when considered in isolation. However, if the F-value is significant, it is highly likely that at least one or more regression coefficients are significant.

If we find that the F-test is significant, we can interpret the model fit by using the $\mathbf{R^2}$. The R^2 (also called the **coefficient of determination**) indicates the degree to which the model, relative to the mean, explains the observed variation in the dependent variable. In Fig. 7.6, we illustrate this graphically by means of a scatter plot. The y-axis relates to the dependent variable *sales* (weekly sales in dollars) and the x-axis to the independent variable *promotion*. In the scatter plot, we see 30 observations of sales and price (note that we use a small sample size for illustration purposes). The horizontal line (at about \$5,000 sales per week) refers to the average sales in all 30 observations. This is also our benchmark. After all, if we were to have no regression line, our best estimate of the weekly sales would also be the average. The sum of all the squared differences between each observation and the average is the total variation or the *total sum of squares* (SS_T). We indicate the total variation in only one observation on the right of the scatter plot.

The straight upward sloping line (starting at the y-axis at about \$2,500 sales per week when there are no promotional activities) is the regression line that OLS estimates. If we want to understand what the regression model adds beyond the average (which is the benchmark for calculating the R^2), we can calculate the difference between the regression line and the line indicating the average. We call this the *regression sum of squares* (SS_R), as it is the variation in the data that the regression analysis explains. The final point we need to understand regarding how well a regression line fits the available data, is the unexplained sum of the squares. This is the difference between the observations (indicated by the dots) and the regression line. The squared sum of these differences refers to the regression error that we discussed previously and which is therefore denoted as the *error sum*

[9]This hypothesis can also be read as that a model with only an intercept is sufficient.

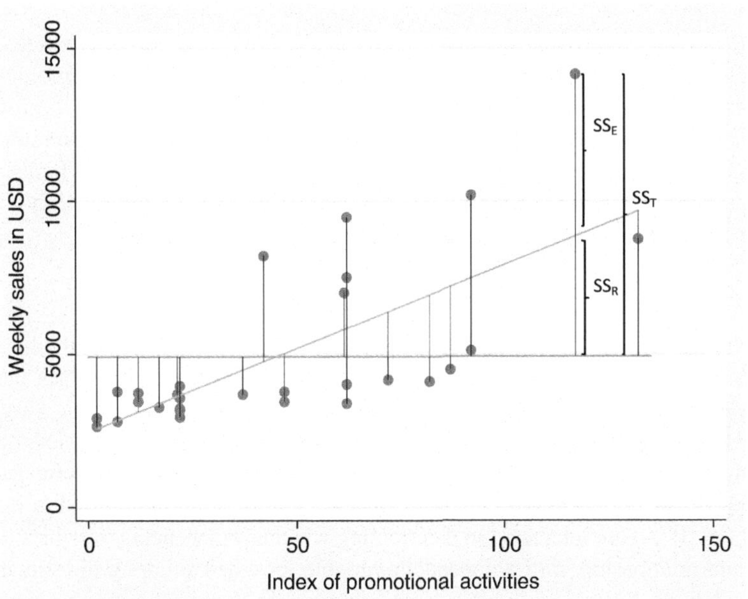

Fig. 7.6 Explanation of the R^2

of squares (SS_E). In more formal terms, we can describe these types of variation as follows:

$$SS_T = SS_R + SS_E$$

This is the same as:

$$\sum_{i=1}^{n} (y_i - \bar{y})^2 = \sum_{i=1}^{n} (\widehat{y}_i - \bar{y})^2 + \sum_{i=1}^{n} (y_i - \widehat{y}_i)^2$$

Here, n describes the number of observations, y_i is the value of the independent variable for observation i, $\widehat{y}_i$ is the predicted value of observation i, and $\bar{y}$ is the mean value of y. As you can see, this description is like the one-way ANOVA we discussed in Chap. 6. A useful regression line should explain a substantial amount of variation (have a high SS_R) relative to the total variation (SS_T):

$$R^2 = \frac{SS_R}{SS_T}$$

The R^2 always lies between 0 and 1, with a higher R^2 indicating a better model fit. When interpreting the R^2, higher values indicate that the variation in x explains more of the variation in y. Therefore, relative to the SS_R, the SS_E is low.

It is difficult to provide rules of thumb regarding what R^2 is appropriate, as this varies from research area to research area. For example, in longitudinal studies, R^2s of 0.90 and higher are common. In cross-sectional designs, values of around 0.30 are common, while values of 0.10 are normal in cross-sectional data in exploratory research. In scholarly research focusing on marketing, R^2 values of 0.50, 0.30, and 0.10 can, as a rough rule of thumb, be respectively described as substantial, moderate, and weak.

If we use the R^2 to compare different regression models (but with the same dependent variable), we run into problems. If we add irrelevant variables that are slightly correlated with the dependent variable, the R^2 will increase. Thus, if we only use the R^2 as the basis for understanding regression model fit, we are biased towards selecting regression models with many independent variables. Selecting a model only based on the R^2 is generally not a good strategy, unless we are interested in making predictions. If we are interested in determining whether independent variables have a significant relationship with a dependent variable, or when we wish to estimate the relative strength of different independent variables' effects, we need regression models that do a good job of explaining the data (which have a low SS_E), but which also have a few independent variables. It is easier to recommend that a management should change a few key variables to improve an outcome than to recommend a long list of somewhat related variables. We also do not want too many independent variables, because they are likely to complicate the insights. Consequently, it is best to rely on simple models when possible. Relevant variables should, of course, always be included. To avoid a bias towards complex models, we can use the **adjusted R^2** to select regression models. The adjusted R^2 only increases if the addition of another independent variable explains a substantial amount of the variance. We calculate the adjusted R^2 as follows:

$$R^2_{adj} = 1 - \left(1 - R^2\right) \cdot \frac{n - 1}{n - k - 1}$$

Here, n describes the number of observations and k the number of independent variables (not counting the constant α). This adjusted R^2 is a relative measure and should be used to compare different but **nested models** with the same dependent variable. Nested means that all of a simpler model's terms are included in a more complex model, as well as additional variables. You should pick the model with the highest adjusted R^2 when comparing regression models. However, do not blindly use the adjusted R^2 as a guide, but also look at each individual variable and see if it is relevant (practically) for the problem you are researching. Furthermore, it is important to note that we cannot interpret the adjusted R^2 as the percentage of explained variance as we can with the regular R^2. The adjusted R^2 is only a measure of how much the model explains while controlling for model complexity.

Because the adjusted R^2 can only compare nested models, there are additional fit indices that can be used to compare models with the same dependent variable, but different independent variables (Treiman 2014). The **Akaike information criterion (AIC)** and the **Bayes information criterion (BIC)** are such measures of model fit. More precisely, AIC and BIC are relative measures indicating the difference in information when a set of candidate models with different independent variables is estimated. For example, we can use these criteria to compare two models where the first regression model explains the *sales* by using two independent variables (e.g., *price* and *promotions*) and the second model adds one more independent variable (e.g., *price*, *promotions*, and *service quality*). We can also use the AIC[10] and BIC when we explain sales by using two different sets of independent variables.

Both the AIC and BIC apply a penalty (the BIC a slightly larger one), as the number of independent variables increases with the sample size (Treiman 2014). Smaller values are better and, when comparing models, a rough guide is that when the more complex model's AIC (or BIC) is 10 lower than that of another model, the former model should be given strong preference (Fabozzi et al. 2014). When the difference is less than 2, the simpler model is preferred. For values between 2 and 10, the evidence shifts towards the more complex model, although a specific cut-off point is hard to recommend. When interpreting these statistics, note that the AIC tends to point towards a more complex model than the BIC.

7.3.4.2 Effects of Individual Variables

Having established that the overall model is significant and that the R^2 is satisfactory, we need to interpret the effects of the various independent variables used to explain the dependent variable. If a regression coefficient's *p*-value is below 0.05, we generally say that the specific independent variable relates significantly to the dependent variable. To be precise, the null and alternative hypotheses tested for an individual parameter (e.g., β_1) are:

$$H_0: \beta_1 = 0$$

$$H_1: \beta_1 \neq 0.$$

If a coefficient is significant (i.e., the *p*-value is below 0.05), we reject the null hypothesis and support the alternative hypothesis, concluding that the parameter differs significantly from zero. For example, if we estimate a regression model on the data shown in Fig. 7.1, the (unstandardized) β_1 coefficient of promotional activities' effect on sales is 54.591, with a *t*-value of 5.35. This *t*-value results in a *p*-value less than 0.05, indicating that the effect is significantly different from zero. If we hypothesize a direction (i.e., smaller or larger than zero) instead of significantly different from zero, we should divide the corresponding *p*-value by

[10]The AIC is specifically calculated as $\text{AIC} = n \cdot \ln(SS_E/n) + 2 \cdot k$, where n is the number of observations and k the number of independent variables, while the BIC is calculated as $\text{BIC} = n \cdot \ln(SS_E/n) + k \cdot \ln(n)$.

two. This is the same as applying the *t*-test for a directional effect, which is explained in Chap. 6.

The next step is to interpret the actual size of the β coefficients, which we can interpret in terms of **unstandardized effects** and **standardized effects**. The unstandardized β coefficient indicates the effect that a one-unit increase in the independent variable (on the scale used to measure the original independent variable) has on the dependent variable. This effect is therefore the partial relationship between a change in a single independent variable and the dependent variable. For example, the unstandardized β_1 coefficient of promotional activities' effect on sales (54.59) indicates that a one-unit change in (the index of) promotional activities increases sales by 54.59 units. Importantly, if we have multiple independent variables, a variable's unstandardized coefficient is the effect of that independent variable's increase by one unit, but keeping the other independent variables constant. While this is a very simple example, we might run a multiple regression in which the independent variables are measured on different scales, such as in dollars, units sold, or on Likert scales. Consequently, the independent variables' effects cannot be directly compared with one another, because their influence also depends on the type of scale used. Comparing the unstandardized β coefficients would, in any case, amount to comparing apples with oranges!

Fortunately, the standardized βs allow us to compare the relative effect of differently measured independent variables by expressing the effect in terms of standard deviation changes from the mean. More precisely, the standardized β coefficient expresses the effect that a single standard deviation change in the independent variable has on the dependent variable. The standardized β is used to compare different independent variables' effects. All we need to do is to find the highest absolute value, which indicates the variable that has the strongest effect on the dependent variable. The second highest absolute value indicates the second strongest effect, etc.

Two further tips: First, only consider significant βs in this respect, as insignificant βs do not (statistically) differ from zero! Second, while the standardized βs are helpful from a practical point of view, standardized βs only allow for comparing the coefficients within and not between models! Even if you just add a single variable to your regression model, the standardized βs may change substantially.

When interpreting (standardized) β coefficients, you should always keep the effect size in mind. If a β coefficient is significant, it merely indicates an effect that differs from zero. This does not necessarily mean that the effect is managerially relevant. For example, we may find a \$0.01 sales effect of spending \$1 more on promotional activities that is statistically significant. Statistically, we could conclude that the effect of a \$1 increase in promotional

(continued)

activities increases sales by an average of $0.01 (just one dollar cent). While this effect differs significantly from zero, we would probably not recommend increasing promotional activities in practice (we would lose money on the margin) as the effect size is just too small.[11]

Another way to interpret the size of individual effects is to use the η^2 (pronounced **eta-squared**), which, similar to the R^2, is a measure of the variance accounted for. There are two types of η^2: The model η^2, which is identical to the R^2, and each variable's partial η^2, which describes how much of the total variance is accounted for by that variable. Just like the R^2, the η^2 can only be used to compare variables within a regression model and cannot be used to compare them between regression models. The η^2 relies on different rules of thumb regarding what are small, medium, and large effect sizes. Specifically, an effect of 0.02 is small, 0.15 is medium, and 0.30 and over is large (Cohen 1992).

There are also situations in which an effect is not constant for all observations, but depends on another variable's values. Researchers can run a **moderation analysis**, which we discuss in Box 7.2, to estimate such effects.

7.3.5 Validate the Regression Results

Having checked for the assumptions of the regression analysis and interpreted the results, we need to assess the regression model's stability. Stability means that the results are stable over time, do not vary across different situations, and are not heavily dependent on the model specification. We can check for a regression model's stability in several ways:

1. We can randomly split the dataset into two parts (called **split-sample validation**) and run the regression model again on each data subset. 70% of the randomly chosen data are often used to estimate the regression model (called **estimation sample**) and the remaining 30% is used for comparison purposes (called **validation sample**). We can only split the data if the remaining 30% still meets the sample size rules of thumb discussed earlier. If the use of the two samples results in similar effects, we can conclude that the model is stable. Note that it is mere convention to use 70% and 30% and there is no specific reason for using these percentages.
2. We can also cross-validate our findings on a new dataset and examine whether these findings are similar to the original findings. Again, similarity in the

[11]Cohen's (1994) classical article "The Earth is Round ($p < 0.05$)" offers an interesting perspective on significance and effect sizes.

Box 7.2 Moderation

The discussion of individual variables' effects assumes that there is only one effect. That is, that only one β parameter represents all observations well. This is often not true. For example, the link between sales and price has been shown to be stronger when promotional activities are higher. In other words, the effect of price (β_1) is not constant, but with the level of promotional activities.

Moderation analysis is one way of testing if such heterogeneity is present. A moderator variable, usually denoted by m, is a variable that changes the strength (or even direction) of the relationship between the independent variable (x_1) and the dependent variable (y). You only need to create a new variable that is the multiplication of x_1 and m (i.e., $x_1 \cdot m$). The regression model then takes the following form:

$$y = \alpha + \beta_1 x_1 + \beta_2 m + \beta_3 x_1 \cdot m + e$$

In words, a moderator analysis requires entering the independent variable x_1, the moderator variable m, and the product $x_1 \cdot m$, which represents the interaction between the independent variable and the moderator. Moderation analysis is therefore also commonly referred to as an analysis of **interaction effects**. After estimating this regression model, you can interpret the significance and sign of the β_3 parameter. A significant effect suggests that:

- when the sign of β_3 is positive, the effect β_1 increases as m increases,
- when the sign of β_3 is negative, the effect β_1 decreases as m increases.

For further details on moderation analysis, please see David Kenny's discussion on moderation (http://www.davidakenny.net/cm/moderation.htm), or the advanced discussion by Aiken and West (1991). Jeremy Dawson's website (http://www.jeremydawson.co.uk/slopes.htm) offers a tool for visualizing moderation effects. An example of a moderation analysis is found in Mooi and Frambach (2009).

findings indicates stability and that our regression model is properly specified. **Cross-validation** does, of course, assume that we have a second dataset.

3. We can add several alternative variables to the model and examine whether the original effects change. For example, if we try to explain weekly supermarket sales, we could use several additional variables, such as the breadth of the assortment or the downtown/non-downtown location in our regression model. If the basic findings we obtained earlier continue to hold even when adding these two new variables, we conclude that the effects are stable. This analysis does, of

course, require us to have more variables available than those included in the original regression model.

7.3.6 Use the Regression Model

When we have found a useful regression model that satisfies regression analysis's assumptions, it is time to use it. Prediction is a key use of regression models. Essentially, prediction entails calculating the values of the dependent variables based on assumed values of the independent variables and their related, previously calculated, unstandardized β coefficients. Let us illustrate this by returning to our opening example. Imagine that we are trying to predict weekly supermarket sales (in dollars) (y) and have estimated a regression model with two independent variables: price (x_1) and an index of promotional activities price (x_2). The regression model is as follows:

$$y = \alpha + \beta_1 x_1 + \beta_2 x_2 + e$$

If we estimate this model on the previously used dataset, the estimated coefficients using regression analysis are 30,304.05 for the intercept, −25,209.86 for price, and 42.27 for promotions. We can use these coefficients to predict sales in different situations. Imagine, for example, that we set the price at $1.10 and the promotional activities at 50. Our expectation of the weekly sales would then be:

$$\widehat{y} = 30,304.05 - 25,209.86 \cdot \$1.10 + 42.27 \cdot 50 = \$4,686.70.$$

We could also build several scenarios to plan for different situations by, for example, increasing the price to $1.20 and reducing the promotional activities to 40. By using regression models like this, one can, for example, automate stocking and logistical planning, or develop strategic marketing plans.

Regression can also help by providing insight into variables' specific effects. For example, if the effect of promotions is not significant, it may tell managers that the supermarket's sales are insensitive to promotions. Alternatively, if there is some effect, the strength and direction of promotional activities' effect may help managers understand whether they are useful.

Table 7.2 summarizes (on the left side) the major theoretical decisions we need to make if we want to run a regression model. On the right side, these decisions are then "translated" into Stata actions.

Table 7.2 Steps involved in carrying out a regression analysis

Theory	Action
Consider the regression analysis data requirements	
Sufficient sample size	Check if sample size is 104+k, where k indicates the number of independent variables. If the expected effects are weak (the R^2 is .10 or lower), use at least $30 \cdot k$ observations per independent variable.
	This can be done easily by calculating the correlation matrix. Note the number of observations (**obs**=. . .) immediately under the correlate command to determine the sample size available for regression.
	`correlate commitment s9 s10 s19 s21 s23 status age gender`
Do the dependent and independent variables show variation?	Calculate the standard deviation of the variables by going to ▶ Statistics ▶ Summaries, tables, and tests ▶ Summary and descriptive statistics ▶ Summary statistics (enter the dependent and independent variables). At the very least, the standard deviation (indicated by **Std. Dev.** in the output) should be greater than 0.
	`summarize commitment s9 s10 s19 s21 s23 i.status age i.gender`
Is the dependent variable interval or ratio scaled?	See Chap. 3 to determine the measurement level.
Is (multi)collinearity present?	The presence of (multi)collinearity can only be assessed after the regression analysis has been conducted (to run a regression model; ▶ Statistics ▶ Linear models and related ▶ Linear regression. Under **Dependent variable** enter the dependent variable and add all the independent variables under the box **Independent variables** and click on **OK**).
	Check the VIF: ▶ Statistics ▶ Postestimation ▶ Specification, diagnostic, and goodness-of-fit analysis ▶ Variance inflation factors. Then click on **Launch** and **OK**. The VIF should be below 10 (although it can be higher, or lower, in some cases; see Sect. 7.3.1.4 for specifics).
	`vif`
Specify and estimate the regression model	
Model specification	1. Pick distinct variables
	2. Try to build a robust model
	3. Consider the variables that are needed to give advice
	4. Consider whether the number of independent variables is in relation to the sample size
Estimate the regression model	▶ Statistics ▶ Linear models and related ▶ Linear regression. Under **Dependent variable** enter the dependent variable and add all the independent

(continued)

Table 7.2 (continued)

Theory	Action
	variables under **Independent variables** and click on **OK**.
	`regress commitment s9 s10 s19 s21 s23 i.status age gender`
	Use robust regression when heteroskedasticity is present:
	`regress commitment s9 s10 s19 s21 s23 i.status age gender, robust`
Test the regression analysis assumptions	
Can the regression model be specified linearly?	Consider whether you can write the regression model as: $y = \alpha + \beta_1 \cdot x_1 + \beta_2 \cdot x_2 + \ldots + e$
Is the relationship between the independent and dependent variables linear?	Plot the dependent variable against the independent variable using a scatterplot matrix to see if the relation (if any) appears to be linear. ► Graphics ► Scatterplot matrix. Then add all the variables and click on **Marker properties** where, under **Symbol**, you can choose **Point** for a clearer matrix. Note that you cannot add variables that start with i. (i.e., categorical variables). Then click on **OK**.
	`graph matrix commitment s9 s10 s19 s21 s23 status age gender, msymbol(point)`
	Conduct Ramsey's RESET test to test for non-linearities. Go to ► Statistics ► Postestimation ► Specification, diagnostic, and goodness-of-fit analysis ► Ramsey regression specification-error test for omitted variables. Then click on **Launch** and **OK**.
	`estat ovtest`
Is the expected mean error of the regression model zero?	Choice made on theoretical grounds.
Are the errors constant (homoscedastic)?	Breusch-Pagan test: This can only be checked right after running a regression model. Go to ► Statistics ► Postestimation ► Specification, diagnostic, and goodness-of-fit analysis ► Tests for heteroskedasticity (hettest). Then click on **Launch** and then **OK**. Check that the Breusch-Pagan / Cook-Weisberg test for heteroskedasticity is not significant. If it is, you can use robust regression to remedy this.
	`estat hettest`
	White's test: This can only be checked right after running a regression model. Go to ► Statistics ► Postestimation ► Specification, diagnostic, and goodness-of-fit analysis ► information matrix test (imtest). Then click on **Launch** and then **OK**.
	`estat imtest`
Are the errors correlated (autocorrelation)?	This can only be checked after running a regression model and by declaring the time aspect. This means

(continued)

Table 7.2 (continued)

Theory	Action
	you need a variable that indicates how the variables are organized over time. This variable, for example, *week*, should be declared in Stata using the `tsset` command, for example, `tsset week`. Then conduct the Durbin–Watson test. You can select this test by going to ▶ Statistics ▶ Postestimation ▶ Specification, diagnostic, and goodness-of-fit analysis ▶ Durbin-Watson statistic to test for first-order serial correlation. Click on **Launch** and then **OK**. The Durbin-Watson test for first-order serial correlation should not be significant. The critical values can be found on the website accompanying this book (⭳ Web Appendix → Downloads).
	`tsset week`
	`estat dwatson`
Are the errors normally distributed?	This can only be checked after running a regression model. You should first save the errors by going to ▶ Statistics ▶ Postestimation ▶ Predictions ▶ Predictions and their SEs, leverage statistics, distance statistics, etc. Then click on **Launch**. Enter the name of the error variable (we use *error* in this chapter), making sure **Residuals (equation-level scores)** is ticked, and click on **OK**.
	You should calculate the Shapiro-Wilk test to test the normality of the errors. To select the Shapiro-Wilk test, go to Statistics ▶ Summaries, tables, and tests ▶Distributional plots and tests ▶ Shapiro-Wilk normality test. Under **Variables** enter *error* and click on **OK**. Check if the Shapiro-Wilk test under **Prob>z** reports a *p*-value greater than 0.05.
	To visualize, create a histogram of the errors containing a standard normal curve: ▶ Graphics ▶ Histogram and enter ***error***. Under ▶ Density plots, tick **Add normal-density plot**.
	`predict error, res`
	`swilk error`
	`histogram error, normal`
Interpret the regression model	
Consider the overall model fit	Check the R^2 and significance of the F-value.
Consider the effects of the independent variables separately	Check the (standardized) β. Also check the sign of the β. Consider the significance of the *t*-value (under `P>|t|` **in the regression table).**
To compare models	Calculate the AIC and BIC ▶ Statistics ▶ Postestimation ▶ Specification, diagnostic, and goodness-of-fit analysis ▶ Information criteria – AIC and BIC. Click on **Launch** and then **OK**.
	Check the AIC and BIC, and ascertain if the simpler model has AIC or BIC values that are at least 2, but

(continued)

Table 7.2 (continued)

Theory	Action
	preferably 10, lower than that of the more complex model.
	`estat ic`
Calculate the standardized effects	Check **Standardized beta coefficients** under the **Reporting** tab of the regression dialog box, which can be found under ▶ Statistics ▶ Linear models and related ▶ Linear regression ▶ Reporting
	Determine, sequentially, the highest absolute values
Calculate the effect size	Make sure you have used OLS regression (and not robust regression). Then go to ▶ Statistics ▶ Postestimation ▶ Specification, diagnostic, and goodness-of-fit analysis ▶ Eta-squared and omega-squared effect sizes. Then click on **Launch** and **OK**.
	Interpret each eta squared as the percentage of variance explained (i.e., as that variable's R^2). An effect of individual variables of 0.02 is small, 0.15 is medium, and 0.30 and greater is large.
Validate the model	
Are the results robust?	This can only be done easily using the command window. First create a random variable.
	`set seed 12345`
	`gen validate = runiform() < 0.7`
	Then run the regression model where you first select 70% and then last 30% of the cases. Do this by going to ▶ Statistics ▶ Linear models and related ▶ Linear regression. Then click on **by/if/in** and under **If: (expression)** enter *validate==1*. Then repeat and enter *validate==0*.
	`regress commitment s9 s10 s19 s21 s23 i.status age gender, robust if validate==1`
	`regress commitment s9 s10 s19 s21 s23 i.status age gender, robust if validate==0`
	Compare the model results to ensure they are equal.

7.4 Example

Let's go back to the Oddjob Airways case study and run a regression analysis on the data. Our aim is to explain commitment—the customer's intention to continue the relationship. This variable is formed from three items in the dataset: *com1* ("I am very committed to Oddjob Airways"), *com2* ("My relationship with Oddjob

Airways means a lot to me"), and *com3* ("If Oddjob Airways would not exist any longer, it would be a hard loss for me"). Specifically, it is formed by taking the mean of these three variables.[12]

Our task is to identify which variables relate to commitment to Oddjob Airways. Regression analysis can help us determine which variables relate significantly to commitment, while also identifying the relative strength of the different independent variables.

The Oddjob Airways dataset (⤓ Web Appendix → Downloads) offers several variables that may explain commitment (*commitment*). Based on prior research and discussions with Oddjob Airway's management, the following variables have been identified as promising candidates:

- Oddjob Airways gives you a sense of safety (*s9*),
- The condition of Oddjob Airways' aircraft is immaculate (*s10*),
- Oddjob Airways also pays attention to its service delivery's details. (*s19*),
- Oddjob Airways makes traveling uncomplicated (*s21*), and
- Oddjob Airways offers great value for money (*s23*).

As additional variables, we add the following three categories to the model: the respondent's status (*status*), age (*age*), and gender (*gender*).

7.4.1 Check the Regression Analysis Data Requirements

Before we start, let's see if we have a sufficient sample size. The easiest way to do this is to correlate all the dependent and independent variables (see Chap. 5) by entering all the variables we intend to include in the regression analysis. To do so go to ► Statistics ► Summaries, tables, and tests ► Summary and descriptive statistics ► Correlations and covariances. In the dialog box, enter each variable separately (i.e., *commitment, s9 s10,* etc.) and click on **OK**.

As is indicated by the first line in Table 7.3, the number of observations is 973 (**obs=973**). Green's (1991) rule of thumb suggests that we need at least $104 + k$ observations, where k is the number of independent variables. Since we have 9 independent variables, we satisfy this criterion. Note that if we treat *status* as the categorical variable, which it is, we need to estimate two parameters for *status*—one for the *Silver* and one for the *Gold* category—where *Blue* is the baseline (and cannot be estimated). We thus estimate 10 parameters in total (still satisfying this criterion). In fact, even if we apply VanVoorhis and Morgan's (2007) more stringent criteria of 30 observations per variable, we still have a sufficient sample size. In Table 7.3, we can also examine the pairwise correlations to get an idea of which independent variables relate to the dependent variable and which of them might be collinear.

---

[12]Using the Stata command `egen commitment=rowmean(com1 com2 com3)`

Table 7.3 Correlation matrix to determine sample size

```
correlate commitment s9 s10 s19 s21 s23 status age gender
(obs=973)

             | commit~t       s9      s10      s19      s21      s23   status      age   gender
-------------+---------------------------------------------------------------------------------
  commitment |   1.0000
          s9 |   0.3857   1.0000
         s10 |   0.3740   0.6318   1.0000
         s19 |   0.4655   0.5478   0.5673   1.0000
         s21 |   0.4951   0.5342   0.5135   0.6079   1.0000
         s23 |   0.4618   0.4528   0.5076   0.5682   0.5367   1.0000
      status |   0.0388   0.0114  -0.0237  -0.0042  -0.0149  -0.1324   1.0000
         age |   0.1552   0.1234   0.1738   0.0965   0.1083   0.1445   0.0187   1.0000
      gender |  -0.0743   0.0203   0.0388   0.0186  -0.0043  -0.0337   0.2128  -0.0047   1.0000
```

Table 7.4 Descriptive statistics to determine variation

```
summarize commitment s9 s10 s19 s21 s23 i.status age i.gender

    Variable |        Obs        Mean    Std. Dev.       Min        Max
-------------+---------------------------------------------------------
  commitment |      1,065    4.163693    1.739216          1          7
          s9 |      1,036    72.23359    20.71326          1        100
         s10 |      1,025    64.53854    21.40811          1        100
         s19 |      1,013    57.21027    21.66066          1        100
         s21 |      1,028    58.96498    22.68369          1        100
-------------+---------------------------------------------------------
         s23 |      1,065    48.93521    22.71068          1        100
             |
      status |
     Silver  |      1,065    .2300469    .4210604          0          1
       Gold  |      1,065    .1342723    .3411048          0          1
             |
         age |      1,065    50.41972    12.27464         19        101
             |
      gender |
       male  |      1,065    .7370892    .4404212          0          1
```

Next we should ascertain if our variables display some variation. This is very easy in Stata when using the `summarize` command (as described in Chap. 5) by going to ► Statistics ► Summaries, tables, and tests ► Summary and descriptive statistics ► Summary statistics and by entering the variables in the **Variables** box. This, as shown in Table 7.4, results in output indicating the number of observations per variable and their means and standard deviations, along with their minimum and maximum values. Note that the number of observations in Table 7.3 is **973** and indicates the number of cases in which we fully observe all the variables in the set. However, each variable has a larger number of non-missing observations, which is shown in Table 7.4.

When working with categorical variables, we check whether all observations fall into one category. For example, *status* is coded using thee labels, *Blue*, *Silver*, and *Gold*, representing the values 1, 2, and 3. Having information on two categories makes information on the last category redundant (i.e., knowing that an observation does not fall into the *Silver* or *Gold* category implies it is *Blue*); Stata will therefore only show you one less than the total number of categories. The lowest value (here *Blue*) is removed by default. Categories can be easier to use when the data are nominal or ordinal. Note that each category is coded 0 when absent and 1 when present. For example, looking at Table 7.4, we can see that **.2300469** or 23% of the respondents fall into the *Silver* category and **.1342723** or 13% into the *Gold* category (implying that 64% fall into the *Blue* category). We also did this with the *gender* variable. You can tell Stata to show categories rather than the actual values by using `i.` in front of the variable's name.

The scale of the dependent variable is interval or ratio scaled. Specifically, three 7-point Likert scales create the mean of three items that form commitment. Most researchers would consider this to be interval or ratio scaled, which meets the OLS regression data assumptions.

We should next check for collinearity. While having no collinearity is important, we can only check this assumption after having run a regression analysis. To do so, go to ► Statistics ► Linear models and related ► Linear regression. In the dialog box that follows (Fig. 7.7), enter the dependent variable *commitment* under **Dependent variable** and *s9 s10 s19 s21 s23 i.status age gender* under **Independent variables**. Note that because *status* has multiple levels, using *i.* is necessary to tell you how many observations fall into each category, but be aware that the first level is not shown. Then click on **OK**.

Stata will show the regression output (Table 7.5). However, as the task is to check for collinearity, and not to interpret the regression results, we proceed by going to ► Statistics ► Postestimation ► Specification, diagnostic, and goodness-of-fit analysis ► Variance inflation factors. Then click on **Launch** and **OK**.

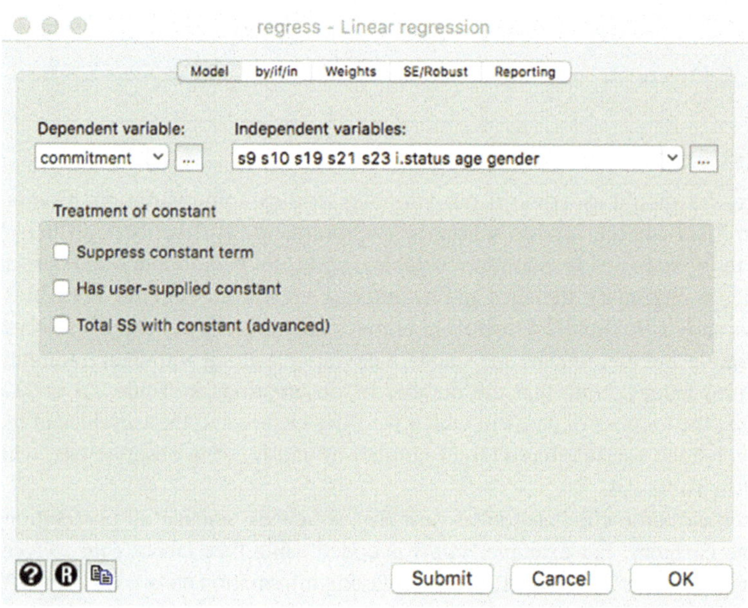

Fig. 7.7 The regression dialog box

Table 7.5 Regression output

```
regress commitment s9 s10 s19 s21 s23 i.status age gender

      Source |       SS           df       MS      Number of obs   =       973
-------------+----------------------------------   F(9, 963)       =     54.59
       Model |  966.268972         9  107.363219   Prob > F        =    0.0000
    Residual |  1893.84455       963  1.96660909   R-squared       =    0.3378
-------------+----------------------------------   Adj R-squared   =    0.3317
       Total |  2860.11352       972  2.94250363   Root MSE        =    1.4024

------------------------------------------------------------------------------
  commitment |      Coef.   Std. Err.      t    P>|t|     [95% Conf. Interval]
-------------+----------------------------------------------------------------
          s9 |   .0051594   .0029967     1.72   0.085    -.0007213    .0110401
         s10 |   .0006685   .0029835     0.22   0.823    -.0051865    .0065235
         s19 |   .0122601   .0030111     4.07   0.000      .006351    .0181691
         s21 |   .0186644   .0027365     6.82   0.000     .0132942    .0240345
         s23 |   .0157612   .0026255     6.00   0.000     .0106088    .0209135
             |
      status |
      Silver |    .183402   .1117365     1.64   0.101    -.0358732    .4026771
        Gold |   .4277363   .1377598     3.10   0.002     .1573922    .6980804
             |
         age |   .0102835   .0038561     2.67   0.008     .0027162    .0178509
      gender |  -.3451731   .1050914    -3.28   0.001    -.5514077   -.1389385
       _cons |   1.198751   .2998922     4.00   0.000     .6102336    1.787269
------------------------------------------------------------------------------
```

Table 7.6 Calculation of the variance inflation factors

```
vif

    Variable |       VIF       1/VIF
-------------+----------------------
          s9 |      1.92    0.521005
         s10 |      2.01    0.497696
         s19 |      2.07    0.483267
         s21 |      1.88    0.532030
         s23 |      1.75    0.570861
      status |
           2 |      1.11    0.902297
           3 |      1.12    0.896848
         age |      1.05    0.951353
      gender |      1.06    0.946307
-------------+----------------------
    Mean VIF |      1.55
```

As you can see in Table 7.6, the highest VIF value is **2.07**, which is below 10 and no reason for concern. Note that the individual VIF values are important and not the mean VIF, as individual variables might be problematic even though, on average, collinearity is not a concern. Note that Stata shows the two *status* levels *Silver* and *Gold* as **2** and **3**.

Having met all the described requirements for a regression analysis, our next task is to interpret the regression analysis results. Since we already had to specify a regression model to check the requirement of no collinearity, we know what this regression model will look like!

7.4.2 Specify and Estimate the Regression Model

We know exactly which variables to select for this model: *commitment*, as the dependent variable, and *s9*, *s10*, *s19*, *s21*, *s23*, *status*, *age*, and *gender* as the independent variables. Run the regression analysis again by going to ► Statistics ► Linear models and related ► Linear regression. Having entered the dependent and independent variables in the corresponding boxes, click on the **SE/Robust** tab.

Stata will show you several estimation options (Fig. 7.8). You should maintain the **Default standard errors**, which is identical to **Ordinary least squares (OLS)**. However, when heteroskedasticity is present, use **Robust** standard errors.

Next, click on the **Reporting** tab (Fig. 7.9). Under this tab, you find several options, including reporting the **Standardized beta coefficients**. You can also change the confidence level to 0.90 or 0.99, as discussed in Chap. 6. Under **Set table formats**, you can easily select how you want the regression results to be reported, for example, the number of decimals, US or European notation (1,000.00 vs. 1.000,00), and whether you want to see leading zeros (0.00 vs .00).

Next click on **OK**. This produces the same output as in Table 7.5. Before we interpret the output, let's first consider the assumptions.

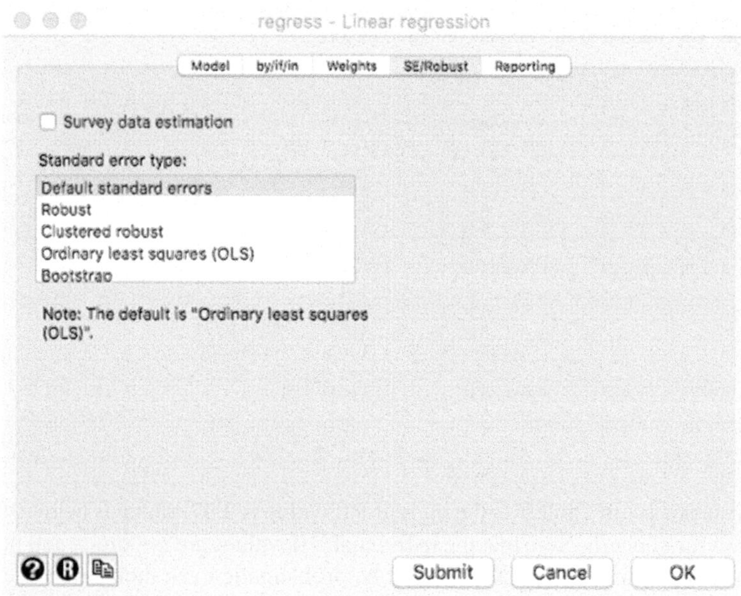

Fig. 7.8 The SE/Robust tab

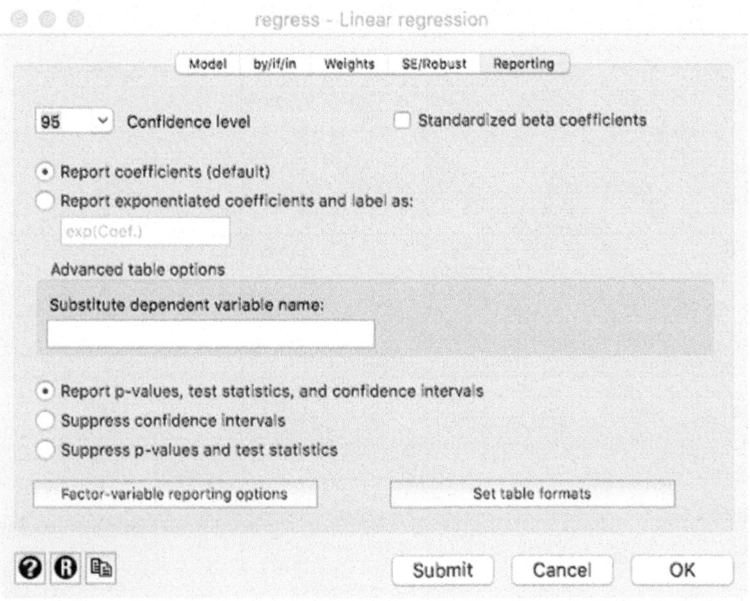

Fig. 7.9 The Reporting tab

7.4.3 Test the Regression Analysis Assumptions

The first assumption is whether the regression model can be expressed linearly. Since no variable transformations occurred, with the exception of categorizing the *status* variable, we meet this assumption, because we can write the regression model linearly as:

$$commitment = \alpha + \beta_1 s9 + \beta_2 s10 + \beta_3 s19 + \beta_4 s21 + \beta_5 s23 + \beta_6 status\_Silver \\ + \beta_7 status\_Gold + \beta_8 age + \beta_9 gender + e$$

Note that because *status* has three levels, two (three minus one) variables are used to estimate the effect of *status*; consequently, we have two βs.

Separately, we also check whether the relationships between the independent and dependent variables are linear. To do this, create a scatterplot matrix of the dependent variable against all the independent variables. This matrix is a combination of all scatterplots (in Chap. 5). To do this, go to ► Graphics ► Scatterplot matrix. Then add all the variables and click on **Marker properties**, where, under **Symbol**, you can choose **Point** for a clearer matrix. Note that you cannot add variables that start with *i.* (i.e., categorical variables) and we therefore just enter *status* (and not *i.status*). Then click on **OK**, after which Stata produces a graph similar to Fig. 7.10. To interpret this graph, look at the first cell, which reads **commitment**. All the scatterplots in the first row (and not the column, as this shows the transpose) show the relationship between the dependent variable and each

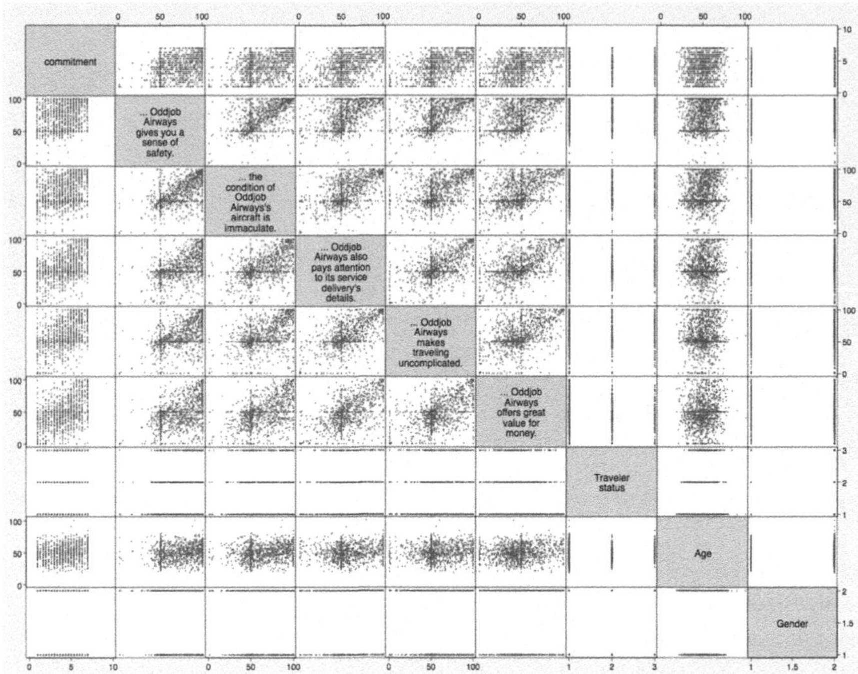

Fig. 7.10 A scatterplot matrix of the dependent variable against all the independent variables

independent variable. The large number of dots makes it difficult to see whether the relationships are linear. However, linearity is also not clearly rejected. Note that the cells **Traveler status** and **Gender** show three (two) distinct bands. This is because these variables take on three distinct values (*Blue*, *Silver*, and *Gold*) for *status* and two (*female* and *male*) for *gender*. When an independent variable has a small number of categories, linearity is not important, although you can still see the form that the relationship might take.

We can also test for the presence of nonlinear relationships between the independent and dependent variable by means of Ramsey's RESET test. Go to ► Statistics ► Postestimation ► Specification, diagnostic, and goodness-of-fit analysis ► Ramsey regression specification-error test for omitted variables. Then click on **Launch** and **OK**.

The results in Table 7.7 of this test under **Prob > F** suggest no non-linearities are present, as the *p*-value (**0.3270**) is greater than 0.05. Bear in mind, however, that this test does not consider all forms of non-linearities.

To check the second assumption, we should assess whether the regression model's expected mean error is zero. Remember, this choice is made on theoretical grounds and there is no empirical test for this. We have a randomly drawn sample from the population and the model is similar in specification to other models

Table 7.7 Ramsey's RESET test

```
estat ovtest

Ramsey RESET test using powers of the fitted values of commitment
       Ho:  model has no omitted variables
                 F(3, 960) =      1.15
                  Prob > F =      0.3270
```

Table 7.8 Breusch-Pagan test for heteroskedasticity

```
estat hettest

Breusch-Pagan / Cook-Weisberg test for heteroskedasticity
         Ho: Constant variance
         Variables: fitted values of commitment

         chi2(1)      =     8.59
         Prob > chi2  =   0.0034
```

explaining commitment. This makes it highly likely that the regression model's expected mean error is zero.

The third assumption is that of homoscedasticity. To test for this, use the Breusch-Pagan test and go to ► Statistics ► Postestimation ► Specification, diagnostic, and goodness-of-fit analysis ► Tests for heteroskedasticity. Then click on **Launch** and **OK**.

The output in Table 7.8 shows the results of the Breusch-Pagan / Cook-Weisberg test for heteroskedasticity. With a *p*-value (**Prob** > **chi2**) of **0.0034**, we should reject the null hypothesis that the error variance is constant, thus suggesting that the error variance is not constant.

To test for heteroskedasticity by means of White's test, go to ► Statistics ► Postestimation ► Specification, diagnostic, and goodness-of-fit analysis ► Information matrix test (imtest). Then click on **Launch** and then **OK**.

As you can see in Table 7.9, the output consists of four tests. Under **Heteroskedasticity**, the first element is the most important, as it gives an indication of whether heteroskedasticity is present. The null hypothesis is that there is no heteroskedasticity. In Table 7.9, this null hypothesis is rejected and the findings suggest that heteroskedasticity is present. Both the Breusch-Pagan and White's test are in agreement. Consequently, we should use a robust estimator; this option is shown in Fig. 7.8. as **Robust**. Please note that because we now use a robust estimator, Stata no longer shows the adjusted R^2 and we can only use the AIC and BIC to compare the models.

If we had data with a time component, we would also perform the Durbin-Watson test to check for potential autocorrelation (fourth assumption). This

Table 7.9 White's test for heteroskedasticity

```
estat imtest

Cameron & Trivedi's decomposition of IM-test

---------------------------------------------------
              Source |       chi2     df      p
---------------------+-----------------------------
  Heteroskedasticity |      78.04     50    0.0068
            Skewness |      30.73      9    0.0003
            Kurtosis |       6.70      1    0.0097
---------------------+-----------------------------
               Total |     115.47     60    0.0000
---------------------------------------------------
```

requires us to first specify a time component, which is absent in the dataset; however, if we had access to a time variable, say *week*, we could time-set the data by using `tsset week`. We can then use the command `estat dwatson` to calculate the Durbin-Watson d-statistic and check whether autocorrelation is present. However, since the data do not include any time component, we should not conduct this test.

Lastly, we should explore how the errors are distributed. We first need to save the errors to do so by going to ► Statistics ► Postestimation ► Predictions ► Predictions and their SEs, leverage statistics, distance statistics, etc. Then click on **Launch**. In the dialog box that opens (Fig. 7.11), enter *error* under **New variable name** and tick **Residuals (equation-level scores)**, which is similar to what we discussed at the beginning of this chapter. There are also several other types of variables that Stata can save. The first option, **Linear prediction (xb)**, saves the predicted values. The other options are more advanced and discussed in detail in the Stata Manual (StataCorp 2015).

Next, click on **OK**. Stata now saves the errors so that they can be used to test and visualize whether they are normally distributed. To test for normality, you should run the Shapiro-Wilk test (Chap. 6) by going go to ► Statistics ► Summaries, tables, and tests ► Distributional plots and tests ► Shapiro-Wilk normality test. In the dialog box that opens (Fig. 7.12), enter *error* under **Variables** and click on **OK**. The output in Table 7.10 indicates a *p*-value (**Prob** > **z**) of **0.08348**, suggesting that the errors are approximately normally distributed. Hence, we can interpret the regression parameters' significance by using *t*-tests.

We also create a histogram of the errors comprising a standard normal curve. To do so, go to ► Graphics ► Histogram and enter *error* under **Variable**. Click on the **Density plots** tab and tick **Add normal-density plot**. The chart in Fig. 7.13 also suggests that our data are normally distributed, as the bars indicating the frequency of the errors generally follow a normal curve.

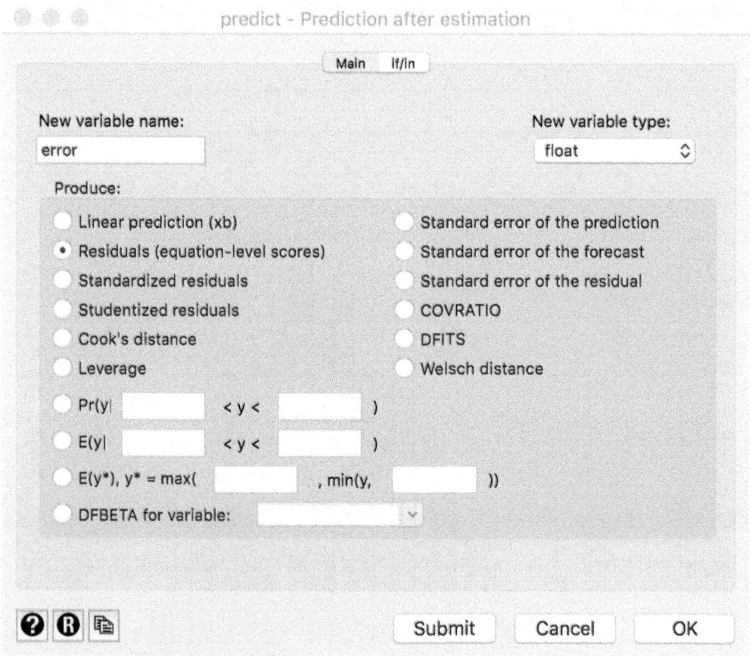

Fig. 7.11 Saving the predicted errors

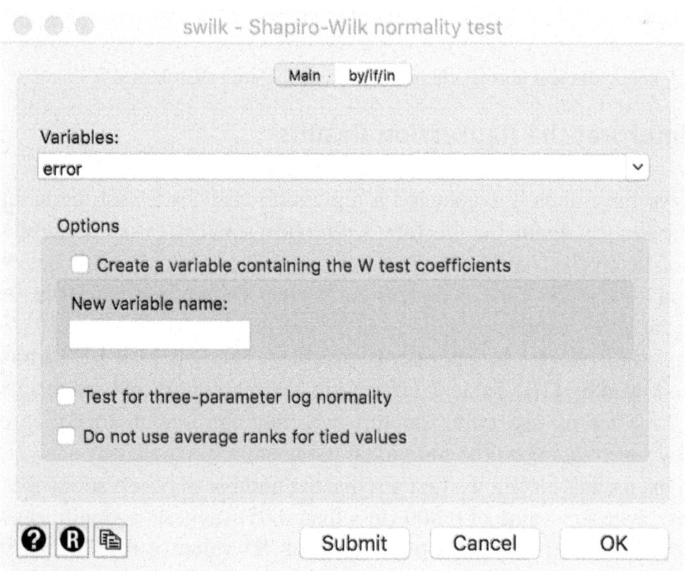

Fig. 7.12 Test for the errors' normality

Table 7.10 The Shapiro-Wilk test for normality

```
swilk error

                   Shapiro-Wilk W test for normal data

    Variable |       Obs       W           V         z       Prob>z
-------------+------------------------------------------------------
       error |       973    0.99716      1.748     1.382    0.08348
```

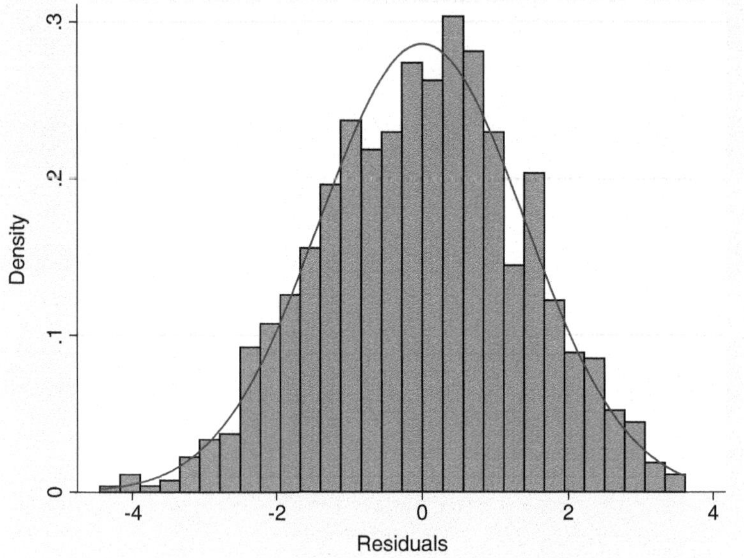

Fig. 7.13 A histogram and normal curve to visualize the error distribution

7.4.4 Interpret the Regression Results

Although we have already conducted a regression analysis to test the assumptions, let's run the analysis again, but this time with robust standard errors because we found evidence of heteroskedasticity. To run the regression, go to ▶ Statistics ▶ Linear models and related ▶ Linear regression. Under **Dependent variable**, enter the dependent variable *commitment* and add all the independent variables *s9 s10 s19 s21 s23 i.status age gender* under **Independent variables**. Then click on **SE/Robust**, select **Robust**, followed by **OK**. Table 7.11 presents the regression analysis results.

Table 7.11 has of two parts; on top, you find the overall model information followed by information on the individual parameters (separated by - - - - -). In the section on the overall model, we first see that the number of observations is 973. Next is the F-test, whose p-value of **0.000** (less than 0.05) suggests a significant model.[13] Further down, we find that the model yields an R^2 value of **0.3378**, which seems satisfactory and is above the value of 0.30 that is common for cross-sectional research.

[13]Note that a p-value is never exactly zero, but has values different from zero in later decimal places.

Table 7.11 Regression output

```
regress commitment s9 s10 s19 s21 s23 i.status age gender, robust

Linear regression                               Number of obs     =        973
                                                F(9, 963)         =      80.20
                                                Prob > F          =     0.0000
                                                R-squared         =     0.3378
                                                Root MSE          =     1.4024

------------------------------------------------------------------------------
             |               Robust
  commitment |      Coef.   Std. Err.      t    P>|t|     [95% Conf. Interval]
-------------+----------------------------------------------------------------
          s9 |   .0051594   .0031804     1.62   0.105    -.0010819    .0114007
         s10 |   .0006685   .0032532     0.21   0.837    -.0057157    .0070527
         s19 |   .0122601   .0032491     3.77   0.000      .005884    .0186361
         s21 |   .0186644    .002807     6.65   0.000     .0131558     .024173
         s23 |   .0157612   .0027849     5.66   0.000     .0102961    .0212263
             |
      status |
     Silver  |    .183402   .1152451     1.59   0.112    -.0427585    .4095624
       Gold  |   .4277363   .1333499     3.21   0.001     .1660463    .6894263
             |
         age |   .0102835    .003807     2.70   0.007     .0028125    .0177546
      gender |  -.3451731   .1029958    -3.35   0.001    -.5472952   -.143051
       _cons |   1.198751    .289718     4.14   0.000     .6301998    1.767303
------------------------------------------------------------------------------
```

In the section on the individual parameters, we find, from left to right, information on the included variables (with the dependent *commitment* listed on top), the coefficients, the robust standard errors, the *t*-values and associated *p*-values (indicated as **P** > **|t|**), and the confidence intervals. First, you should look at the individual coefficients. For example, for *s19*, we get an effect of **0.0122601**, suggesting that when variable *s19* moves up by one unit, the dependent variable *commitment* goes up by **.0122601** units. Under **P** > **|t|**, we find that the regression coefficient of *s19* is significant, as the *p*-value is smaller **0.105** is greater than 0.05. Conversely, *s9* has no significant effect on commitment, as the corresponding *p*-value of **0.105** is greater than 0.05. Further analyzing the output, we find that the variables *s21*, *s23*, *status Gold*, *age*, and *gender* have significant effects. Note that of all the *status* variables, only the coefficient of the tier *status Gold* is significant, whereas the coefficient of *status Silver* is not. A particular issue with these categorical variables is that if you change the base category to, for example, *Silver*, neither *Gold* nor *Blue* will be significant. Always interpret significant findings of categorical variables in relation to their base category. That is, you can claim that *Gold* status travelers have a significantly higher commitment than those of a *Blue* status.[14] The coefficient of *gender* is significant and negative. Because the variable *gender* is scaled 0 (female) to 1 (male), this implies that males (the higher value) show less commitment to Oddjob Airways than females. Note that because the *gender* variable is measured binary (i.e., it is a *dummy variable*; see Chap. 5), it is always relative to the other gender and therefore always significant. Only when there are 3 or more categories does the interpretation issue, which we saw regarding *status*, occur. Specifically, on average, a male customer shows −**.3451731** units less commitment.

In this example, we estimate one model as determined by prior research and the company management input. However, in other instances, we might have alternative models, which we wish to compare in terms of their fit. In Box 7.3, we describe how to do this by using the relative fit statistics AIC and BIC.

Next, we should check the standardized coefficients and effect sizes to get an idea of which variables are most important. This cannot be read from the *t*-values or *p*-values! To calculate the standardized β coefficients, return to ► Statistics ► Linear models and related ► Linear regression. In the **Reporting** tab, check the **Standardized beta coefficients** and click on **OK**. This will produce the output in Table 7.13. To interpret the standardized β coefficents, look at the largest absolute number, which is **.4277363** for the variable *status Gold*. The second highest value relates to *gender* (−**.3451731**) and is binary. While the third-highest is *s21* ("Oddjob Airways makes traveling uncomplicated"). These variables contribute the most in this order.[15] Note, however, that *gender* is not a variable that marketing

Box 7.3 Model Comparison Using AIC and BIC

When comparing different models with the same dependent variable (e.g., *commitment*), but with different independent variables, we can compare the models' adequacy by means of the AIC and BIC statistics. To do this, go to Statistics ► Postestimation ► Specification, diagnostic, and goodness-of-fit analysis ► Information criteria – AIC and BIC. Click on **Launch** and then **OK**. Stata will then show the output as in Table 7.12. The AIC and BIC are respectively listed as 3429.253 and 3478.057. Remember that the AIC and BIC can be used to compare different models. For example, we can drop *age* and *gender* from the previous model and calculate the AIC and BIC again. Although we do not show this output, the resultant AIC and BIC would then respectively be 3443.283 and 3482.326, which are higher, indicating worse fit and suggesting that our original specification is better.

(continued)

[14]Note that it is possible to show all categories for regression tables by typing `set showbaselevels on`. This can be made permanent by typing `set showbaselevels on, permanent`.

[15]Note that while the constant has the highest value (1.19), this is not a coefficient and should not be interpreted as an effect size.

Box 7.3 (continued)

Table 7.12 Relative measures of fit

```
estat ic

Akaike's information criterion and Bayesian information criterion

-----------------------------------------------------------------------------
       Model |        Obs  ll(null)  ll(model)      df         AIC        BIC
-------------+---------------------------------------------------------------
           . |        973 -1905.187  -1704.627      10    3429.253   3478.057
-----------------------------------------------------------------------------
```

Table 7.13 Regression output

```
regress commitment s9 s10 s19 s21 s23 i.status age gender, vce(robust) beta

Linear regression                               Number of obs     =        973
                                                F(9, 963)         =      80.20
                                                Prob > F          =     0.0000
                                                R-squared         =     0.3378
                                                Root MSE          =     1.4024

------------------------------------------------------------------------------
             |               Robust
  commitment |      Coef.   Std. Err.      t    P>|t|                     Beta
-------------+----------------------------------------------------------------
          s9 |   .0051594   .0031804     1.62   0.105                 .0625475
         s10 |   .0006685   .0032532     0.21   0.837                  .008328
         s19 |   .0122601   .0032491     3.77   0.000                 .1535821
         s21 |   .0186644    .002807     6.65   0.000                 .2451992
         s23 |   .0157612   .0027849     5.66   0.000                 .2083433
             |
      status |
     Silver  |    .183402   .1152451     1.59   0.112                 .0453108
       Gold  |   .4277363   .1333499     3.21   0.001                 .0859729
             |
         age |   .0102835    .003807     2.70   0.007                 .0716953
      gender |  -.3451731   .1029958    -3.35   0.001                -.0885363
       _cons |   1.198751    .289718     4.14   0.000                        .
------------------------------------------------------------------------------
```

managers can change and the number of people flying determines the *status*. Making flying with Oddjob airways less complicated may be something marketing managers can influence.

To obtain a better understanding of the effect sizes, we can calculate the η^2. Effect sizes can only be calculated when OLS regression and not robust regression is used. Therefore, run the regression model without the `, robust` option and go to ► Statistics ► Postestimation ► Specification, diagnostic, and goodness-of-fit analysis ► Eta-squared and omega-squared effect sizes. Then click on **Launch** and **OK**. Stata will calculate the effect sizes, as shown in Table 7.14. These effect

Table 7.14 Effect sizes

```
estat esize

Effect sizes for linear models

-----------------------------------------------------------------
            Source |  Eta-Squared     df     [95% Conf. Interval]
-------------------+---------------------------------------------
             Model |   .3378429        9      .286636    .3754686
                   |
                s9 |   .0030688        1            .    .0138597
               s10 |   .0000521        1            .    .0041106
               s19 |   .0169237        1     .0045586    .0364154
               s21 |   .0460813        1     .0236554    .0743234
               s23 |   .0360722        1      .016498    .0619021
            status |   .0107431        2     .0010116    .0259672
               age |   .0073311        1     .0005013     .021729
            gender |   .0110784        1     .0017974    .0277598
-----------------------------------------------------------------
```

sizes can be interpreted as the R^2, but for each individual variable in that specific model. First, the overall η^2 of **.3378429** is identical to the model R^2 as shown in Table 7.11. We should consider the largest value of the individual variables (*s21*) as the most important variable, because it contributes the most to the explained variance (4.6% or, specifically, **.0460813**). Although this is the largest value, Cohen's (1992) rules of thumb suggest this is a small effect size.[16]

7.4.5 Validate the Regression Results

Next, we need to validate the model. Let's first split-validate our model. This can only be done by means of easy instructions in the command window. First, we need to create a variable that helps us select two samples. A uniform distribution is very useful for this purpose, which we can make easily by typing `set seed 12345` in the command window (press enter), followed by `gen validate=runiform() < 0.7`. The first command tells Stata to use random numbers, but since we fix the "seed," we can replicate these numbers later.[17] The second part makes a new variable called *validate*, which has the values zero and one. We can use this variable to help Stata select a random 70% and 30% of cases. This requires us to run the regression model again and selecting the first 70% and the last 30% of the cases. Let's first estimate our model over the 70% of cases. Do this by going to ► Statistics ► Linear models and related ► Linear regression. Then click on **by/if/**

[16]Please note that only Stata 13 or above feature built-in routines to calculate η^2.

[17]The seed specifies the initial value of the random-number generating process such that it can be replicated later.

Table 7.15 Assessing robustness

```
regress commitment s9 s10 s19 s21 s23 i.status age gender if validate==1, vce(robust)

Linear regression                               Number of obs     =        687
                                                F(9, 677)         =      53.40
                                                Prob > F          =     0.0000
                                                R-squared         =     0.3330
                                                Root MSE          =     1.3969

------------------------------------------------------------------------------
             |               Robust
  commitment |      Coef.   Std. Err.      t    P>|t|     [95% Conf. Interval]
-------------+----------------------------------------------------------------
          s9 |   .0037881    .003764     1.01   0.315    -.0036024    .0111786
         s10 |   .0051292   .0038798     1.32   0.187    -.0024886     .012747
         s19 |   .0095681   .0037821     2.53   0.012     .0021421     .016994
         s21 |   .0176205   .0033582     5.25   0.000     .0110269    .0242142
         s23 |   .0155433   .0033412     4.65   0.000      .008983    .0221036
             |
      status |
       Blue  |          0  (base)
      Silver |   .1400931   .1376231     1.02   0.309    -.1301264    .4103126
       Gold  |   .3958519   .1634049     2.42   0.016     .0750106    .7166931
             |
         age |   .0099883   .0044713     2.23   0.026      .001209    .0187676
      gender |  -.2638042    .124435    -2.12   0.034    -.5081291   -.0194794
       _cons |   1.177727   .3408402     3.46   0.001     .5084958    1.846958
------------------------------------------------------------------------------

regress commitment s9 s10 s19 s21 s23 i.status age gender if validate==0, vce(robust)

Linear regression                               Number of obs     =        286
                                                F(9, 276)         =      34.58
                                                Prob > F          =     0.0000
                                                R-squared         =     0.3694
                                                Root MSE          =     1.4105

------------------------------------------------------------------------------
             |               Robust
  commitment |      Coef.   Std. Err.      t    P>|t|     [95% Conf. Interval]
-------------+----------------------------------------------------------------
          s9 |   .0070651   .0060363     1.17   0.243     -.004818    .0189482
         s10 |  -.0091891   .0058937    -1.56   0.120    -.0207915    .0024132
         s19 |   .0186433   .0063824     2.92   0.004     .0060788    .0312078
         s21 |   .0223386   .0048513     4.60   0.000     .0127884    .0318888
         s23 |   .0157948    .005125     3.08   0.002     .0057057    .0258839
             |
      status |
       Blue  |          0  (base)
      Silver |   .2740726   .2103388     1.30   0.194    -.1399995    .6881447
       Gold  |   .5126804   .2389762     2.15   0.033     .0422327    .9831281
             |
         age |   .0099737   .0071877     1.39   0.166    -.0041761    .0241234
      gender |  -.5085486   .1843601    -2.76   0.006    -.8714793   -.1456179
       _cons |   1.245528   .5731065     2.17   0.031     .1173124    2.373743
------------------------------------------------------------------------------
```

in and under **IF:(expression)** enter *validate==1* and click on **OK**. Stata will now estimate the regression model using 70% of the observations (i.e., the estimation sample). When repeating the process for the remaining 30%, enter *validate==0* under **IF:(expression)** (i.e., the validation sample). Table 7.15 shows the results of these two model estimations. As we can see, the models are quite similar (also when compared to the original model), but the effect of *age* is not significant in the second model, although the coefficient is very similar. This suggests that the effects are robust.

As we have no second dataset available, we cannot re-run the analysis to compare. We do, however, have access to other variables such as *country*. If we add this variable, all the variables that were significant at $p < 0.05$ remain significant, while *country* is not significant. Based on this result, we can conclude that the results are stable.

7.5 Farming with AgriPro (Case Study)

AgriPro (http://www.agriprowheat.com) is a firm based in Colorado, USA, which does research on and produces genetically modified wheat seed. Every year, AgriPro conducts thousands of experiments on different varieties of wheat seeds in different USA locations. In these experiments, the agricultural and economic characteristics, regional adaptation, and yield potential of different varieties of wheat seeds are investigated. In addition, the benefits of the wheat produced, including the milling and baking quality, are examined. If a new variety of wheat seed with superior characteristics is identified, AgriPro produces and markets it throughout the USA and parts of Canada.

AgriPro's product is sold to farmers through their distributors, known in the industry as growers. Growers buy wheat seed from AgriPro, grow wheat, harvest the seeds, and sell the seed to local farmers, who plant them in their fields. These growers also provide the farmers, who buy their seeds, with expert local knowledge about management and the environment.

AgriPro sells its products to these growers in several geographically defined markets. These markets are geographically defined, because the different local conditions (soil, weather, and local plant diseases) force AgriPro to produce different products. One of these markets, the heartland region of the USA, is an important AgriPro market, but the company has been performing below the management expectations in it. The heartland region includes the states of Ohio, Indiana, Missouri, Illinois, and Kentucky.

To help AgriPro understand more about farmers in the heartland region, it commissioned a marketing research project involving the farmers in these states. AgriPro, together with a marketing research firm, designed a survey, which included questions regarding what farmers planting wheat find important, how they obtain information on growing and planting wheat, what is important for their purchasing decision, and their loyalty to and satisfaction with the top five wheat suppliers (including AgriPro). In addition, questions were asked about how many acres of farmland the respondents farm, how much wheat they planted, how old they were, and their level of education.

This survey was mailed to 650 farmers from a commercial list that includes nearly all farmers in the heartland region. In all, 150 responses were received, resulting in a 23% response rate. The marketing research firm also assisted AgriPro to assign variable names and labels. They did not delete any questions or observations due to nonresponse to items.

Your task is to analyze the dataset further and, based on the dataset, provide the AgriPro management with advice. This dataset is labeled *agripro.dta* and is available in the ⭳ Web Appendix (→ Chap. 7 → Downloads). Note that the dataset contains the variable names and labels matching those in the survey. In the Web Appendix (⭳ Web Appendix → Downloads), we also include the original survey.[18] To help you with this task, AgriPro has prepared several questions that it would like to see answered:

1. What do these farmers find important when growing wheat? Please describe the variables *import1* ("Wheat fulfills my rotational needs"), *import2* ("I double crop soybeans"), *import3* ("Planting wheat improves my corn yield"), *import4* ("It helps me break disease and pest cycles"), and *import5* ("It gives me summer cash flow") and interpret.
2. What drives how much wheat these farmers grow (*wheat*)? Agripro management is interested in whether *import1*, *import2*, *import3*, *import4*, and *import5* can explain *wheat*. Please run this regression model and test the assumptions. Can you report on this model to AgriPro's management? Please discuss.
3. Please calculate the AIC and BIC for the model discussed in question 2. Then add the variables *acre* and *age*. Calculate the AIC and BIC. Which model is better? Should we present the model with or without *acre* and *age* to our client?
4. AgriPro expects that farmers who are more satisfied with their products devote a greater percentage of their total number of acres to wheat (*wheat*). Please test this assumption by using regression analysis. The client has requested that you control for the number of acres of farmland (*acre*), the age of the respondent (*age*), the quality of the seed (*var3*), and the availability of the seed (*var4*), and check the assumptions of the regression analysis. Note that a smaller sample size is available for this analysis, which means the sample size requirement cannot be met. Proceed with the analysis nevertheless. Are all of the other assumptions satisfied? If not, is there anything we can do about this, or should we ignore the assumptions if they are not satisfied?
5. Agripro wants you to consider which customers are most loyal to its biggest competitor Pioneer (*loyal5*). Use the number of acres (*acre*), number of acres planted with wheat (*wheat*), the age of the respondent (*age*), and this person's education. Use the `i.` operator for education to gain an understanding of the group differences. Does this regression model meet the requirements and assumptions?
6. As an AgriPro's consultant, and based on this study's empirical findings, what marketing advice do you have for AgriPro's marketing team? Using bullet points, provide four or five carefully thought through suggestions.

---

[18]We would like to thank Dr. D.I. Gilliland and AgriPro for making the data and case study available.

7.6 Review Questions

1. Explain what regression analysis is in your own words.
2. Imagine you are asked to use regression analysis to explain the profitability of new supermarket products, such as the introduction of a new type of jam or yoghurt, during the first year of their launch. Which independent variables would you use to explain these new products' profitability?
3. Imagine you have to present the findings of a regression model to a client. The client believes that the regression model is a "black box" and that anything can be made significant. What would your reaction be?
4. I do not care about the assumptions—just give me the results! Please evaluate this statement in the context of regression analysis. Do you agree?
5. Are all regression assumptions equally important? Please discuss.
6. Using standardized βs, we can compare effects between different variables. Can we compare apples and oranges after all? Please discuss.
7. Try adding or deleting variables from the regression model in the Oddjob Airways example and use the adjusted R^2, as well as AIC and BIC statistics, to assess if these models are better.

7.7 Further Readings

Hair, J. F., Black, W. C., Babin, B. J., & Anderson, R. E. (2013). *Multivariate data analysis. A global perspective* (7th ed.). Upper Saddle River: Pearson Prentice Hall.

This is an excellent book which, in a highly accessible way, discusses many statistical terms from a theoretical perspective.

Nielsen at http://www.nielsen.com

This is the website for Nielsen, one of the world's biggest market research companies. They publish many reports that use regression analysis.

The Food Marketing Institute at http://www.fmi.org

This website contains data, some of which can be used for regression analysis.

Treiman, D. J. (2014). *Quantitative data analysis: Doing social research to test ideas*. Hoboken: Wiley.

This is a very good introduction to single and multiple regression. It discusses categorical independent variables in great detail while using Stata.

http://www.ats.ucla.edu/stat/stata/topics/regression.htm

This is an excellent and detailed website dealing with more advanced regression topics in Stata.

References

Aiken, L. S., & West, S. G. (1991). *Multiple regression: Testing and interpreting interactions.* Thousand Oaks: Sage.

Baum, C. F. (2006). *An introduction to modern econometrics using Stata.* College Station: Stata Press.

Breusch, T. S., & Pagan, A. R. (1980). The Lagrange multiplier test and its applications to model specification in econometrics. *Review of Economic Studies, 47*(1), 239–253.

Cameron, A.C. & Trivedi, P.K. (1990). *The information matrix test and its implied alternative hypotheses.* (Working Papers from California Davis – Institute of Governmental Affairs, pp. 1–33).

Cameron, A. C., & Trivedi, P. K. (2010). *Microeconometrics using stata* (Revised ed.). College Station: Stata Press.

Cohen, J. (1992). A power primer. *Psychological Bulletin, 112*(1), 155–159.

Cohen, J. (1994). The earth is round (p < .05). *The American Psychologist, 49*(912), 997–1003.

Cook, R. D., & Weisberg, S. (1983). Diagnostics for heteroscedasticity in regression. *Biometrika, 70*(1), 1–10.

Durbin, J., & Watson, G. S. (1951). Testing for serial correlation in least squares regression, II. *Biometrika, 38*(1–2), 159–179.

Fabozzi, F. J., Focardi, S. M., Rachev, S. T., & Arshanapalli, B. G. (2014). *The basics of financial econometrics: Tools, concepts, and asset management applications.* Hoboken: Wiley.

Green, S. B. (1991). How many subjects does it take to do a regression analysis? *Multivariate Behavioral Research, 26*(3), 499–510.

Greene, W. H. (2011). *Econometric analysis* (7th ed.). Upper Saddle River: Prentice Hall.

Hair, J. F., Jr., Black, W. C., Babin, B. J., & Anderson, R. E. (2013). *Multivariate data analysis.* Upper Saddle River: Pearson.

Hill, C., Griffiths, W., & Lim, G. C. (2008). *Principles of econometrics* (3rd ed.). Hoboken: Wiley.

Kelley, K., & Maxwell, S. E. (2003). Sample size for multiple regression: Obtaining regression coefficients that are accurate, not simply significant. *Psychological Methods, 8*(3), 305–321.

Mason, C. H., & Perreault, W. D., Jr. (1991). Collinearity, power, and interpretation of multiple regression analysis. *Journal of Marketing Research, 28*, 268–280.

Mooi, E. A., & Frambach, R. T. (2009). A stakeholder perspective on buyer–supplier conflict. *Journal of Marketing Channels, 16*(4), 291–307.

O'brien, R. M. (2007). A caution regarding rules of thumb for variance inflation factors. *Quality and Quantity, 41*(5), 673–690.

Ramsey, J. B. (1969). Test for specification errors in classical linear least-squares regression analysis. *Journal of the Royal Statistical Society, Series B, 31*(2), 350–371.

Sin, C., & White, H. (1996). Information criteria for selecting possibly misspecified parametric models. *Journal of Econometrics, 71*(1–2), 207–225.

StataCorp. (2015). *Stata 14 base reference manual.* College Station: Stata Press.

Treiman, D. J. (2014). *Quantitative data analysis: Doing social research to test ideas.* Hoboken: Wiley.

VanVoorhis, C. R. W., & Morgan, B. L. (2007). Understanding power and rules of thumb for determining sample sizes. *Tutorial in Quantitative Methods for Psychology, 3*(2), 43–50.

White, H. (1980). A heteroskedasticity-consistent covariance matrix estimator and a direct test for heteroskedasticity. *Econometrica: Journal of the Econometric Society, 48*(4), 817–838.

Principal Component and Factor Analysis 8

Keywords

Akaike Information Criterion (AIC) • Anti-image • Bartlett method • Bayes Information Criterion (BIC) • Communality • Components • Confirmatory factor analysis • Correlation residuals • Covariance-based structural equation modeling • Cronbach's alpha • Eigenvalue • Eigenvectors • Exploratory factor analysis • Factor analysis • Factor loading • Factor rotation • Factor scores • Factor weights • Factors • Heywood cases • Internal consistency reliability • Kaiser criterion • Kaiser–Meyer–Olkin criterion • Latent root criterion • Measure of sampling adequacy • Oblimin rotation • Orthogonal rotation • Oblique rotation • Parallel analysis • Partial least squares structural equation modeling • Path diagram • Principal axis factoring • Principal components • Principal component analysis • Principal factor analysis • Promax rotation • Regression method • Reliability analysis • Scree plot • Split-half reliability • Structural equation modeling • Test-retest reliability • Uniqueness • Varimax rotation

Learning Objectives

After reading this chapter, you should understand:

- The basics of principal component and factor analysis.
- The principles of exploratory and confirmatory factor analysis.
- Key terms, such as communality, eigenvalues, factor loadings, factor scores, and uniqueness.
- What rotation is.
- The principles of exploratory and confirmatory factor analysis.
- How to determine whether data are suitable for carrying out an exploratory factor analysis.
- How to interpret Stata principal component and factor analysis output.

E. Mooi et al., *Market Research*, Springer Texts in Business and Economics,
DOI 10.1007/978-981-10-5218-7_8

- The principles of reliability analysis and its execution in Stata.
- The concept of structural equation modeling.

8.1 Introduction

Principal component analysis (**PCA**) and **factor analysis** (also called **principal factor analysis** or **principal axis factoring**) are two methods for identifying structure within a set of variables. Many analyses involve large numbers of variables that are difficult to interpret. Using PCA or factor analysis helps find interrelationships between variables (usually called items) to find a smaller number of unifying variables called **factors**. Consider the example of a soccer club whose management wants to measure the satisfaction of the fans. The management could, for instance, measure fan satisfaction by asking how satisfied the fans are with the (1) assortment of merchandise, (2) quality of merchandise, and (3) prices of merchandise. It is likely that these three items together measure satisfaction with the merchandise. Through the application of PCA or factor analysis, we can determine whether a single factor represents the three satisfaction items well. Practically, PCA and factor analysis are applied to understand much larger sets of variables, tens or even hundreds, when just reading the variables' descriptions does not determine an obvious or immediate number of factors.

PCA and factor analysis both explain patterns of correlations within a set of observed variables. That is, they identify sets of highly correlated variables and infer an underlying factor structure. While PCA and factor analysis are very similar in the way they arrive at a solution, they differ fundamentally in their assumptions of the variables' nature and their treatment in the analysis. Due to these differences, the methods follow different research objectives, which dictate their areas of application. While the PCA's objective is to *reproduce* a data structure, as well as possible only using a few factors, factor analysis aims to *explain* the variables' correlations by means of factors (e.g., Hair et al. 2013; Matsunaga 2010; Mulaik 2009).[1] We will discuss these differences and their implications in this chapter.

Both PCA and factor analysis can be used for exploratory or confirmatory purposes. What are exploratory and confirmatory factor analyses? Comparing the left and right panels of Fig. 8.1 shows us the difference. **Exploratory factor analysis**, often simply referred to as EFA, does not rely on previous ideas on the factor structure we may find. That is, there may be relationships (indicated by the arrows) between each factor and each item. While some of these relationships may be weak (indicated by the dotted arrows), others are more pronounced, suggesting that these items represent an underlying factor well. The left panel of Fig. 8.1 illustrates this point. Thus, an exploratory factor analysis reveals the number of factors and the items belonging to a specific factor. In a **confirmatory factor**

[1]Other methods for carrying out factor analyses include, for example, unweighted least squares, generalized least squares, or maximum likelihood. However, these are statistically complex and inexperienced users should not consider them.

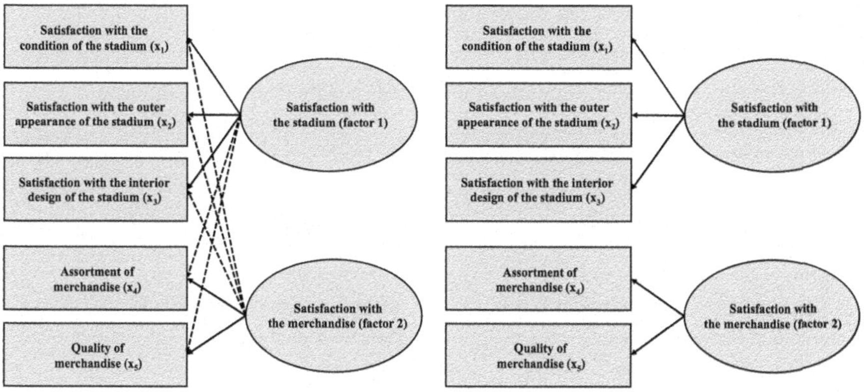

Fig. 8.1 Exploratory factor analysis (*left*) and confirmatory factor analysis (*right*)

analysis, usually simply referred to as CFA, there may only be relationships between a factor and specific items. In the right panel of Fig. 8.1, the first three items relate to factor 1, whereas the last two items relate to factor 2. Different from the exploratory factor analysis, in a confirmatory factor analysis, we have clear expectations of the factor structure (e.g., because researchers have proposed a scale that we want to adapt for our study) and we want to test for the expected structure.

In this chapter, we primarily deal with exploratory factor analysis, as it conveys the principles that underlie all factor analytic procedures and because the two techniques are (almost) identical from a statistical point of view. Nevertheless, we will also discuss an important aspect of confirmatory factor analysis, namely **reliability analysis**, which tests the consistency of a measurement scale (see Chap. 3). We will also briefly introduce a specific confirmatory factor analysis approach called **structural equation modeling** (often simply referred to as SEM). Structural equation modeling differs statistically and practically from PCA and factor analysis. It is not only used to evaluate how well observed variables relate to factors but also to analyze hypothesized relationships between factors that the researcher specifies prior to the analysis based on theory and logic.

8.2 Understanding Principal Component and Factor Analysis

8.2.1 Why Use Principal Component and Factor Analysis?

Researchers often face the problem of large questionnaires comprising many *items*. For example, in a survey of a major German soccer club, the management was particularly interested in identifying and evaluating performance features that relate to soccer fans' satisfaction (Sarstedt et al. 2014). Examples of relevant features include the stadium, the team composition and their success, the trainer, and the

Table 8.1 Items in the soccer fan satisfaction study

Satisfaction with...	
Condition of the stadium	Public appearances of the players
Interior design of the stadium	Number of stars in the team
Outer appearance of the stadium	Interaction of players with fans
Signposting outside the stadium	Volume of the loudspeakers in the stadium
Signposting inside the stadium	Choice of music in the stadium
Roofing inside the stadium	Entertainment program in the stadium
Comfort of the seats	Stadium speaker
Video score boards in the stadium	Newsmagazine of the stadium
Condition of the restrooms	Price of annual season ticket
Tidiness within the stadium	Entry fees
Size of the stadium	Offers of reduced tickets
View onto the playing field	Design of the home jersey
Number of restrooms	Design of the away jersey
Sponsors' advertisements in the stadium	Assortment of merchandise
Location of the stadium	Quality of merchandise
Name of the stadium	Prices of merchandise
Determination and commitment of the players	Pre-sale of tickets
Current success regarding matches	Online-shop
Identification of the players with the club	Opening times of the fan-shops
Quality of the team composition	Accessibility of the fan-shops
Presence of a player with whom fans can identify	Behavior of the sales persons in the fan shops

management. The club therefore commissioned a questionnaire comprising 99 previously identified items by means of literature databases and focus groups of fans. All the items were measured on scales ranging from 1 ("very dissatisfied") to 7 ("very satisfied"). Table 8.1 shows an overview of some items considered in the study.

As you can imagine, tackling such a large set of items is problematic, because it provides quite complex data. Given the task of identifying and evaluating performance features that relate to soccer fans' satisfaction (measured by "Overall, how satisfied are you with your soccer club"), we cannot simply compare the items on a pairwise basis. It is far more reasonable to consider the factor structure first. For example, satisfaction with the condition of the stadium (x_1), outer appearance of the stadium (x_2), and interior design of the stadium (x_3) cover similar aspects that relate to the respondents' satisfaction with the stadium. If a soccer fan is generally very satisfied with the stadium, he/she will most likely answer all three items positively. Conversely, if a respondent is generally dissatisfied with the stadium, he/she is most likely to be rather dissatisfied with all the performance aspects of the stadium, such as the outer appearance and interior design. Consequently, these three items are likely to be highly correlated—they cover related aspects of the respondents' overall satisfaction with the stadium. More precisely, these items can be interpreted as manifestations of the factor capturing the "joint meaning" of the items related to it. The arrows pointing from the factor to the items in Fig. 8.1 indicate this point. In our example, the "joint meaning" of the three items could be described as *satisfaction with the stadium*, since the items represent somewhat different, yet related, aspects of the stadium. Likewise, there is a second factor that relates to the two items x_4 and x_5, which, like the first factor, shares a common meaning, namely *satisfaction with the merchandise*.

PCA and factor analysis are two statistical procedures that draw on item correlations in order to find a small number of factors. Having conducted the analysis, we can make use of few (uncorrelated) factors instead of many variables, thus significantly reducing the analysis's complexity. For example, if we find six factors, we only need to consider six correlations between the factors and overall satisfaction, which means that the recommendations will rely on six factors.

8.2.2 Analysis Steps

Like any multivariate analysis method, PCA and factor analysis are subject to certain requirements, which need to be met for the analysis to be meaningful. A crucial requirement is that the variables need to exhibit a certain degree of correlation. In our example in Fig. 8.1, this is probably the case, as we expect increased correlations between x_1, x_2, and x_3, on the one hand, and between x_4 and x_5 on the other. Other items, such as x_1 and x_4, are probably somewhat correlated, but to a lesser degree than the group of items x_1, x_2, and x_3 and the pair x_4 and x_5. Several methods allow for testing whether the item correlations are sufficiently high.

Both PCA and factor analysis strive to reduce the overall item set to a smaller set of factors. More precisely, PCA extracts factors such that they account for variables' variance, whereas factor analysis attempts to explain the correlations between the variables. Whichever approach you apply, using only a few factors instead of many items reduces its precision, because the factors cannot represent all the information included in the items. Consequently, there is a trade-off between simplicity and accuracy. In order to make the analysis as simple as possible, we want to extract only a few factors. At the same time, we do not want to lose too much information by having too few factors. This trade-off has to be addressed in any PCA and factor analysis when deciding how many factors to extract from the data.

Once the number of factors to retain from the data has been identified, we can proceed with the interpretation of the factor solution. This step requires us to produce a label for each factor that best characterizes the joint meaning of all the variables associated with it. This step is often challenging, but there are ways of facilitating the interpretation of the factor solution. Finally, we have to assess how well the factors reproduce the data. This is done by examining the solution's goodness-of-fit, which completes the standard analysis. However, if we wish to continue using the results in further analyses, we need to calculate the factor scores.

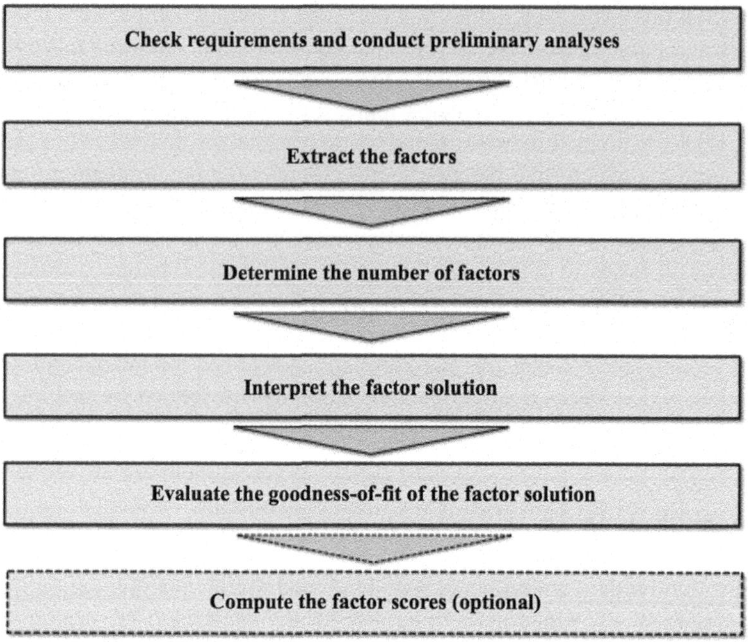

Fig. 8.2 Steps involved in a PCA

Factor scores are linear combinations of the items and can be used as variables in follow-up analyses.

Figure 8.2 illustrates the steps involved in the analysis; we will discuss these in more detail in the following sections. In doing so, our theoretical descriptions will focus on the PCA, as this method is easier to grasp. However, most of our descriptions also apply to factor analysis. Our illustration at the end of the chapter also follows a PCA approach but uses a Stata command (`factor, pcf`), which blends the PCA and factor analysis. This blending has several advantages, which we will discuss later in this chapter.

8.3 Principal Component Analysis

8.3.1 Check Requirements and Conduct Preliminary Analyses

Before carrying out a PCA, we have to consider several requirements, which we can test by answering the following questions:

- Are the measurement scales appropriate?
- Is the sample size sufficiently large?
- Are the observations independent?
- Are the variables sufficiently correlated?

Are the measurement scales appropriate?
For a PCA, it is best to have data measured on an interval or ratio scale. In practical applications, items measured on an ordinal scale level have become common. Ordinal scales can be used if:

- the scale points are equidistant, which means that the difference in the wording between scale steps is the same (see Chap. 3), *and*
- there are five or more response categories.

Is the sample size sufficiently large?
Another point of concern is the sample size. As a rule of thumb, the number of (valid) observations should be at least ten times the number of items used for analysis. This only provides a rough indication of the necessary sample size. Fortunately, researchers have conducted studies to determine minimum sample size requirements, which depend on other aspects of the study. MacCallum et al. (1999) suggest the following:

- When all communalities (we will discuss this term in Sect. 8.3.2.4) are above 0.60, small sample sizes of below 100 are adequate.
- With communalities around 0.50, sample sizes between 100 and 200 are sufficient.
- When communalities are consistently low, with many or all under 0.50, a sample size between 100 and 200 is adequate if the number of factors is small and each of these is measured with six or more indicators.
- When communalities are consistently low and the factors numbers are high or are measured with only few indicators (i.e., 3 or less), 300 observations are recommended.

Are the observations independent?
We have to ensure that the observations are independent. This means that the observations need to be completely unrelated (see Chap. 3). If we use dependent observations, we would introduce "artificial" correlations, which are not due to an underlying factor structure, but simply to the same respondents answered the same questions multiple times.

Are the variables sufficiently correlated?
As indicated before, PCA is based on correlations between items. Consequently, conducting a PCA only makes sense if the items correlate sufficiently. The problem is deciding what "sufficient" actually means.

An obvious step is to examine the correlation matrix (Chap. 5). Naturally, we want the correlations between different items to be as high as possible, but they will not always be. In our previous example, we expect high correlations between x_1, x_2, and x_3, on the one hand, and x_4 and x_5 on the other. Conversely, we might expect lower correlations between, for example, x_1 and x_4 and between x_3 and x_5. Thus, not all of the correlation matrix's elements need to have high values. The PCA depends on the *relative* size of the correlations. Therefore, if single correlations are very low, this is not necessarily problematic! Only when all the correlations are around zero is PCA no longer useful. In addition, the statistical significance of each correlation coefficient helps decide whether it differs significantly from zero.

There are additional measures to determine whether the items correlate sufficiently. One is the **anti-image**. The anti-image describes the portion of an item's variance that is independent of another item in the analysis. Obviously, we want all items to be highly correlated, so that the anti-images of an item set are as small as possible. Initially, we do not interpret the anti-image values directly, but use a measure based on the anti-image concept: **The Kaiser–Meyer–Olkin (KMO)** statistic. The KMO statistic, also called the **measure of sampling adequacy (MSA)**, indicates whether the other variables in the dataset can explain the correlations between variables. Kaiser (1974), who introduced the statistic, recommends a set of distinctively labeled threshold values for KMO and MSA, which Table 8.2 presents.

To summarize, the correlation matrix with the associated significance levels provides a first insight into the correlation structures. However, the final decision of whether the data are appropriate for PCA should be primarily based on the KMO statistic. If this measure indicates sufficiently correlated variables, we can continue the analysis of the results. If not, we should try to identify items that correlate only weakly with the remaining items and remove them. In Box 8.1, we discuss how to do this.

Table 8.2 Threshold values for KMO and MSA

KMO/MSA value	Adequacy of the correlations
Below 0.50	Unacceptable
0.50–0.59	Miserable
0.60–0.69	Mediocre
0.70–0.79	Middling
0.80–0.89	Meritorious
0.90 and higher	Marvelous

Box 8.1 Identifying Problematic Items

Examining the correlation matrix and the significance levels of correlations allows identifying items that correlate only weakly with the remaining items. An even better approach is examining the variable-specific MSA values, which are interpreted like the overall KMO statistic (see Table 8.2). In fact, the KMO statistic is simply the overall mean of all variable-specific MSA values. Consequently, all the MSA values should also lie above the threshold

(continued)

Box 8.1 (continued)
level of 0.50. If this is not the case, consider removing this item from the analysis. An item's communality or uniqueness (see next section) can also serve as a useful indicators of how well the factors extracted represent an item. However, communalities and uniqueness are mostly considered when evaluating the solution's goodness-of-fit.

8.3.2 Extract the Factors

8.3.2.1 Principal Component Analysis vs. Factor Analysis

Factor analysis assumes that each variable's variance can be divided into common variance (i.e., variance shared with all the other variables in the analysis) and unique variance (Fig. 8.3), the latter of which can be further broken down into specific variance (i.e., variance associated with only one specific variable) and error variance (i.e., variance due to measurement error). The method, however, can only reproduce common variance. Thereby factor analysis explicitly recognizes the presence of error. Conversely, PCA assumes that all of each variable's variance is common variance, which factor extraction can explain fully (e.g., Preacher and MacCallum 2003). These differences entail different interpretations of the analysis's outcomes. PCA asks:

Which umbrella term can we use to summarize a set of variables that loads highly on a specific factor?

Conversely, factor analysis asks:

What is the common reason for the strong correlations between a set of variables?

From a theoretical perspective, the assumption that there is a unique variance for which the factors cannot fully account, is generally more realistic, but simultaneously more restrictive. Although theoretically sound, this restriction can sometimes lead to complications in the analysis, which have contributed to the widespread use of PCA, especially in market research practice.

Researchers usually suggest using PCA when data reduction is the primary concern; that is, when the focus is to extract a minimum number of factors that account for a maximum proportion of the variables' total variance. In contrast, if the primary concern is to identify latent dimensions represented in the variables, factor analysis should be applied. However, prior research has shown that both approaches arrive at essentially the same result when:

- more than 30 variables are used, *or*
- most of the variables' communalities exceed 0.60.

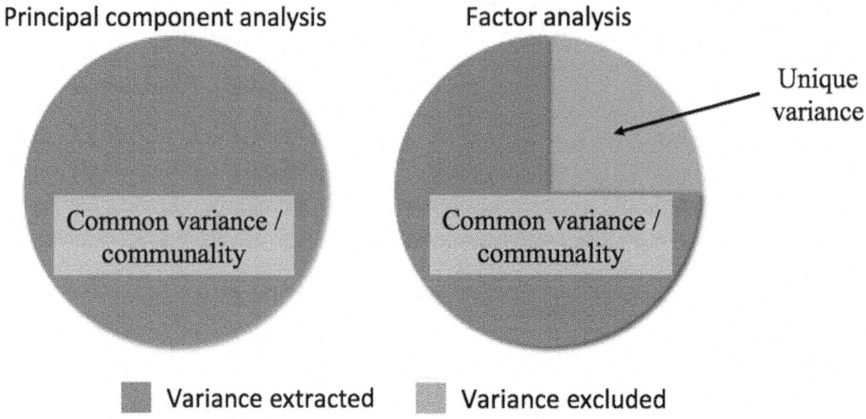

Fig. 8.3 Principal component analysis vs. factor analysis

With 20 or fewer variables and communalities below 0.40—which are clearly undesirable in empirical research—the differences are probably pronounced (Stevens 2009).

Apart from these conceptual differences in the variables' nature, PCA and factor analysis differ in the aim of their analysis. Whereas the goal of factor analysis is to explain the correlations between the variables, PCA focuses on explaining the variables' variances. That is, the PCA's objective is to determine the linear combinations of the variables that retain as much information from the original variables as possible. Strictly speaking, PCA does not extract factors, but **components**, which are labeled as such in Stata.

Despite these differences, which have very little relevance in many common research settings in practice, PCA and factor analysis have many points in common. For example, the methods follow very similar ways to arrive at a solution and their interpretations of statistical measures, such as KMO, eigenvalues, or factor loadings, are (almost) identical. In fact, Stata blends these two procedures in its `factor, pcf` command, which runs a factor analysis but rescales the estimates such that they conform to a PCA. That way, the analysis assumes that the entire variance is common but produces (rotated) loadings (we will discuss factor rotation later in this chapter), which facilitate the interpretation of the factors. In contrast, if we would run a standard PCA, Stata would only offer us eigenvectors whose (unrotated) weights would not allow for a concluding interpretation of the factors. In fact, in many PCA analyses, researchers are not interested in the interpretation of the extracted factors but merely use the method for data reduction. For example, in sensory marketing research, researchers routinely use PCA to summarize a large set of *sensory variables* (e.g., haptic, smell, taste) to derive a set of factors whose scores are then used as input for cluster analyses (Chap. 9). This approach allows for identifying distinct groups of products from which one or more representative products can then be chosen for a more detailed comparison using qualitative research or further assessment in field experiments (e.g., Carbonell et al. 2008; Vigneau and Qannari 2002).

Despite the small differences of PCA and factor analysis in most research settings, researchers have strong feelings about the choice of PCA or factor analysis. Cliff (1987, p. 349) summarizes this issue well, by noting that proponents of factor analysis "insist that components analysis is at best a common factor analysis with some error added and at worst an unrecognizable hodgepodge of things from which nothing can be determined." For further discussions on this topic, see also Velicer and Jackson (1990) and Widaman (1993).[2]

8.3.2.2 How Does Factor Extraction Work?

PCA's objective is to reproduce a data structure with only a few factors. PCA does this by generating a new set of factors as linear composites of the original variables, which reproduces the original variables' variance as best as possible. These linear composites are called **principal components**, but, for simplicity's sake, we refer to them as factors. More precisely, PCA computes **eigenvectors**. These eigenvectors include so called **factor weights**, which extract the maximum possible variance of all the variables, with successive factoring continuing until a significant share of the variance is explained.

Operationally, the first factor is extracted in such a way that it maximizes the variance accounted for in the variables. We can visualize this easily by examining the vector space illustrated in Fig. 8.4. In this example, we have five variables (x_1–x_5) represented by five vectors starting at the zero point, with each vector's length standardized to one. To maximize the variance accounted for, the first factor F_1 is fitted into this vector space in such a way that the sum of all the angles between this factor and the five variables in the vector space is minimized. We do this to interpret the angle between two vectors as correlations. For example, if the factor's vector and a variable's vector are congruent, the angle between these two is zero,

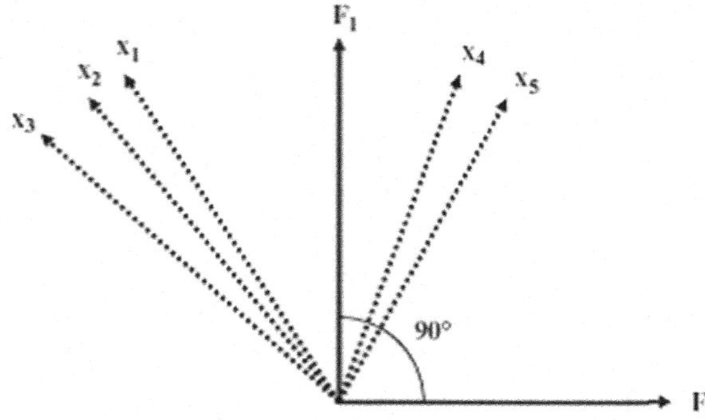

Fig. 8.4 Factor extraction (Note that Fig. 8.4 describes a special case, as the five variables are scaled down into a two-dimensional space. In this set-up, it would be possible for the two factors to explain all five items. However, in real-life, the five items span a five-dimensional vector space.)

[2]Related discussions have been raised in structural equation modeling, where researchers have heatedly discussed the strengths and limitations of factor-based and component-based approaches (e.g., Sarstedt et al. 2016, Hair et al. 2017a, b).

indicating that the factor and the variable correlate perfectly. On the other hand, if the factor and the variable are uncorrelated, the angle between these two is 90°. This correlation between a (unit-scaled) factor and a variable is called the **factor loading**. Note that factor weights and factor loadings essentially express the same thing—the relationships between variables and factors—but they are based on different scales.

After extracting F_1, a second factor (F_2) is extracted, which maximizes the remaining variance accounted for. The second factor is fitted at a 90° angle into the vector space (Fig. 8.4) and is therefore uncorrelated with the first factor.[3] If we extract a third factor, it will explain the maximum amount of variance for which factors 1 and 2 have hitherto not accounted. This factor will also be fitted at a 90° angle to the first two factors, making it independent from the first two factors (we don't illustrate this third factor in Fig. 8.4, as this is a three-dimensional space). The fact that the factors are uncorrelated is an important feature, as we can use them to replace many highly correlated variables in follow-up analyses. For example, using uncorrelated factors as independent variables in a regression analysis helps solve potential collinearity issues (Chap. 7).

The Explained Visually webpage offers an excellent illustration of two- and three-dimensional factor extraction, see http://setosa.io/ev/principal-component-analysis/

An important PCA feature is that it works with standardized variables (see Chap. 5 for an explanation of what standardized variables are). Standardizing variables has important implications for our analysis in two respects. First, we can assess each factor's eigenvalue, which indicates how much a specific factor extracts all of the variables' variance (see next section). Second, the standardization of variables allows for assessing each variable's communality, which describes how much the factors extracted capture or reproduce each variable's variance. A related concept is the uniqueness, which is 1–communality (see Sect. 8.3.2.4).

8.3.2.3 What Are Eigenvalues?

To understand the concept of **eigenvalues**, think of the soccer fan satisfaction study (Fig. 8.1). In this example, there are five variables. As all the variables are standardized prior to the analysis, each has a variance of 1. In a simplified way, we could say that the overall information (i.e., variance) that we want to reproduce by means of factor extraction is 5 units. Let's assume that we extract the two factors presented above.

The first factor's eigenvalue indicates how much variance of the total variance (i.e., 5 units) this factor accounts for. If this factor has an eigenvalue of, let's say

[3]Note that this changes when oblique rotation is used. We will discuss factor rotation later in this chapter.

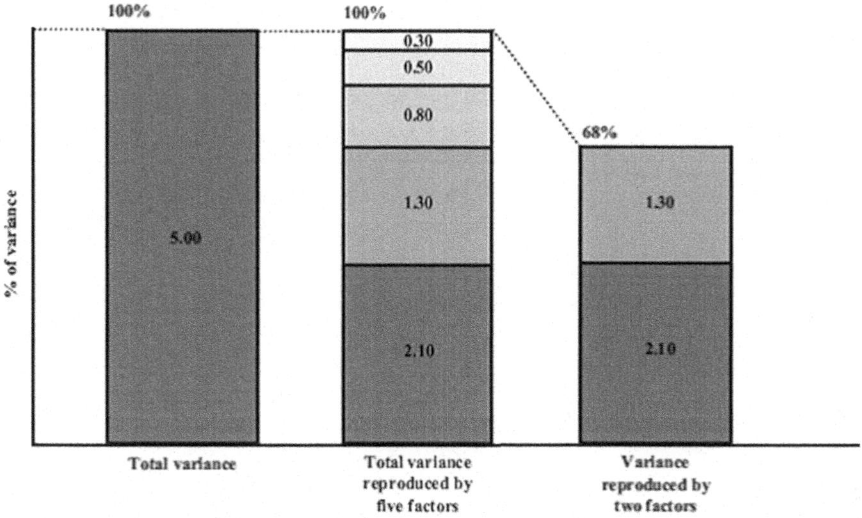

Fig. 8.5 Total variance explained by variables and factors

2.10, it covers the information of 2.10 variables or, put differently, accounts for 2.10/5.00 = 42% of the overall variance (Fig. 8.5).

Extracting a second factor will allow us to explain another part of the remaining variance (i.e., 5.00 – 2.10 = 2.90 units, Fig. 8.5). However, the eigenvalue of the second factor will always be smaller than that of the first factor. Assume that the second factor has an eigenvalue of 1.30 units. The second factor then accounts for 1.30/5.00 = 26% of the overall variance. Together, these two factors explain (2.10 + 1.30)/5.00 = 68% of the overall variance. Every additional factor extracted increases the variance accounted for until we have extracted as many factors as there are variables. In this case, the factors account for 100% of the overall variance, which means that the factors reproduce the complete variance.

Following the PCA approach, we assume that factor extraction can reproduce each variable's entire variance. In other words, we assume that each variable's variance is common; that is, the variance is shared with other variables. This differs in factor analysis, in which each variable can also have a unique variance.

8.3.2.4 What Are Communality and Uniqueness?

Whereas the eigenvalue tells us how much variance each factor is accounts for, the **communality** indicates how much variance of each variable, factor extraction can reproduce. There is no commonly agreed threshold for a variable's communality, as this depends strongly on the complexity of the analysis at hand. However, generally, the extracted factors should account for at least 50% of a variable's variance. Thus, the communalities should be above 0.50. Note that Stata does not indicate each variable's communality but its **uniqueness**, which is 1–communality. Hence, uniqueness gives the proportion of a variable's variance that the factors do *not* capture. For uniqueness the same threshold as for communality applies. Thus, the uniqueness values should be below 0.50. Every additional factor extracted will increase the explained variance, and if we extract as many factors as there are items (in our example five), each variable's communality would be 1.00 and its uniqueness equal to 0. The factors extracted would then fully explain each variable; that is, the first factor will explain a certain amount of each variable's variance, the second factor another part, and so on.

However, since our overall objective is to reduce the number of variables through factor extraction, we should extract only a few factors that account for a high degree of overall variance. This raises the question of how to decide on the number of factors to extract from the data, which we discuss in the following section.

8.3.3 Determine the Number of Factors

Determining the number of factors to extract from the data is a crucial and challenging step in any PCA. Several approaches offer guidance in this respect, but most researchers do not pick just one method, but determine the number of factors resulting from the application of multiple methods. If multiple methods suggest the same number of factors, this leads to greater confidence in the results.

8.3.3.1 The Kaiser Criterion

An intuitive way to decide on the number of factors is to extract all the factors with an eigenvalue greater than 1. The reason for this is that each factor with an eigenvalue greater than 1 accounts for more variance than a single variable (remember, we are looking at standardized variables, which is why each variable's variance is exactly 1). As the objective of PCA is to reduce the overall number of variables, each factor should of course account for more variance than a single variable can. If this occurs, then this factor is useful for reducing the set of variables. Extracting all the factors with an eigenvalue greater than 1 is frequently called the **Kaiser criterion** or **latent root criterion** and is commonly used to determine the number of factors. However, the Kaiser criterion is well known for overspecifying the number of factors; that is, the criterion suggests more factors than it should (e.g., Russell 2002; Zwick and Velicer 1986).

8.3.3.2 The Scree Plot

Another popular way to decide on the number of factors to extract is to plot each factor's eigenvalue (y-axis) against the factor with which it is associated (x-axis). This results in a **scree plot**, which typically has a distinct break in it, thereby showing the "correct" number of factors (Cattell 1966). This distinct break is called the "elbow." It is generally recommended that all factors should be retained *above* this break, as they contribute most to the explanation of the variance in the dataset. Thus, we select one factor less than indicated by the elbow. In Box 8.2, we introduce a variant of the scree plot. This variant is however only available when using the `pca` command instead of the `factor, pcf` command, which serves as the basis for our case study illustration.

Box 8.2 Confidence Intervals in the Scree Plot

A variant of the scree plot includes the confidence interval (Chap. 6, Sect. 6.6.7) of each eigenvalue. These confidence intervals allow for identifying factors with an eigenvalue significantly greater than 1. If a confidence interval's lower bound is above the 1 threshold, this suggests that the factor should be extracted. Conversely, if an eigenvalue's confidence interval overlaps with the 1 threshold or falls completely below, this factor should not be extracted.

8.3.3.3 Parallel Analysis

A large body of review papers and simulation studies has produced a prescriptive consensus that Horn's (1965) **parallel analysis** is the best method for deciding how many factors to extract (e.g., Dinno 2009; Hayton et al. 2004; Henson and Roberts 2006; Zwick and Velicer 1986). The rationale underlying parallel analysis is that factors from real data with a valid underlying factor structure should have larger eigenvalues than those derived from randomly generated data (actually pseudorandom deviates) with the same sample size and number of variables.

Parallel analysis involves several steps. First, a large number of datasets are randomly generated; they have the same number of observations and variables as the original dataset. Parallel PCAs are then run on each of the datasets (hence, parallel analysis), resulting in many slightly different sets of eigenvalues. Using these results as input, parallel analysis derives two relevant cues.

First, parallel analysis adjusts the original eigenvalues for sampling error-induced collinearity among the variables to arrive at adjusted eigenvalues (Horn 1965). Analogous to the Kaiser criterion, only factors with adjusted eigenvalues larger than 1 should be retained.

Second, we can compare the randomly generated eigenvalues with those from the original analysis. Only factors whose original eigenvalues are larger than the 95th percentile of the randomly generated eigenvalues should be retained (Longman et al. 1989).

8.3.3.4 Expectations

When, for example, replicating a previous market research study, we might have a priori information on the number of factors we want to find. For example, if a previous study suggests that a certain item set comprises five factors, we should extract the same number of factors, even if statistical criteria, such as the scree plot, suggest a different number. Similarly, theory might suggest that a certain number of factors should be extracted from the data.

Strictly speaking, these are confirmatory approaches to factor analysis, which blur the distinction between these two factor analysis types. Ultimately however, we should not fully rely on the data, but keep in mind that the research results should be interpretable and actionable for market research practice.

When using factor analysis, Stata allows for estimating two further criteria called the **Akaike Information Criterion** (**AIC**) and the **Bayes Information Criterion** (**BIC**). These criteria are relative measures of goodness-of-fit and are used to compare the adequacy of solutions with different numbers of factors. "Relative" means that these criteria are not scaled on a range of, for example, 0 to 1, but can generally take any value. Compared to an alternative solution with a different number of factors, smaller AIC or BIC values indicate a better fit. Stata computes solutions for different numbers of factors (up to the maximum number of factors specified before). We therefore need to choose the appropriate solution by looking for the smallest value in each criterion. When using these criteria, you should note that AIC is well known for overestimating the "correct" number of factors, while BIC has a slight tendency to underestimate this number.

Whatever combination of approaches we use to determine the number of factors, the factors extracted should account for at least 50% of the total variance explained (75% or more is recommended). Once we have decided on the number of factors to retain from the data, we can start interpreting the factor solution.

8.3.4 Interpret the Factor Solution

8.3.4.1 Rotate the Factors

To interpret the solution, we have to determine which variables relate to each of the factors extracted. We do this by examining the *factor loadings*, which represent the correlations between the factors and the variables and can take values ranging from −1 to +1. A high factor loading indicates that a certain factor represents a variable well. Subsequently, we look for high *absolute* values, because the correlation between a variable and a factor can also be negative. Using the highest absolute factor loadings, we "assign" each variable to a certain factor and then produce a label for each factor that best characterizes the joint meaning of all the variables associated with it. This labeling is subjective, but a key PCA step. An example of a label is the respondents' satisfaction with the stadium, which represents the items referring to its condition, outer appearance, and interior design (Fig. 8.1).

We can make use of **factor rotation** to facilitate the factors' interpretation.[4] We do not have to rotate the factor solution, but it will facilitate interpreting the findings, particularly if we have a reasonably large number of items (say six or more). To understand what factor rotation is all about, once again consider the factor structure described in Fig. 8.4. Here, we see that both factors relate to the

[4]Note that factor rotation primarily applies to factor analysis rather than PCA—see Preacher and MacCallum (2003) for details. However, our illustration draws on the `factor, pcf` command, which uses the factor analysis algorithm to compute PCA results for which rotation applies.

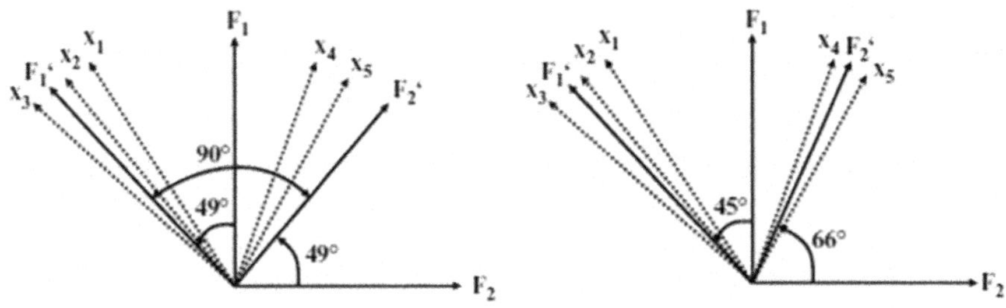

Fig. 8.6 Orthogonal and oblique factor rotation

variables in the set. However, the first factor appears to generally correlate more strongly with the variables, whereas the second factor only correlates weakly with the variables (to clarify, we look for small angles between the factors and variables). This implies that we "assign" all variables to the first factor without taking the second into consideration. This does not appear to be very meaningful, as we want both factors to represent certain facets of the variable set. Factor rotation can resolve this problem. By rotating the factor axes, we can create a situation in which a set of variables loads highly on only one specific factor, whereas another set loads highly on another. Figure 8.6 illustrates the factor rotation graphically.

On the left side of the figure, we see that both factors are orthogonally rotated 49°, meaning that a 90° angle is maintained between the factors during the rotation procedure. Consequently, the factors remain uncorrelated, which is in line with the PCA's initial objective. By rotating the first factor from F_1 to F_1', it is now strongly related to variables x_1, x_2, and x_3, but weakly related to x_4 and x_5. Conversely, by rotating the second factor from F_2 to F_2' it is now strongly related to x_4 and x_5, but weakly related to the remaining variables. The assignment of the variables is now much clearer, which facilitates the interpretation of the factors significantly.

Various **orthogonal rotation** methods exist, all of which differ with regard to their treatment of the loading structure. The **varimax rotation** (the default option for orthogonal rotation in Stata) is the best-known one; this procedure aims at maximizing the dispersion of loadings within factors, which means a few variables will have high loadings, while the remaining variables' loadings will be considerably smaller (Kaiser 1958).

Alternatively, we can choose between several **oblique rotation** techniques. In oblique rotation, the 90° angle between the factors is not maintained during rotation, and the resulting factors are therefore correlated. Figure 8.6 (right side) illustrates an example of an oblique factor rotation. **Promax rotation** (the default option for oblique rotation in Stata) is a commonly used oblique rotation technique. The Promax rotation allows for setting an exponent (referred to as *Promax power* in Stata) that needs to be greater than 1. Higher values make the loadings even more extreme (i.e., high loadings are amplified and weak loadings are reduced even further), which is at the cost of stronger correlations between the factors and less total variance explained (Hamilton 2013). The default value of 3 works well for most applications. **Oblimin rotation** is a popular alternative oblique rotation type.

Oblimin is a class of rotation procedures whereby the degree of obliqueness can be specified. This degree is the *gamma*, which determines the level of the correlation allowed between the factors. A gamma of zero (the default) ensures that the factors are—if at all—only moderately correlated, which is acceptable for most analyses.[5]

Oblique rotation is used when factors are possibly related. It is, for example, very likely that the respondents' satisfaction with the stadium is related to their satisfaction with other aspects of the soccer club, such as the number of stars in the team or the quality of the merchandise. However, relinquishing the initial objective of extracting uncorrelated factors can diminish the factors' interpretability. We therefore recommend using the varimax rotation to enhance the interpretability of the results. Only if the results are difficult to interpret, should an oblique rotation be applied. Among the oblique rotation methods, researchers generally recommend the promax (Gorsuch 1983) or oblimin (Kim and Mueller 1978) methods but differences between the rotation types are typically marginal (Brown 2009).

8.3.4.2 Assign the Variables to the Factors

After rotating the factors, we need to interpret them and give each factor a name, which has some descriptive value. Interpreting factors involves assigning each variable to a specific factor based on the highest *absolute* (!) loading. For example, if a variable has a 0.60 loading with the first factor and a 0.20 loading with the second, we would assign this variable to the first factor. Loadings may nevertheless be very similar (e.g., 0.50 for the first factor and 0.55 for the second one), making the assignment ambiguous. In such a situation, we could assign the variable to another factor, even though it does not have the highest loading on this specific factor. While this step can help increase the results' face validity (see Chap. 3), we should make sure that the variable's factor loading with the designated factor is above an acceptable level. If very few factors have been extracted, the loading should be at least 0.50, but with a high number of factors, lower loadings of above 0.30 are acceptable. Alternatively, some simply ignore a certain variable if it does not fit with the factor structure. In such a situation, we should re-run the analysis without variables that do not load highly on a specific factor. In the end, the results should be interpretable and actionable, but keep in mind that this technique is, first and foremost, exploratory!

8.3.5 Evaluate the Goodness-of-Fit of the Factor Solution

8.3.5.1 Check the Congruence of the Initial and Reproduced Correlations

While PCA focuses on explaining the variables' variances, checking how well the method approximates the correlation matrix allows for assessing the quality of the

[5]When the gamma is set to 1, this is a special case, because the value of 1 represents orthogonality. The result of setting gamma to 1 is effectively a varimax rotation.

solution (i.e., the goodness-of-fit) (Graffelman 2013). More precisely, to assess the solution's goodness-of-fit, we can make use of the differences between the correlations in the data and those that the factors imply. These differences are also called **correlation residuals** and should be as small as possible.

In practice, we check the proportion of correlation residuals with an absolute value higher than 0.05. Even though there is no strict rule of thumb regarding the maximum proportion, a proportion of more than 50% should raise concern. However, high residuals usually go hand in hand with an unsatisfactory KMO measure; consequently, this problem already surfaces when testing the assumptions.

8.3.5.2 Check How Much of Each Variable's Variance Is Reproduced by Means of Factor Extraction

Another way to check the solution's goodness-of-fit is by evaluating how much of each variable's variance the factors reproduce (i.e., the communality). If several communalities exhibit low values, we should consider removing these variables. Considering the variable-specific MSA measures could help us make this decision. If there are more variables in the dataset, communalities usually become smaller; however, if the factor solution accounts for less than 50% of a variable's variance (i.e., the variable's communality is less than 0.50), it is worthwhile reconsidering the set-up. Since Stata does not provide communality but uniqueness values, we have to make this decision in terms of the variance that the factors do *not* reproduce. That is, if several variables exhibit uniqueness values larger than 0.50, we should reconsider the analysis.

8.3.6 Compute the Factor Scores

After the rotation and interpretation of the factors, we can compute the **factor scores**, another element of the analysis. Factor scores are linear combinations of the items and can be used as separate variables in subsequent analyses. For example, instead of using many highly correlated independent variables in a regression analysis, we can use few uncorrelated factors to overcome collinearity problems.

The simplest ways to compute factor scores for each observation is to sum all the scores of the items assigned to a factor. While easy to compute, this approach neglects the potential differences in each item's contribution to each factor (Sarstedt et al. 2016).

Drawing on the eigenvectors that the PCA produces, which include the factor weights, is a more elaborate way of computing factor scores (Hershberger 2005). These weights indicate each item's relative contribution to forming the factor; we simply multiply the standardized variables' values with the weights to get the factor scores. Factor scores computed on the basis of eigenvectors have a zero mean. This means that if a respondent has a value greater than zero for a certain factor, he/she scores above the above average in terms of the characteristic that this factor describes. Conversely, if a factor score is below zero, then this respondent exhibits the characteristic below average.

Different from the PCA, a factor analysis does not produce determinate factor scores. In other words, the factor is indeterminate, which means that part of it remains an arbitrary quantity, capable of taking on an infinite range of values (e.g., Grice 2001; Steiger 1979). Thus, we have to rely on other approaches to computing factor scores such as the **regression method**, which features prominently among factor analysis users. This method takes into account (1) the correlation between the factors and variables (via the item loadings), (2) the correlation between the variables, and (3) the correlation between the factors if oblique rotation has been used (DiStefano et al. 2009). The regression method *z*-standardizes each factor to zero mean and unit standard deviation.[6] We can therefore interpret an observation's score in relation to the mean and in terms of the units of standard deviation from this mean. For example, an observation's factor score of 0.79 implies that this observation is 0.79 standard deviations above the average with regard to the corresponding factor.

Another popular approach is the **Bartlett method**, which is similar to the regression method. The method produces factor scores with zero mean and standard deviations larger than one. Owing to the way they are estimated, the factor scores that the Bartlett method produces are considered are considered more accurate (Hershberger 2005). However, in practical applications, both methods yield highly similar results. Because of the *z*-standardization of the scores, which facilitates the comparison of scores across factors, we recommend using the regression method.

In Table 8.3 we summarize the main steps that need to be taken when conducting a PCA in Stata. Our descriptions draw on Stata's `factor, pcf` command, which carries out a factor analysis but rescales the resulting factors such that the results conform to a standard PCA. This approach has the advantage that it follows the fundamentals of PCA, while allowing for analyses that are restricted to factor analysis (e.g., factor rotation, use of AIC and BIC).

8.4 Confirmatory Factor Analysis and Reliability Analysis

Many researchers and practitioners acknowledge the prominent role that exploratory factor analysis plays in exploring data structures. Data can be analyzed without preconceived ideas of the number of factors or how these relate to the variables under consideration. Whereas this approach is, as its name implies, exploratory in nature, the *confirmatory factor analysis* allows for testing hypothesized structures underlying a set of variables.

In a confirmatory factor analysis, the researcher needs to first specify the constructs and their associations with variables, which should be based on previous measurements or theoretical considerations.

[6]Note that this is not the case when using factor analysis if the standard deviations are different from one (DiStefano et al. 2009).

Table 8.3 Steps involved in carrying out a PCA in Stata

Theory	Stata
Check Assumptions and Carry Out Preliminary Analyses	
Select variables that should be reduced to a set of underlying factors (PCA) or should be used to identify underlying dimensions (factor analysis)	▶ Statistics ▶ Multivariate analysis ▶ Factor and principal component analysis ▶ Factor analysis. Enter the variables in the **Variables** box.
Are the variables interval or ratio scaled?	Determine the measurement level of your variables (see Chap. 3). If ordinal variables are used, make sure that the scale steps are equidistant.
Is the sample size sufficiently large?	Check MacCallum et al.'s (1999) guidelines for minimum sample size requirements, dependent on the variables' communality. For example, if all the communalities are above 0.60, small sample sizes of below 100 are adequate. With communalities around 0.50, sample sizes between 100 and 200 are sufficient.
Are the observations independent?	Determine whether the observations are dependent or independent (see Chap. 3).
Are the variables sufficiently correlated?	Check whether at least some of the variable correlations are significant. ▶ Statistics ▶ Summaries, tables, and tests ▶ Summary and descriptive statistics ▶ Pairwise correlations. Check **Print number of observations for each entry** and **Print significance level for each entry**. Select **Use Bonferroni-adjusted significance level** to maintain the familywise error rate (see Chap. 6).
	`pwcorr s1 s2 s3 s4 s5 s6 s7 s8, obs sig bonferroni`
	Is the KMO ≥ 0.50? ▶ Statistics ▶ Postestimation ▶ Factor analysis reports and graphs ▶ Kaiser-Meyer-Olkin measure of sample adequacy. Then click on **Launch** and **OK**. Note that this analysis can only be run after the PCA has been conducted.
	`estat kmo`
Extract the factors	
Choose the method of factor analysis	▶ Statistics ▶ Multivariate analysis ▶ Factor and principal component analysis ▶ Factor analysis. Click on the **Model 2** tab and select **Principal component factor**.
	`factor s1 s2 s3 s4 s5 s6 s7 s8, pcf`
Determine the number of factors	
Determine the number of factors	Kaiser criterion: ▶ Statistics ▶ Multivariate analysis ▶ Factor and principal component analysis ▶ Factor analysis. Click on the **Model 2** tab and enter **1** under **Minimum value of eigenvalues to be retained**.

(continued)

Table 8.3 (continued)

Theory	Stata
	`factor s1 s2 s3 s4 s5 s6 s7 s8 pcf mineigen (1)`
	Parallel analysis: Download and install *paran* (`help paran`) and enter `paran s1 s2 s3 s4 s5 s6 s7 s8, centile (95) q all graph`
	Extract factors (1) with adjusted eigenvalues greater than 1, and (2) whose adjusted eigenvalues are greater than the random eigenvalues.
	Scree plot: ► Statistics ► Postestimation ► Factor analysis reports and graphs ► Scree plot of eigenvalues. Then click on **Launch** and **OK**.
	`screeplot`
	Pre-specify the number of factors based on a priori information: ► Statistics ► Multivariate analysis ► Factor and principal component analysis ► Factor analysis. Under the **Model 2** tab, tick **Maximum number of factors to be retained** and specify a value in the box below (e.g., **2**).
	`factor s1 s2 s3 s4 s5 s6 s7 s8, factors(2)`
	Make sure that the factors extracted account for at least 50% of the total variance explained (75% or more is recommended): Check the Cumulative column in the PCA output.
Interpret the Factor Solution	
Rotate the factors	Use the varimax procedure or, if necessary, the promax procedure with gamma set to 3 (both with Kaiser normalization): ► Statistics ► Postestimation ► Principal component analysis reports and graphs ► Rotate factor loadings. Select the corresponding option in the menu.
	Varimax: `rotate, kaiser`
	Promax: `rotate, promax(3) oblique Kaiser`
Assign variables to factors	Check the Factor loadings (pattern matrix) table in the output of the rotated solution. Assign each variable to a certain factor based on the highest absolute loading. To facilitate interpretation, you may also assign a variable to a different factor, but check that the loading is at an acceptable level (0.50 if only a few factors are extracted, 0.30 if many factors are extracted).
	Consider making a loadings plot: ► Statistics ► Postestimation ► Factor analysis reports

(continued)

Table 8.3 (continued)

Theory	Stata
	and graphs ▶ Plot of factor loadings. Under **Plot all combinations of the following**, indicate the number of factors for which you want to plot. Check which items load highly on which factor.
Compute factor scores	Save factor scores as new variables: ▶ Statistics ▶ Postestimation ▶ Predictions ▶ Regression and Bartlett scores. Under **New variable names or variable stub*** enter *factor**. Select **Factors scored by the regression scoring method**
	`predict score*, regression`
Evaluate the Goodness-of-fit of the Factor Solution	
Check the congruence of the initial and reproduced correlations	Create a reproduced correlation matrix: ▶ Statistics ▶ Postestimation ▶ Factor analysis reports and graphs ▶ Matrix of correlation residuals. Is the proportion of residuals greater than 0.05 ≤ 50%?
	`estat residuals`
Check how much of each variable's variance is reproduced by means of factor extraction	Check the **Uniqueness** column in the PCA output. Are all the values lower than 0.50?

Instead of allowing the procedure to determine the number of factors, as is done in an exploratory factor analysis, a confirmatory factor analysis tells us how well the actual data fit the pre-specified structure. Reverting to our introductory example, we could, for example, assume that the construct *satisfaction with the stadium* can be measured by means of the three items x_1 (condition of the stadium), x_2 (appearance of the stadium), and x_3 (interior design of the stadium). Likewise, we could hypothesize that *satisfaction with the merchandise* can be adequately measured using the items x_4 and x_5. In a confirmatory factor analysis, we set up a theoretical model linking the items with the respective construct (note that in confirmatory factor analysis, researchers generally use the term construct rather than factor). This process is also called operationalization (see Chap. 3) and usually involves drawing a visual representation (called a **path diagram**) indicating the expected relationships.

Figure 8.7 shows a path diagram—you will notice the similarity to the diagram in Fig. 8.1. Circles or ovals represent the constructs (e.g., Y_1, satisfaction with the stadium) and boxes represent the items (x_1 to x_5). Other elements include the relationships between the constructs and respective items (the loadings l_1 to l_5), the error terms (e_1 to e_5) that capture the extent to which a construct does not explain a specific item, and the correlations between the constructs of interest (r_{12}).

Having defined the individual constructs and developed the path diagram, we can estimate the model. The relationships between the constructs and items (the loadings l_1 to l_5) and the item correlations (not shown in Fig. 8.7) are of particular

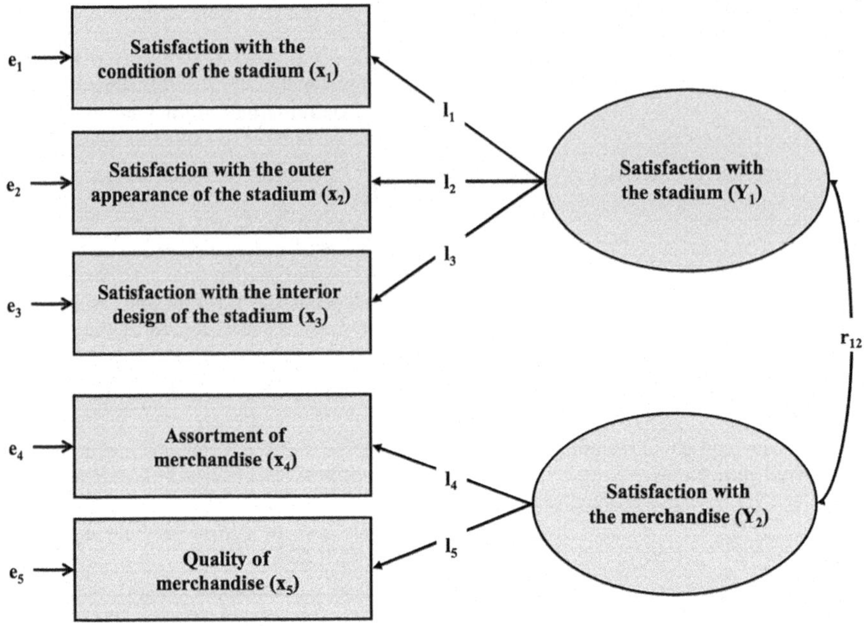

Fig. 8.7 Path diagram (confirmatory factor analysis)

interest, as they indicate whether the construct has been reliably and validly measured.

Reliability analysis is an important element of a confirmatory factor analysis and essential when working with measurement scales. The preferred way to evaluate reliability is by taking two independent measurements (using the same subjects) and comparing these by means of correlations. This is also called **test-retest reliability** (see Chap. 3). However, practicalities often prevent researchers from surveying their subjects a second time.

An alternative is to estimate the **split-half reliability**. In the split-half reliability, scale items are divided into halves and the scores of the halves are correlated to obtain an estimate of reliability. Since all items should be consistent regarding what they indicate about the construct, the halves can be considered approximations of alternative forms of the same scale. Consequently, instead of looking at the scale's test-retest reliability, researchers consider the scale's equivalence, thus showing the extent to which two measures of the same general trait agree. We call this type of reliability the **internal consistency reliability**.

In the example of *satisfaction with the stadium*, we compute this scale's split-half reliability manually by, for example, splitting up the scale into x_1 on the one side and x_2 and x_3 on the other. We then compute the sum of x_2 and x_3 (or calculate the items' average) to form a total score and correlate this score with x_1. A high correlation indicates that the two subsets of items measure related aspects of the same underlying construct and, thus, suggests a high degree of internal consistency.

However, with many indicators, there are many different ways to split the variables into two groups.

Cronbach (1951) proposed calculating the average of all possible split-half coefficients resulting from different ways of splitting the scale items. The **Cronbach's Alpha** coefficient has become by far the most popular measure of internal consistency. In the example above, this would comprise calculating the average of the correlations between (1) x_1 and $x_2 + x_3$, (2) x_2 and $x_1 + x_3$, as well as (3) x_3 and $x_1 + x_2$. The Cronbach's Alpha coefficient generally varies from 0 to 1, whereas a generally agreed lower limit for the coefficient is 0.70. However, in exploratory studies, a value of 0.60 is acceptable, while values of 0.80 or higher are regarded as satisfactory in the more advanced stages of research (Hair et al. 2011). In Box 8.3, we provide more advice on the use of Cronbach's Alpha. We illustrate a reliability analysis using the standard Stata module in the example at the end of this chapter.

Box 8.3 Things to Consider When Calculating Cronbach's Alpha

When calculating Cronbach's Alpha, ensure that all items are formulated in the same direction (positively or negatively worded). For example, in psychological measurement, it is common to use both negatively and positively worded items in a questionnaire. These need to be reversed prior to the reliability analysis. In Stata, this is done automatically when an item is negatively correlated with the other items. It is possible to add the option`,` `reverse (variable name)` to force Stata to reverse an item or you can stop Stata from reversing items automatically by using`, asis`. Furthermore, we have to be aware of potential subscales in our item set. Some multi-item scales comprise subsets of items that measure different facets of a multidimensional construct. For example, soccer fan satisfaction is a multidimensional construct that includes aspects such as satisfaction with the stadium, the merchandise (as described above), the team, and the coach, each measured with a different item set. It would be inappropriate to calculate one Cronbach's Alpha value for all 99 items. Cronbach's Alpha is always calculated over the items belonging to one construct and not all the items in the dataset!

8.5 Structural Equation Modeling

Whereas a confirmatory factor analysis involves testing if and how items relate to specific constructs, *structural equation modeling* involves the estimation of relations between these constructs. It has become one of the most important methods in social sciences, including marketing research.

There are broadly two approaches to structural equation modeling: **Covariance-based structural equation modeling** (e.g., Jöreskog 1971) and **partial least**

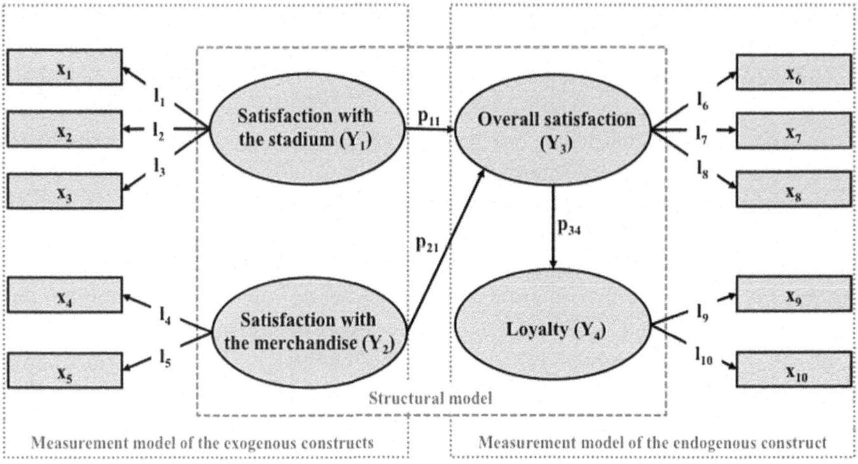

Fig. 8.8 Path diagram (structural equation modeling)

squares structural equation modeling (e.g., Wold 1982), simply referred to as **CB-SEM** and **PLS-SEM**. Both estimation methods are based on the idea of an underlying model that allows the researcher to test relationships between multiple items and constructs.

Figure 8.8 shows an example path diagram with four constructs (represented by circles or ovals) and their respective items (represented by boxes).[7] A path model incorporates two types of constructs: (1) exogenous constructs (here, satisfaction with the stadium (Y_1) and satisfaction with the merchandise (Y_2)) that do not depend on other constructs, and (2) endogenous constructs (here, overall satisfaction (Y_3) and loyalty (Y_4)) that depend on one or more exogenous (or other endogenous) constructs. The relations between the constructs (indicated with p) are called path coefficients, while the relations between the constructs and their respective items (indicated with l) are the indicator loadings. One can distinguish between the structural model that incorporates the relations between the constructs and the (exogenous and endogenous) measurement models that represent the relationships between the constructs and their related items. Items that measure constructs are labeled x.

In the model in Fig. 8.8, we assume that the two exogenous constructs *satisfaction with the stadium* and *satisfaction with the merchandise* relate to the endogenous construct *overall satisfaction* and that *overall satisfaction* relates to *loyalty*. Depending on the research question, we could of course incorporate additional exogenous and endogenous constructs. Using empirical data, we could then test this model and, thus, evaluate the relationships between all the constructs and between each construct and its indicators. We could, for example, assess which of the two

[7]Note that we omitted the error terms for clarity's sake.

constructs, Y_1 or Y_2, exerts the greater influence on Y_3. The result would guide us when developing marketing plans in order to increase overall satisfaction and, ultimately, loyalty by answering the research question whether we should rather concentrate on increasing the fans' satisfaction with the stadium or with the merchandise.

The evaluation of a path model analysis requires several steps that include the assessment of both measurement models and the structural model. Diamantopoulos and Siguaw (2000) and Hair et al. (2013) provide a thorough description of the covariance-based structural equation modeling approach and its application. Acock (2013) provides a detailed explanation of how to conduct covariance-based structural equation modeling analyses in Stata. Hair et al. (2017a, b, 2018) provide a step-by-step introduction on how to set up and test path models using partial least squares structural equation modeling.

8.6 Example

In this example, we take a closer look at some of the items from the Oddjob Airways dataset (⤓ Web Appendix → Downloads). This dataset contains eight items that relate to the customers' experience when flying with Oddjob Airways. For each of the following items, the respondents had to rate their degree of agreement from 1 ("completely disagree") to 100 ("completely agree"). The variable names are included below:

- with Oddjob Airways you will arrive on time (*s1*),
- the entire journey with Oddjob Airways will occur as booked (*s2*),
- in case something does not work out as planned, Oddjob Airways will find a good solution (*s3*),
- the flight schedules of Oddjob Airways are reliable (*s4*),
- Oddjob Airways provides you with a very pleasant travel experience (*s5*),
- Oddjob Airways's on board facilities are of high quality (*s6*),
- Oddjob Airways's cabin seats are comfortable (*s7*), and
- Oddjob Airways offers a comfortable on-board experience (*s8*).

Our aim is to reduce the complexity of this item set by extracting several factors. Hence, we use these items to run a PCA using the `factor, pcf` procedure in Stata.

8.6.1 Principal Component Analysis

8.6.1.1 Check Requirements and Conduct Preliminary Analyses

All eight variables are interval scaled from 1 ("very unsatisfied") to 100 ("very satisfied"), therefore meeting the requirements in terms of the measurement scale.

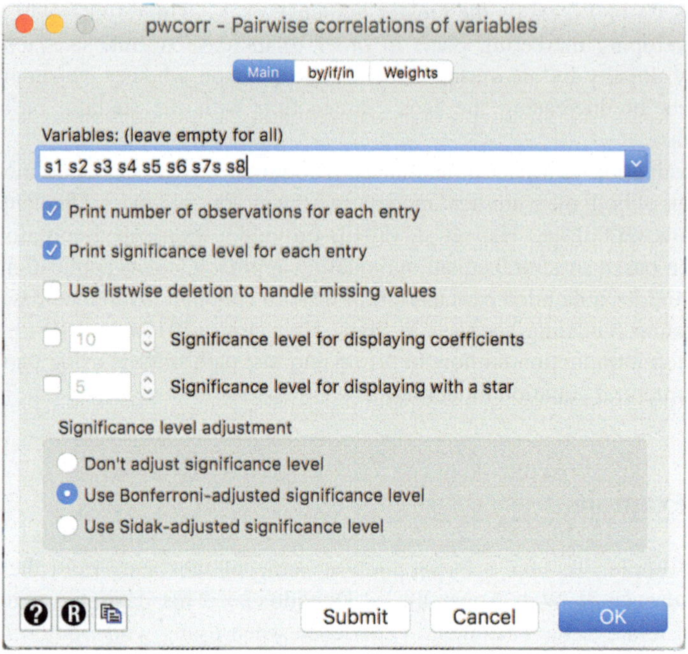

Fig. 8.9 Pairwise correlations of variables

With 1,065 independent observations, the sample size requirements are clearly met, even if the analysis yields very low communality values.

Determining if the variables are sufficiently correlated is easy if we go to ▶ Statistics ▶ Summaries, tables, and tests ▶ Summary and descriptive statistics ▶ Pairwise correlations. In the dialog box shown in Fig. 8.9, either enter each variable separately (i.e., *s1 s2 s3* etc.) or simply write *s1-s8* as the variables appear in this order in the dataset. Then also tick **Print number of observations for each entry**, **Print significance levels for each entry**, as well as **Use Bonferroni-adjusted significance level**. The latter option corrects for the many tests we execute at the same time and is similar to what we discussed in Chap. 6.

Table 8.4 shows the resulting output. The values in the diagonal are all **1.000**, which is logical, as this is the correlation between a variable and itself! The off-diagonal cells correspond to the pairwise correlations. For example, the pairwise correlation between *s1* and *s2* is **0.7392**. The value under it denotes the *p*-value (**0.000**), indicating that the correlation is significant. To determine an absolute minimum standard, check if at least one correlation in all the off-diagonal cells is significant. The last value of **1037** indicates the sample size for the correlation between *s1* and *s2*.

The correlation matrix in Table 8.4 indicates that there are several pairs of highly correlated variables. For example, not only *s1* is highly correlated with *s2* (correlation = **0.7392**), but also *s3* is highly correlated with *s1* (correlation = **0.6189**), just

Table 8.4 Pairwise correlation matrix

```
pwcorr s1-s8, obs sig bonferroni
```

	s1	s2	s3	s4	s5	s6	s7	s8
s1	1.0000							
	1038							
s2	0.7392	1.0000						
	0.0000							
	1037	1040						
s3	0.6189	0.6945	1.0000					
	0.0000	0.0000						
	952	952	954					
s4	0.7171	0.7655	0.6447	1.0000				
	0.0000	0.0000	0.0000					
	1033	1034	951	1035				
s5	0.5111	0.5394	0.5593	0.4901	1.0000			
	0.0000	0.0000	0.0000	0.0000				
	1026	1027	945	1022	1041			
s6	0.4898	0.4984	0.4972	0.4321	0.8212	1.0000		
	0.0000	0.0000	0.0000	0.0000	0.0000			
	1025	1027	943	1022	1032	1041		
s7	0.4530	0.4555	0.4598	0.3728	0.7873	0.8331	1.0000	
	0.0000	0.0000	0.0000	0.0000	0.0000	0.0000		
	1030	1032	947	1027	1038	1037	1048	
s8	0.5326	0.5329	0.5544	0.4822	0.8072	0.8401	0.7773	1.0000
	0.0000	0.0000	0.0000	0.0000	0.0000	0.0000	0.0000	
	1018	1019	937	1014	1029	1025	1032	1034

like *s4* (correlation = **0.7171**). As these variables' correlations with the remaining ones are less pronounced, we suspect that these four variables constitute one factor. As you can see by just looking at the correlation matrix, we can already identify the factor structure that might result.

However, at this point of the analysis, we are more interested in checking whether the variables are sufficiently correlated to conduct a PCA. When we examine Table 8.4, we see that all correlations have *p*-values below 0.05. This result indicates that the variables are sufficiently correlated. However, for a concluding evaluation, we need to take the anti-image and related statistical measures into account. Most importantly, we should also check if the KMO values are at least 0.50. As we can only do this *after* the actual PCA, we will discuss this point later.

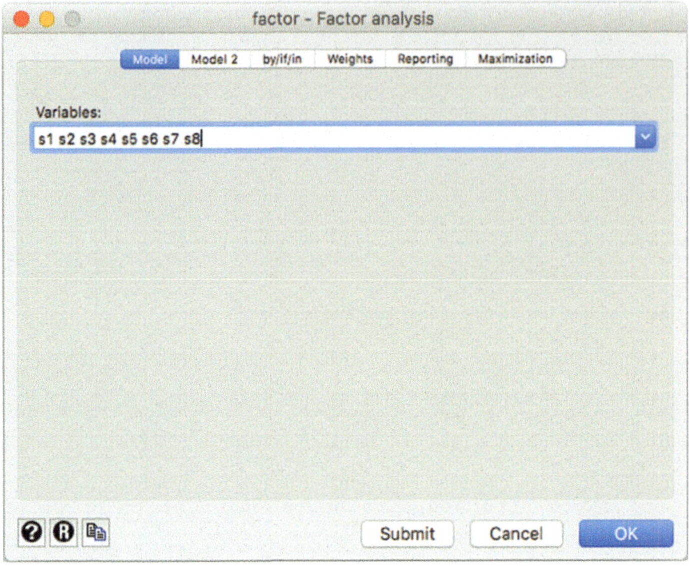

Fig. 8.10 Factor analysis dialog box

8.6.1.2 Extract the Factors and Determine the Number of Factors

To run the PCA, click on ► Statistics ► Multivariate analysis ► Factor and principal component analysis ► Factor analysis, which will open a dialog box similar to Fig. 8.10. Next, enter *s1 s2 s3 s4 s5 s6 s7 s8* in the **Variables** box. Alternatively, you can also simply enter *s1*-*s8* in the box as the dataset includes these eight variables in consecutive order.

Under the **Model 2** tab (see Fig. 8.11), we can choose the method for factor extraction, the maximum factors to be retained, as well as the minimum value for the eigenvalues to be retained. As the aim of our analysis is to reproduce the data structure, we choose **Principal-component factor**, which initiates the PCA based on Stata's `factor, pcf` procedure. By clicking on **Minimum value of eigenvalues to be retained** and entering **1** in the box below, we specify the Kaiser criterion. If we have a priori information on the factor structure, we can specify the number of factors manually by clicking on **Maximum number of factors to be retained**. Click on **OK** and Stata will display the results (Table 8.5).

Before we move to the further interpretation of the PCA results in Table 8.5, let's first take a look at the KMO. Although we need to know if the KMO is larger than 0.50 to interpret the PCA results with confidence, we can only do this after having run the PCA. We can calculate the KMO values by going to ► Statistics ► Postestimation ► Principal component analysis reports and graphs ► Kaiser-Meyer-Olkin measure of sample adequacy (Fig. 8.12). In the dialog box that opens, simply click on **OK** to initiate the analysis.

The analysis result at the bottom of Table 8.6 reveals that the KMO value is **0.9073**, which is "marvelous" (see Table 8.2). Likewise, the variable-specific MSA

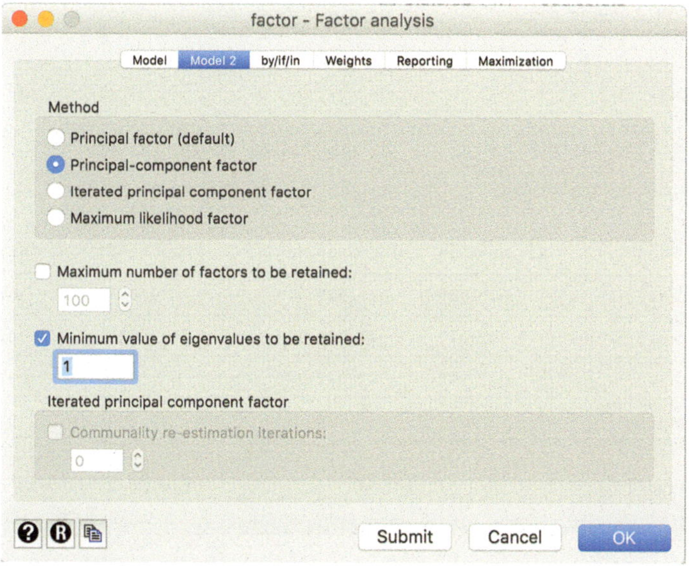

Fig. 8.11 Factor analysis dialog box (options)

values in the table are all above the threshold value of 0.50. For example, *s1* has an MSA value of **0.9166**.

The output in Table 8.5 shows three blocks. On the top right of the first block, Stata shows the number of observations used in the analysis (**Number of obs = 921**) and indicates that the analysis yielded two factors (**Retained factors = 2**). In the second block, Stata indicates the eigenvalues for each factor. With an eigenvalue of **5.24886**, the first factor extracts a large amount of variance, which accounts for 5.24886/8 = 65.61% of the total variance (see column: **Proportion**). With an eigenvalue of **1.32834**, factor two extracts less variance (16.60%). Using the **Kaiser criterion** (i.e., eigenvalue >1), we settle on two factors, because the third factor's eigenvalue is clearly lower than 1 (**0.397267**). The **Cumulative** column indicates the cumulative variance extracted. The two factors extract **0.8222** or 82.22% of the variance, which is highly satisfactory. The next block labeled **Factor loadings (pattern matrix) and unique variances** shows the factor loadings along with the **Uniqueness**, which indicates the amount of each variable's variance that the factors cannot reproduce (i.e., 1-communality) and is therefore lost in the process. All uniqueness values are very low, indicating that the factors reproduce the variables' variance well. Specifically, with a value of **0.3086**, *s3* exhibits the highest uniqueness value, which suggests a communality of 1–0.3086 = 0.6914 and is clearly above the 0.50 threshold.

Beside the Kaiser criterion, the scree plot helps determine the number of factors. To create a scree plot, go to ► Statistics ► Postestimation ► Principal component analysis reports and graphs ► Scree plot of eigenvalues. Then click on **Launch** and

Table 8.5 PCA output

```
factor s1 s2 s3 s4 s5 s6 s7 s8, pcf mineigen(1)

(obs=921)

Factor analysis/correlation                      Number of obs    =        921
    Method: principal-component factors          Retained factors =          2
    Rotation: (unrotated)                        Number of params =         15

    --------------------------------------------------------------------------
         Factor  |   Eigenvalue   Difference        Proportion   Cumulative
    -------------+------------------------------------------------------------
        Factor1  |      5.24886      3.92053            0.6561       0.6561
        Factor2  |      1.32834      0.93107            0.1660       0.8222
        Factor3  |      0.39727      0.13406            0.0497       0.8718
        Factor4  |      0.26321      0.03202            0.0329       0.9047
        Factor5  |      0.23119      0.03484            0.0289       0.9336
        Factor6  |      0.19634      0.00360            0.0245       0.9582
        Factor7  |      0.19274      0.05067            0.0241       0.9822
        Factor8  |      0.14206            .            0.0178       1.0000
    --------------------------------------------------------------------------
    LR test: independent vs. saturated:  chi2(28) = 6428.36 Prob>chi2 = 0.0000

Factor loadings (pattern matrix) and unique variances

    ------------------------------------------------
        Variable |  Factor1   Factor2 |   Uniqueness
    -------------+--------------------+-------------
              s1 |   0.7855    0.3995 |      0.2235
              s2 |   0.8017    0.4420 |      0.1619
              s3 |   0.7672    0.3206 |      0.3086
              s4 |   0.7505    0.5090 |      0.1777
              s5 |   0.8587   -0.3307 |      0.1533
              s6 |   0.8444   -0.4203 |      0.1104
              s7 |   0.7993   -0.4662 |      0.1439
              s8 |   0.8649   -0.3291 |      0.1436
    ------------------------------------------------
```

OK. Stata will produce a graph as shown in Fig. 8.13. There is an “elbow” in the line at three factors. As the number of factors that the scree plot suggests is one factor less than the elbow indicates, we conclude that two factors are appropriate. This finding supports the conclusion based on the Kaiser criterion.

Stata also allows for plotting each eigenvalue’s confidence interval. However, to display such a scree plot requires running a PCA with a different command in combination with a postestimation command. The following syntax produces a scree plot for our example with a 95% confidence interval (heteroskedastic), as well as a horizontal reference line at the 1 threshold.

```
pca s1 s2 s3 s4 s5 s6 s7 s8, mineigen(1)
screeplot, recast(line) ci(heteroskedastic) yline(1)
```

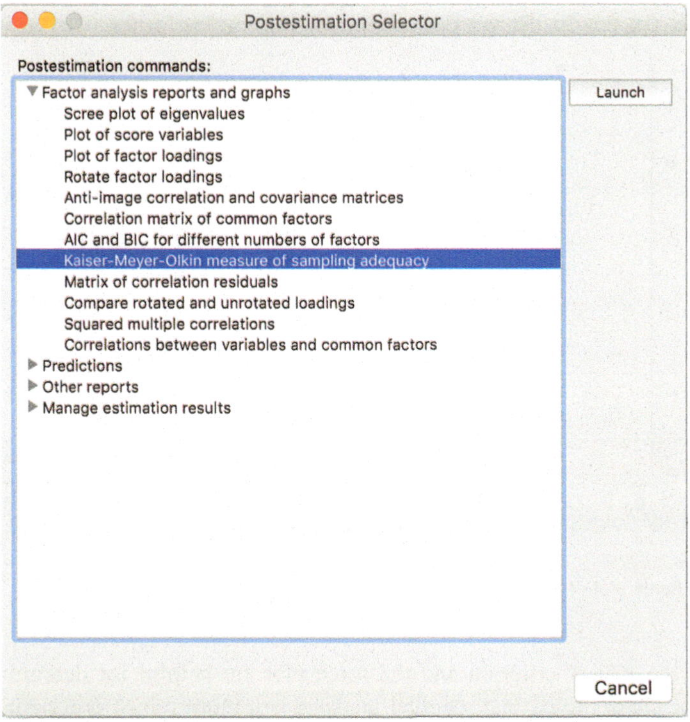

Fig. 8.12 Postestimation dialog box

Table 8.6 The KMO statistic

```
estat kmo

Kaiser-Meyer-Olkin measure of sampling adequacy

    -----------------------
        Variable |     kmo
    -------------+---------
              s1 |  0.9166
              s2 |  0.8839
              s3 |  0.9427
              s4 |  0.8834
              s5 |  0.9308
              s6 |  0.8831
              s7 |  0.9036
              s8 |  0.9180
    -------------+---------
         Overall |  0.9073
```

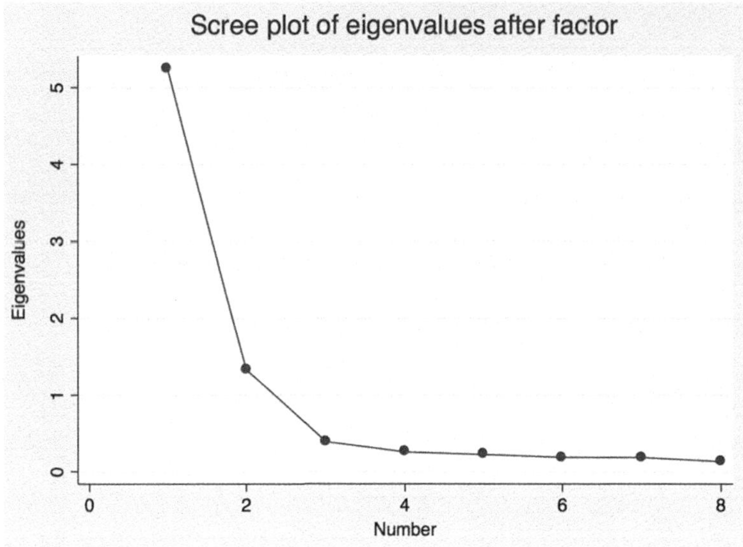

Fig. 8.13 Scree plot of eigenvalues

While the Kaiser criterion and the scree plot are helpful for determining the number of factors to extract, parallel analysis is a more robust criterion. Parallel analysis can only be accessed through the free add-on package paran. To install the package, type in `help paran` in the command window and follow the instructions to install the package. Having installed the package, type in `paran s1 s2 s3 s4 s5 s6 s7 s8, centile(95) q all graph` in the command window and Stata will produce output similar to Table 8.7 and Fig. 8.14.

Table 8.7 contains two rows of eigenvalues, with the first column (**Adjusted Eigenvalue**) indicating the sampling error-adjusted eigenvalues obtained by parallel analysis. Note that your results will look slightly different, as parallel analysis draws on randomly generated datasets. The second column (**Unadjusted Eigenvalue**) contains the eigenvalues as reported in the PCA output (Table 8.5). Analogous to the original analysis, the two factors exhibit adjusted eigenvalues larger than 1, indicating a two-factor solution. The scree plot in Fig. 8.14 also supports this result, as the first two factors exhibit adjusted eigenvalues larger than the randomly generated eigenvalues. Conversely, the random eigenvalue of the third factor is clearly larger than the adjusted one.

Finally, we can also request the model selection statistics AIC and BIC for different numbers of factors (see Fig. 8.15). To do so, go to ► Statistics ► Postestimation ► AIC and BIC for different numbers of factors. As our analysis draws on eight variables, we restrict the number of factors to consider to 4 (**Specify the maximum number of factors to include in summary table: 4**). Table 8.8 shows the results of our analysis. As can be seen, AIC has the smallest value (**53.69378**) for a four-factor solution, whereas BIC's minimum value occurs for a

Table 8.7 Parallel analysis output

```
paran s1 s2 s3 s4 s5 s6 s7 s8, centile (95) q all graph

Computing: 10% 20% 30% 40% 50% 60% 70% 80% 90% 100%


Results of Horn's Parallel Analysis for principal components
240 iterations, using the p95 estimate

--------------------------------------------------
Component   Adjusted     Unadjusted     Estimated
or Factor   Eigenvalue   Eigenvalue     Bias
--------------------------------------------------
 1          5.1571075    5.2488644      .09175694
 2          1.2542191    1.3283371      .07411802
 3          .3409911     .39726683      .05627573
 4          .23833934    .26320842      .02486908
 5          .22882924    .23118517      .00235593
 6          .23964682    .19634077      -.04330605
 7          .2775334     .19273589      -.0847975
 8          .26333341    .14206139      -.12127203
--------------------------------------------------
Criterion: retain adjusted components > 1
```

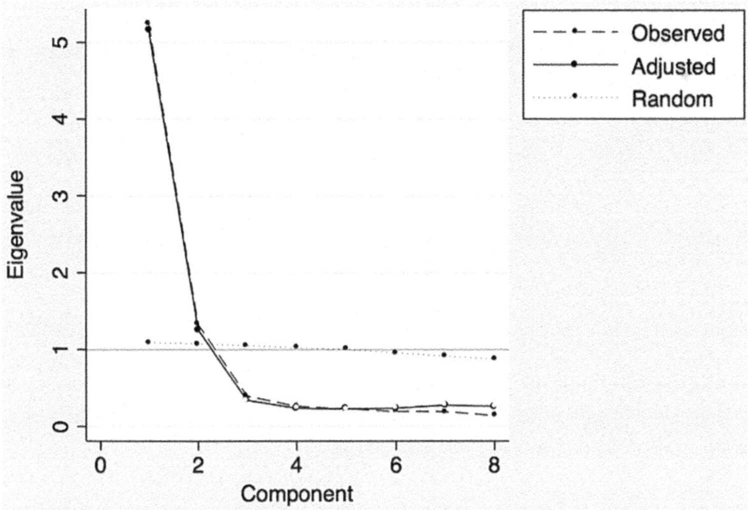

Fig. 8.14 Scree plot of parallel analysis

two-factor solution (**141.8393**). However, as AIC is well known to overspecify the number of factors, this result gives confidence that the two-factor solution as indicated by the BIC is appropriate. Note that Table 8.8 says no Heywood cases encountered.

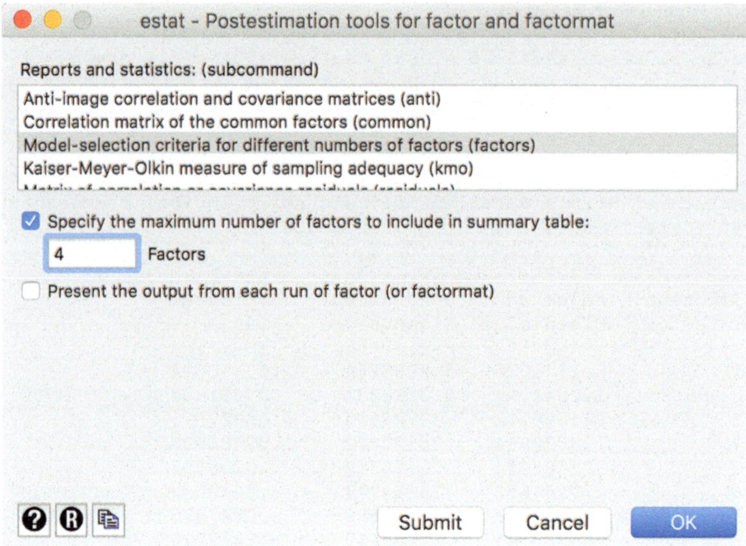

Fig. 8.15 AIC and BIC

Table 8.8 Factor rotation output

```
estat factors, factors(4)

Factor analysis with different numbers of factors (maximum likelihood)

    ----------------------------------------------------------
    #factors |     loglik   df_m    df_r         AIC       BIC
    ---------+------------------------------------------------
           1 |  -771.1381      8      20    1558.276   1596.88
           2 |  -19.72869     15      13    69.45738  141.8393
           3 |  -10.09471     21       7    62.18943  163.5241
           4 |  -.8468887     26       2    53.69378  179.1557
    ----------------------------------------------------------
    no Heywood cases encountered
```

Heywood cases indicate negative estimates of variances or correlation estimates greater than one in absolute value. In the Stata output, they are typically noted as `Beware: solution is a Heywood case`.

8.6.1.3 Interpret the Factor Solution

To facilitate the interpretation of the factor solution, we need to rotate the factors. To initiate this analysis, go to ► Statistics ► Postestimation ► Factor analysis reports and graphs ► Rotate factor loadings. In the dialog box that opens, select **Varimax (default)** under **Orthogonal rotation** and click on **Apply the Kaiser normalization**, followed by **OK**. Table 8.9 shows the resulting output.

Table 8.9 Factor rotation output

```
rotate, kaiser

Factor analysis/correlation                        Number of obs    =        921
    Method: principal-component factors            Retained factors =          2
    Rotation: orthogonal varimax (Kaiser on)       Number of params =         15

    --------------------------------------------------------------------------
         Factor  |     Variance   Difference        Proportion   Cumulative
    -------------+------------------------------------------------------------
        Factor1  |      3.41063      0.24405            0.4263       0.4263
        Factor2  |      3.16657            .            0.3958       0.8222
    --------------------------------------------------------------------------
    LR test: independent vs. saturated:  chi2(28) = 6428.36 Prob>chi2 = 0.0000

Rotated factor loadings (pattern matrix) and unique variances

    ------------------------------------------------
        Variable |  Factor1   Factor2 |   Uniqueness
    -------------+--------------------+-------------
              s1 |   0.2989    0.8290 |      0.2235
              s2 |   0.2817    0.8711 |      0.1619
              s3 |   0.3396    0.7590 |      0.3086
              s4 |   0.1984    0.8848 |      0.1777
              s5 |   0.8522    0.3470 |      0.1533
              s6 |   0.9031    0.2719 |      0.1104
              s7 |   0.9017    0.2076 |      0.1439
              s8 |   0.8557    0.3524 |      0.1436
    ------------------------------------------------

Factor rotation matrix

    --------------------------------
                 | Factor1  Factor2
    -------------+------------------
         Factor1 |  0.7288   0.6847
         Factor2 | -0.6847   0.7288
    --------------------------------
```

The upper part of Table 8.9 is the same as the standard PCA output (Table 8.5), showing that the analysis draws on **921** observations and extracts two factors, which jointly capture 82.22% of the variance. As its name implies, the **Rotated factor loadings** block shows the factor loadings after rotation. Recall that rotation is carried out to facilitate the interpretation of the factor solution. To interpret the factors, we first "assign" each variable to a certain factor based on its maximum absolute factor loading. That is, if the highest absolute loading is negative, higher values of a particular variable relate negatively to the assigned factor. After that, we should find an umbrella term for each factor that best describes the set of variables associated with that factor. Looking at Table 8.9, we see that *s1*–*s4* load highly on the second factor, whereas *s5*-*s8* load on the first factor. For example, *s1* has a **0.2989** loading on the first factor, while its loading is much stronger on the second factor (**0.8290**). Finally, note that the uniqueness and, hence, the communality values are unaffected by the rotation (see Table 8.5).

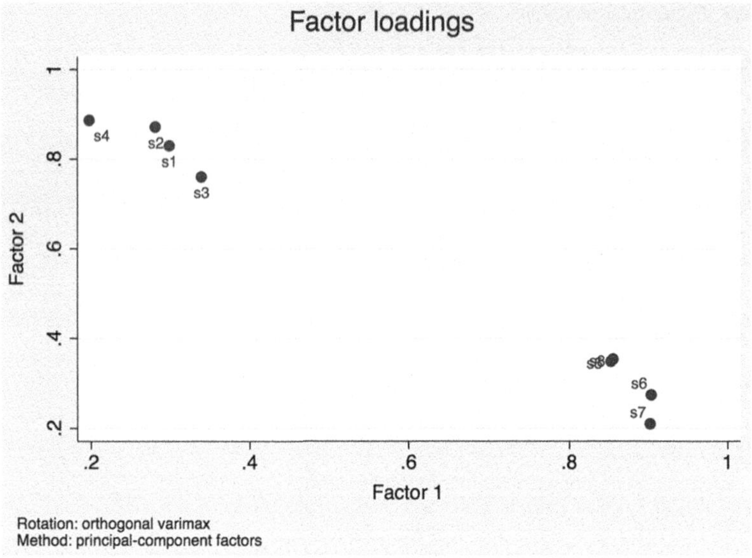

Fig. 8.16 Factor loadings plot

To facilitate the identification of labels, we can plot each item's loading against each factor. To request a factor loadings plot, go to ▶ Statistics ▶ Postestimation ▶ Factor analysis reports and graphs ▶ Plot of factor loadings and click on **Launch**. In the dialog box that follows, retain the default settings and click on **OK**. The resulting plot (Fig. 8.16) shows two cluster of variables, which strongly load on one factor while having low loadings on the other factor. This result supports our previous conclusion in terms of the variable assignments.

Having identified which variables load highly on which factor in the rotated solution, we now need to identify labels for each factor. Looking at the variable labels, we learn that the first set of variables (*s1–s4*) relate to reliability aspects of the journey and related processes, such as the booking. We could therefore label this factor (i.e., factor 2) *reliability*. The second set of variables (*s5–s8*) relate to different aspects of the onboard facilities and the travel experience. Hence, we could label this factor (i.e., factor 1) *onboard experience*. The labeling of factors is of course subjective and you could provide different descriptions.

8.6.1.4 Evaluate the Goodness-of-fit of the Factor Solution

The last step involves assessing the analysis's goodness-of-fit. To do so, we first look at the residuals (i.e., the differences between observed and reproduced correlations) in the reproduced correlation matrix. To create this matrix, go to ▶ Statistics ▶ Postestimation ▶ Factor analysis reports and graphs ▶ Matrix of correlation residuals. In the dialog box that opens, select **Matrix of correlation of covariance residuals (residuals)** and click on **OK**. Table 8.10 shows the resulting output.

Table 8.10 Correlation residual matrix

```
estat residuals

Raw residuals of correlations (observed-fitted)

--------------------------------------------------------------------------------------------
    Variable |       s1        s2        s3        s4        s5        s6        s7        s8
-------------+------------------------------------------------------------------------------
          s1 |   0.0000
          s2 |  -0.0525    0.0000
          s3 |  -0.1089   -0.0602    0.0000
          s4 |  -0.0598   -0.0557   -0.0907    0.0000
          s5 |  -0.0174   -0.0063   -0.0029    0.0076    0.0000
          s6 |   0.0046    0.0023   -0.0216    0.0159   -0.0464    0.0000
          s7 |   0.0136    0.0182   -0.0118    0.0042   -0.0529   -0.0393    0.0000
          s8 |  -0.0056   -0.0135   -0.0056    0.0052   -0.0404   -0.0265   -0.0645    0.0000
--------------------------------------------------------------------------------------------
```

When examining the lower part of Table 8.10, we see that there are several residuals with absolute values larger than 0.05. A quick count reveals that 8 out of 29 (i.e., 27.59%) residuals are larger than 0.05. As the percentage of increased residuals is well below 50%, we can presume a good model fit.

Similarly, our previous analysis showed that the two factors reproduce a sufficient amount of each variable's variance. The uniqueness values are clearly below 0.50 (i.e., the communalities are larger than 0.50), indicating that the factors account for more than 50% of the variables' variance (Table 8.5).

8.6.1.5 Compute the Factor Scores

The evaluation of the factor solution's goodness-of-fit completes the standard PCA analysis. However, if we wish to continue using the results for further analyses, we should calculate the factor scores. Go to ▶ Statistics ▶ Postestimation ▶ Predictions ▶ Regression and Bartlett scores and Stata will open a dialog box similar to Fig. 8.17. Under **New variable names or variable stub*** you can enter a prefix name, which Stata uses to name the saved factor scores. For example, specifying *factor**, as shown in Fig. 8.17, will create two variables called *factor1* and *factor2*. Next, select **Factors scored by the regression scoring method** and click on **OK**. Stata will produce an output table showing the scoring coefficients, which are the weights used to compute the factor scores from the standardized data. However, we are more interested in the actual factor scores, which we can access by clicking on the **Data Browser** button in Stata's menu bar. Figure 8.18 shows the scores of *factor1* and *factor2* for the first ten observations.

Being z-standardized, the newly generated variables *factor1* and *factor2* have mean values (approximately) equal to zero and standard deviations equal to 1. Thus, the factor scores are estimated in units of standard deviations from their means. For example, the first observation is about **1.91** standard deviations below average on the *onboard experience* factor (i.e., factor 1) and about **0.91** standard deviations above average on the *reliability* factor (i.e., factor 2). In contrast, the second observation is clearly above average in terms of *reliability* and *onboard experience*. Note that if the original variables include a missing value, the factor score will also be missing (i.e., only a "**.**" (dot) will be recorded).

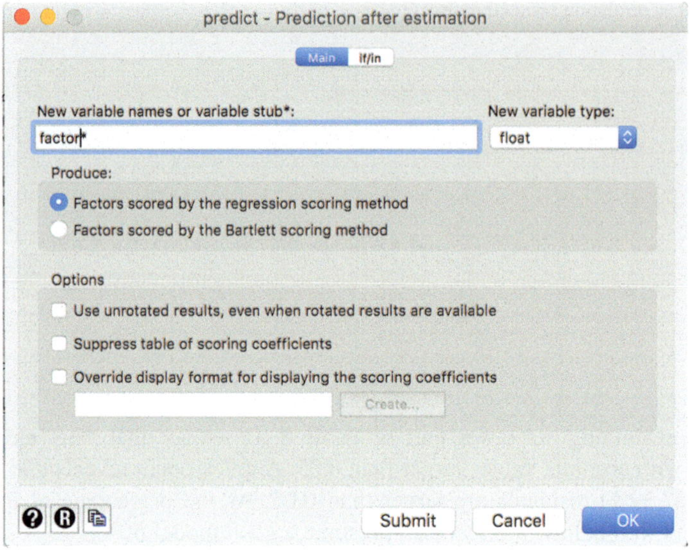

Fig. 8.17 Saving factor scores as new variables

Fig. 8.18 Overview of factor scores for the first ten observations

factor1	factor2
-1.911223	.9144877
1.684422	1.297118
-2.432783	.4436803
.1665416	.8420736
-.1204526	-.85245
-.4704992	.3993561
.8271663	-1.990248
.2047913	.9054558
.4886046	1.043682
.5272735	-.7065509

8.6.2 Reliability Analysis

To illustrate its usage, let's carry out a reliability analysis of the first factor *onboard experience* by calculating Cronbach's Alpha as a function of variables *s5* to *s8*. To run the reliability analysis, click on ► Statistics ► Multivariate analysis ► Cronbach's Alpha. A window similar to Fig. 8.19 will appear. Next, enter variables *s5*–*s8* into the **Variables** box.

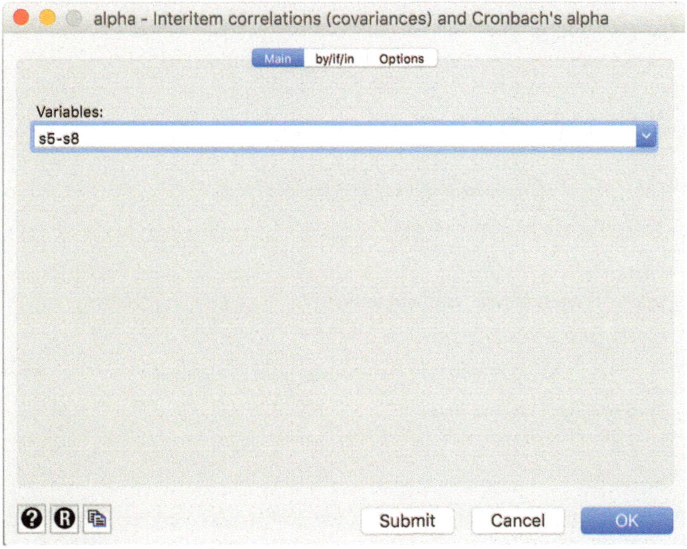

Fig. 8.19 Reliability analysis dialog box

The **Options** tab (Fig. 8.20) provides options for dealing with missing values and requesting descriptive statistics for each item and the entire scale or item correlations. Check **Display item-test and item-rest correlations** and click on **OK**.

The results in Table 8.11 show that the scale exhibits a high degree of internal consistency reliability. With a value of **0.9439** (see row **Test scale**), the Cronbach's Alpha coefficient lies well above the commonly suggested threshold of 0.70. This result is not surprising, since we are simply testing a scale previously established by means of item correlations. Keep in mind that we usually carry out a reliability analysis to test a scale using a different sample—this example is only for illustration purposes! The rightmost column of Table 8.11 indicates what the Cronbach's Alpha would be if we deleted the item indicated in that row. When we compare each of the values with the overall Cronbach's Alpha value, we can see that any change in the scale's set-up would reduce the Cronbach's Alpha value. For example, by removing *s5* from the scale, the Cronbach's Alpha of the new scale comprising only *s6*, *s7*, and *s8* would be reduced to **0.9284**. Therefore, deleting this item (or any others) makes little sense. In the leftmost column of Table 8.11, Stata indicates the number of observations (**Obs**), as well as whether that particular item correlates positively or negatively with the sum of the other items (**Sign**). This information is useful for determining whether reverse-coded items were also identified as such. Reverse-coded items should have a minus sign. The columns **item-test**, **item-rest**, and **average interitem covariance** are not needed for a basic interpretation.

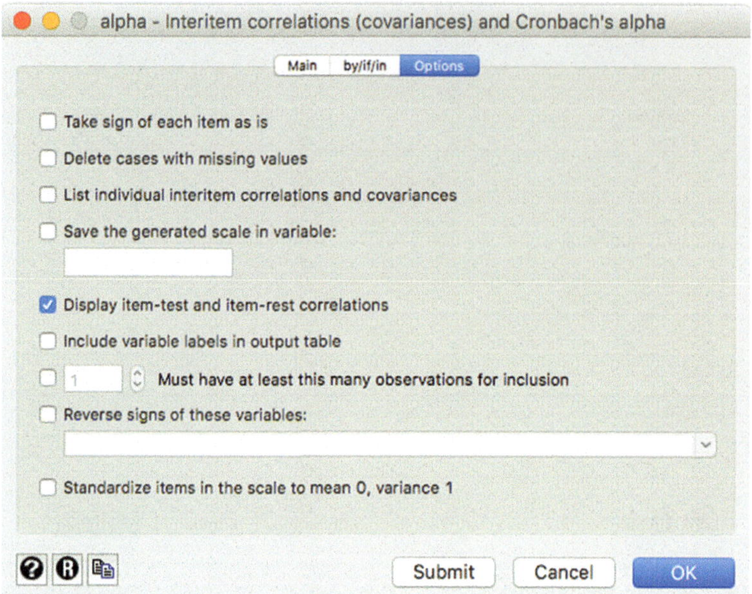

Fig. 8.20 Options tab for Cronbach's alpha

Table 8.11 Reliability statistics

```
alpha s5 s6 s7 s8, item

Test scale = mean(unstandardized items)

                                                                average
                               item-test      item-rest        interitem
Item          |  Obs   Sign   correlation    correlation       covariance       alpha
--------------+--------------------------------------------------------------------
s5            | 1041     +       0.9211         0.8578          421.5723       0.9284
s6            | 1041     +       0.9422         0.8964          412.3225       0.9168
s7            | 1048     +       0.9222         0.8506          399.2565       0.9330
s8            | 1034     +       0.9203         0.8620          434.3355       0.9282
--------------+--------------------------------------------------------------------
Test scale    |                                                 416.8996       0.9439
-----------------------------------------------------------------------------------
```

8.7 Customer Satisfaction at Haver and Boecker (Case Study)

Haver and Boecker (http://www.haverboecker.com) is one of the world's leading and most renowned machine producers in the fields of mineral processing, as well as the storing, conveying, packing, and loading of bulk material. The family-owned group operates through its global network of facilities, with manufacturing units, among others, in Germany, the UK, Belgium, US, Canada, Brazil, China, and India.

The company's relationships with its customers are usually long-term oriented and complex. Since the company's philosophy is to help customers and business partners solve their challenges or problems, they often customize their products and services to meet the buyers' needs. Therefore, the customer is no longer a passive buyer, but an active partner. Given this background, the customers' satisfaction plays an important role in establishing, developing, and maintaining successful customer relationships.

Very early on, the company's management realized the importance of customer satisfaction and decided to commission a market research project in order to identify marketing activities that can positively contribute to the business's overall success. Based on a thorough literature review, as well as interviews with experts, the company developed a short survey to explore their customers' satisfaction with specific performance features and their overall satisfaction. All the items were measured on 7-point scales, with higher scores denoting higher levels of satisfaction. A standardized survey was mailed to customers in 12 countries worldwide, which yielded 281 fully completed questionnaires. The following items (names in parentheses) were listed in the survey:

- Reliability of the machines and systems. (s_1)
- Life-time of the machines and systems. (s_2)
- Functionality and user-friendliness operation of the machines and systems. (s_3)
- Appearance of the machines and systems. (s_4)
- Accuracy of the machines and systems. (s_5)

- Timely availability of the after-sales service. (s_6)
- Local availability of the after-sales service. (s_7)
- Fast processing of complaints. (s_8)
- Composition of quotations. (s_9)
- Transparency of quotations. (s_{10})
- Fixed product prize for the machines and systems. (s_{11})
- Cost/performance ratio of the machines and systems. (s_{12})
- Overall, how satisfied are you with the supplier (*overall*)?

Your task is to analyze the dataset to provide the management of Haver and Boecker with advice for effective customer satisfaction management. The dataset is labeled *haver_and_boecker.dta* (⭳ Web Appendix → Downloads).

1. Determine the factors that characterize the respondents by means a factor analysis. Use items s_1–s_{12} for this. Run a PCA with varimax rotation to facilitate interpretation. Consider the following aspects:
   (a) Are all assumptions for carrying out a PCA met? Pay special attention to the question whether the data are sufficiently correlated.
   (b) How many factors would you extract? Base your decision on the Kaiser criterion, the scree plot, parallel analysis, and the model selection statistics AIC and BIC.
   (c) Find suitable labels for the extracted factors.
   (d) Evaluate the factor solution's goodness-of-fit.
2. Use the factor scores and regress the customers' overall satisfaction (*overall*) on these. Evaluate the strength of the model and compare it with the initial regression. What should Haver and Boecker's management do to increase their customers' satisfaction?
3. Calculate the Cronbach's Alpha over items s_1–s_5 and interpret the results.

For further information on the dataset and the study, see Festge and Schwaiger (2007), as well as Sarstedt et al. (2009).

8.8 Review Questions

1. What is factor analysis? Try to explain what factor analysis is in your own words.
2. What is the difference between exploratory factor analysis and confirmatory factor analysis?
3. What is the difference between PCA and factor analysis?
4. Describe the terms communality, eigenvalue, factor loading, and uniqueness. How do these concepts relate to one another?
5. Describe three approaches used to determine the number of factors.
6. What are the purpose and the characteristic of a varimax rotation? Does a rotation alter eigenvalues or factor loadings?

7. Re-run the Oddjob Airways case study by carrying out a factor analysis and compare the results with the example carried out using PCA. Are there any significant differences?
8. What is reliability analysis and why is it important?
9. Explain the basic principle of structural equation modeling.

8.9 Further Readings

Nunnally, J. C., & Bernstein, I. H. (1993). *Psychometric theory* (3rd ed.). New York: McGraw-Hill.

Psychometric theory is a classic text and the most comprehensive introduction to the fundamental principles of measurement. Chapter 7 provides an in-depth discussion of the nature of reliability and its assessment.

Sarstedt, M., Hair, J. F., Ringle, C. M., Thiele, K. O., & Gudergan, S. P. (2016). Estimation issues with PLS and CBSEM: where the bias lies! *Journal of Business Research, 69*(10), 3998–4010.

This paper discusses the differences between covariance-based and partial least squares structural equation modeling from a measurement perspective. This discussion relates to the differentiation between factor analysis and PCA and the assumptions underlying each approach to measure unobservable phenomena.

Stewart, D. W., (1981). The application and misapplication of factor analysis in marketing research. *Journal of Marketing Research, 18*(1), 51–62.

David Stewart discusses procedures for determining when data are appropriate for factor analysis, as well as guidelines for determining the number of factors to extract, and for rotation.

References

Acock, A. C. (2013). *Discovering structural equation modeling using Stata* (Revised ed.). College Station: Stata Press.

Brown, J. D. (2009). Choosing the right type of rotation in PCA and EFA. *JALT Testing & Evaluation SIG Newsletter, 13*(3), 20–25.

Carbonell, L., Izquierdo, L., Carbonell, I., & Costell, E. (2008). Segmentation of food consumers according to their correlations with sensory attributes projected on preference spaces. *Food Quality and Preference, 19*(1), 71–78.

Cattell, R. B. (1966). The scree test for the number of factors. *Multivariate Behavioral Research, 1*(2), 245–276.

Cliff, N. (1987). *Analyzing multivariate data.* New York: Harcourt Brace Jovanovich.

Cronbach, L. J. (1951). Coefficient alpha and the internal structure of tests. *Psychometrika, 16*(3), 297–334.

Diamantopoulos, A., & Siguaw, J. A. (2000). *Introducing LISREL: A guide for the uninitiated.* London: Sage.

Dinno, A. (2009). Exploring the sensitivity of Horn's parallel analysis to the distributional form of random data. *Multivariate Behavioral Research, 44*(3), 362–388.

DiStefano, C., Zhu, M., & Mîndriă, D. (2009). Understanding and using factor scores: Considerations fort he applied researcher. *Practical Assessment, Research & Evaluation, 14*(20), 1–11.

Festge, F., & Schwaiger, M. (2007). The drivers of customer satisfaction with industrial goods: An international study. *Advances in International Marketing, 18*, 179–207.

Gorsuch, R. L. (1983). *Factor analysis* (2nd ed.). Hillsdale: Lawrence Erlbaum Associates.

Graffelman, J. (2013). Linear-angle correlation plots: New graphs for revealing correlation structure. *Journal of Computational and Graphical Statistics, 22*(1), 92–106.

Grice, J. W. (2001). Computing and evaluating factor scores. *Psychological Methods, 6*(4), 430–450.

Hair, J. F., Black, W. C., Babin, B. J., & Anderson, R. E. (2013). *Multivariate data analysis. A global perspective* (7th ed.). Upper Saddle River: Pearson Prentice Hall.

Hair, J. F., Ringle, C. M., & Sarstedt, M. (2011). PLS-SEM: Indeed a silver bullet. *Journal of Marketing Theory and Practice, 19*(2), 139–151.

Hair, J. F., Hult, G. T. M., Ringle, C. M., & Sarstedt, M. (2017a). *A primer on partial least squares structural equation modeling (PLS-SEM)* (2nd ed.). Thousand Oaks: Sage.

Hair, J. F., Hult, G. T. M., Ringle, C. M., Sarstedt, M., & Thiele, K. O. (2017b). Mirror, mirror on the wall. A comparative evaluation of composite-based structural equation modeling methods. *Journal of the Academy of Marketing Science, 45*(5), 616–632.

Hair, J. F., Sarstedt, M., Ringle, C. M., & Gudergan, S. P. (2018). *Advanced issues in partial least squares structural equation modeling (PLS-SEM)*. Thousand Oaks: Sage.

Hamilton, L. C. (2013), Statistics with Stata: Version 12: Cengage Learning.

Hayton, J. C., Allen, D. G., & Scarpello, V. (2004). Factor retention decisions in exploratory factor analysis: A tutorial on parallel analysis. *Organizational Research Methods, 7*(2), 191–205.

Henson, R. K., & Roberts, J. K. (2006). Use of exploratory factor analysis in published research: Common errors and some comment on improved practice. *Educational and Psychological Measurement, 66*(3), 393–416.

Hershberger, S. L. (2005). Factor scores. In B. S. Everitt & D. C. Howell (Eds.), *Encyclopedia of statistics in behavioral science* (pp. 636–644). New York: John Wiley.

Horn, J. L. (1965). A rationale and test for the number of factors in factor analysis. *Psychometrika, 30*(2), 179–185.

Jöreskog, K. G. (1971). Simultaneous factor analysis in several populations. *Psychometrika, 36*(4), 409–426.

Kaiser, H. F. (1958). The varimax criterion for factor analytic rotation in factor analysis. *Educational and Psychological Measurement, 23*(3), 770–773.

Kaiser, H. F. (1974). An index of factorial simplicity. *Psychometrika, 39*(1), 31–36.

Kim, J. O., & Mueller, C. W. (1978). *Introduction to factor analysis: What it is and how to do it*. Thousand Oaks: Sage.

Longman, R. S., Cota, A. A., Holden, R. R., & Fekken, G. C. (1989). A regression equation for the parallel analysis criterion in principal components analysis: Mean and 95th percentile eigenvalues. *Multivariate Behavioral Research, 24*(1), 59–69.

MacCallum, R. C., Widaman, K. F., Zhang, S., & Hong, S. (1999). Sample size in factor analysis. *Psychological Methods, 4*(1), 84–99.

Matsunga, M. (2010). How to factor-analyze your data right: Do's and don'ts and how to's. *International Journal of Psychological Research, 3*(1), 97–110.

Mulaik, S. A. (2009). *Foundations of factor analysis* (2nd ed.). London: Chapman & Hall.

Preacher, K. J., & MacCallum, R. C. (2003). Repairing Tom Swift's electric factor analysis machine. *Understanding Statistics, 2*(1), 13–43.

Russell, D. W. (2002). In search of underlying dimensions: The use (and abuse) of factor analysis in Personality and Social Psychology Bulletin. *Personality and Social Psychology Bulletin, 28*(12), 1629–1646.

Sarstedt, M., Schwaiger, M., & Ringle, C. M. (2009). Do we fully understand the critical success factors of customer satisfaction with industrial goods? Extending Festge and Schwaiger's model to account for unobserved heterogeneity. *Journal of Business Market Management, 3*(3), 185–206.

Sarstedt, M., Ringle, C. M., Raithel, S., & Gudergan, S. (2014). In pursuit of understanding what drives fan satisfaction. *Journal of Leisure Research, 46*(4), 419–447.

Sarstedt, M., Hair, J. F., Ringle, C. M., Thiele, K. O., & Gudergan, S. P. (2016). Estimation issues with PLS and CBSEM: Where the bias lies! *Journal of Business Research, 69*(10), 3998–4010.

Steiger, J. H. (1979). Factor indeterminacy in the 1930's and the 1970's some interesting parallels. *Psychometrika, 44*(2), 157–167.

Stevens, J. P. (2009). *Applied multivariate statistics for the social sciences* (5th ed.). Hillsdale: Erlbaum.

Velicer, W. F., & Jackson, D. N. (1990). Component analysis versus common factor analysis: Some issues in selecting an appropriate procedure. *Multivariate Behavioral Research, 25*(1), 1–28.

Vigneau, E., & Qannari, E. M. (2002). Segmentation of consumers taking account of external data. A clustering of variables approach. *Food Quality and Preference, 13*(7–8), 515–521.

Widaman, K. F. (1993). Common factor analysis versus principal component analysis: Differential bias in representing model parameters? *Multivariate Behavioral Research, 28*(3), 263–311.

Wold, H. O. A. (1982). Soft modeling: The basic design and some extensions. In K. G. Jöreskog & H. O. A. Wold (Eds.), *Systems under indirect observations: Part II* (pp. 1–54). Amsterdam: North-Holland.

Zwick, W. R., & Velicer, W. F. (1986). Comparison of five rules for determining the number of components to retain. *Psychological Bulletin, 99*(3), 432–442.

Cluster Analysis

9

Keywords

Agglomerative clustering • Average linkage • Canberra distance • Centroid linkage • Chaining effect • Chebychev distance • City-block distance • Clusters • Clustering variables • Complete linkage • Dendrogram • Distance matrix • Divisive clustering • Duda-Hart index • Euclidean distance • Factor-cluster segmentation • Gower's dissimilarity coefficient • Hierarchical clustering methods • Partitioning methods • k-means • k-medians • k-means++ • k-medoids • Label switching • Linkage algorithm • Local optimum • Mahalanobis distance • Manhattan metric • Market segmentation • Matching coefficients • Non-hierarchical clustering methods • Profiling • Russel and Rao coefficient • Single linkage • Simple matching coefficient • Straight line distance • Ties • Variance ration criterion • Ward's linkage • Weighted average linkage

Learning Objectives

After reading this chapter, you should understand:

- The basic concepts of cluster analysis.
- How basic cluster algorithms work.
- How to compute simple clustering results manually.
- The different types of clustering procedures.
- The Stata clustering outputs.

E. Mooi et al., *Market Research*, Springer Texts in Business and Economics,
DOI 10.1007/978-981-10-5218-7_9

9.1 Introduction

Market segmentation is one of the most fundamental marketing activities. Since consumers, customers, and clients have different needs, companies have to divide markets into groups (segments) of consumers, customers, and clients with similar needs and wants. Firms can then target each of these segments by positioning themselves in a unique segment (e.g., Ferrari in the high-end sports car market). Market segmentation "is essential for marketing success: the most successful firms drive their businesses based on segmentation" (Lilien and Rangaswamy 2004, p. 61) and "tools such as segmentation [...] have the largest impact on marketing decisions" (John et al. 2014, p. 127). While market researchers often form market segments based on practical grounds, industry practice and wisdom, cluster analysis uses data to form segments, making segmentation less dependent on subjectivity.

9.2 Understanding Cluster Analysis

Cluster analysis is a method for segmentation and identifies homogenous groups of objects (or cases, observations) called **clusters**. These objects can be individual customers, groups of customers, companies, or entire countries. Objects in a certain cluster should be as similar as possible to each other, but as distinct as possible from objects in other clusters.

Let's try to gain a basic understanding of cluster analysis by looking at a simple example. Imagine that you are interested in segmenting your customer base in order to better target them through, for example, pricing strategies.

The first step is to decide on the characteristics that you will use to segment your customers A to G. In other words, you have to decide which **clustering variables** will be included in the analysis. For example, you may want to segment a market based on customers' price consciousness (x) and brand loyalty (y). These two variables can be measured on a scale from 0 to 100 with higher values denoting a higher degree of price consciousness and brand loyalty. Table 9.1 and the scatter plot in Fig. 9.1 show the values of seven customers (referred to as objects).

The aim of cluster analysis is to identify groups of objects (in this case, customers) that are very similar regarding their price consciousness and brand loyalty, and assign them to clusters. After having decided on the clustering variables (here, price consciousness and brand loyalty), we need to decide on the clustering procedure to form our groups of objects. This step is crucial for the analysis, as different procedures require different decisions prior to analysis. There is an abundance of different approaches and little guidance on which one to use in practice. We will discuss the most popular approaches in market research, including:

- hierarchical methods, and
- partitioning methods (more precisely k-means)

Table 9.1 Data

Customer	A	B	C	D	E	F	G
x	33	82	66	30	79	50	10
y	95	94	80	67	60	33	17

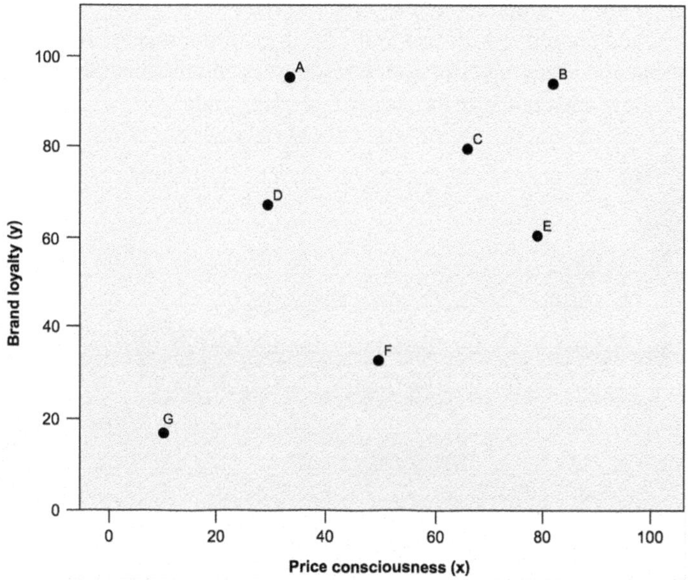

Fig. 9.1 Scatter plot

While the basic aim of these procedures is the same, namely grouping similar objects into clusters, they take different routes, which we will discuss in this chapter. An important consideration before starting the grouping is to determine how similarity should be measured. Most methods calculate measures of (dis) similarity by estimating the distance between pairs of objects. Objects with smaller distances between one another are considered more similar, whereas objects with larger distances are considered more dissimilar. The decision on how many clusters should be derived from the data is a fundamental issue in the application of cluster analysis. This question is explored in the next step of the analysis. In most instances, we do not know the exact number of clusters and then we face a trade-off. On the one hand, we want as few clusters as possible to make the clusters easy to understand and actionable. On the other hand, having many clusters allows us to identify subtle differences between objects.

Megabus is a hugely successful bus line in the US. They completely rethought the nature of their customers and concentrated on three specific segments of the market: College kids, women travelling in groups, and active seniors. To meet these customer segments' needs, Megabus reimagined the entire driving experience by developing double-decker buses with glass roofs and big windows, and equipped with fast WiFi. In light of the success of Megabus's segmenting and targeting efforts, practitioners even talk about the "Megabus Effect"—how one company has shaped an entire industry.

In the final step, we need to interpret the clustering solution by defining and labeling the obtained clusters. We can do so by comparing the mean values of the clustering variables across the different clusters, or by identifying explanatory variables to profile the clusters. Ultimately, managers should be able to identify customers in each cluster on the basis of easily measurable variables. This final step also requires us to assess the clustering solution's stability and validity. Figure 9.2 illustrates the steps associated with a cluster analysis; we will discuss these steps in more detail in the following sections.

9.3 Conducting a Cluster Analysis

9.3.1 Select the Clustering Variables

At the beginning of the clustering process, we have to select appropriate variables for clustering. Even though this choice is critical, it is rarely treated as such. Instead, a mixture of intuition and data availability guide most analyses in marketing

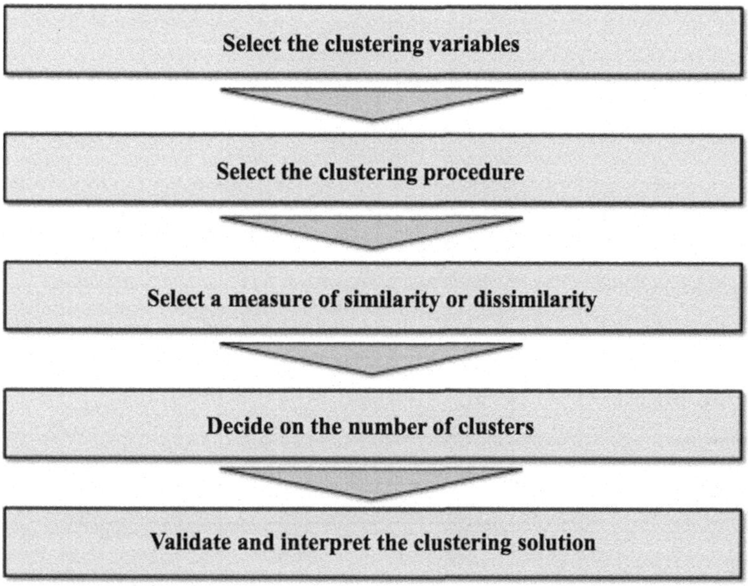

Fig. 9.2 Steps in a cluster analysis

practice. However, faulty assumptions may lead to improper market segmentation and, consequently, to deficient marketing strategies. Thus, great care should be taken when selecting the clustering variables! There are several types of clustering variables, as shown in Fig. 9.3. Sociodemographic variables define clusters based on people's demographic (e.g., age, ethnicity, and gender), geographic (e.g., residence in terms of country, state, and city), and socioeconomic (e.g., education, income, and social class) characteristics. Psychometric variables capture unobservable character traits such as people's personalities or lifestyles. Finally, behavioral clustering variables typically consider different facets of consumer behavior, such as the way people purchase, use, and dispose of products. Other behavioral clustering variables capture specific benefits which different groups of consumers look for in a product.

The types of variables used for cluster analysis provide different solutions and, thereby, influence targeting strategies. Over the last decades, attention has shifted from more traditional sociodemographic clustering variables towards behavioral and psychometric variables. The latter generally provide better guidance for decisions on marketing instruments' effective specification. Generally, clusters based on psychometric variables are more homogenous and these consumers respond more consistently to marketing actions (e.g., Wedel and Kamakura 2000). However, consumers in these clusters are frequently hard to identify as such variables are not easily measured. Conversely, clusters determined by sociodemographic variables are easy to identify but are also more heterogeneous, which complicates targeting efforts. Consequently, researchers frequently combine

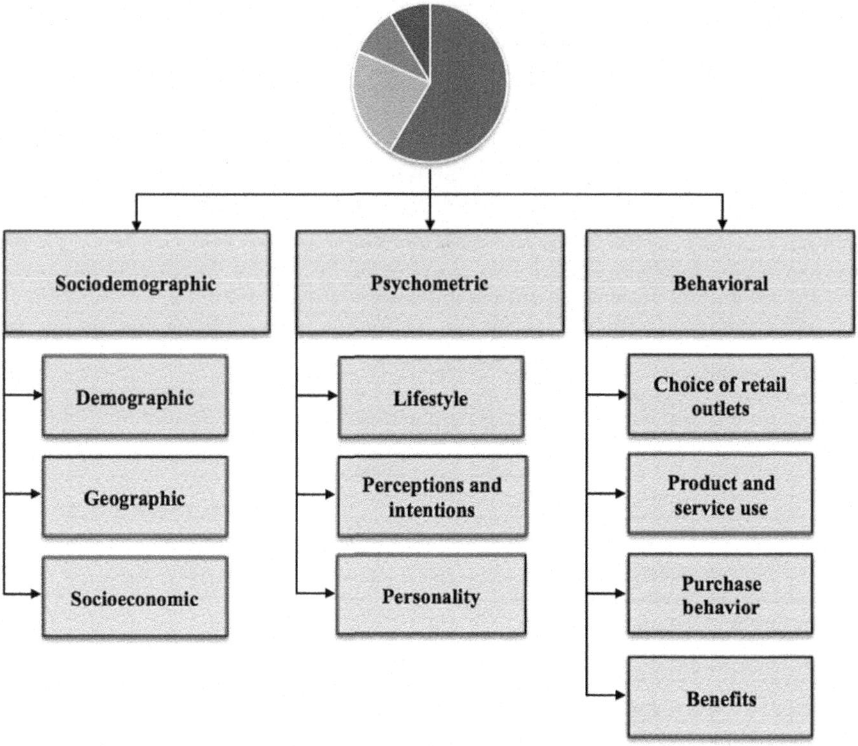

Fig. 9.3 Types of clustering variables

different variables such as lifestyle characteristics and demographic variables, benefiting from each one's strengths.

In some cases, the choice of clustering variables is apparent because of the task at hand. For example, a managerial problem regarding corporate communications will have a fairly well defined set of clustering variables, including contenders such as awareness, attitudes, perceptions, and media habits. However, this is not always the case and researchers have to choose from a set of candidate variables. But how do we make this decision? To facilitate the choice of clustering variables, we should consider the following guiding questions:

- Do the variables differentiate sufficiently between the clusters?
- Is the relation between the sample size and the number of clustering variables reasonable?
- Are the clustering variables highly correlated?
- Are the data underlying the clustering variables of high quality?

Do the variables differentiate sufficiently between the clusters?

It is important to select those clustering variables that provide a clear-cut differentiation between the objects.[1] More precisely, criterion validity is of special interest; that is, the extent to which the "independent" clustering variables are associated with one or more criterion variables not included in the analysis. Such criterion variables generally relate to an aspect of behavior, such as purchase intention or willingness-to-pay. Given this relationship, there should be significant differences between the criterion variable(s) across the clusters (e.g., consumers in one cluster exhibit a significantly higher willingness-to-pay than those in other clusters). These associations may or may not be causal, but it is essential that the clustering variables distinguish significantly between the variable(s) of interest.

Is the relation between the sample size and the number of clustering variables reasonable?

When choosing clustering variables, the sample size is a point of concern. First and foremost, this relates to issues of managerial relevance as the cluster sizes need to be substantial to ensure that the targeted marketing programs are profitable. From a statistical perspective, every additional variable requires an over-proportional increase in observations to ensure valid results. Unfortunately, there is no generally accepted guideline regarding minimum sample sizes or the relationship between the objects and the number of clustering variables used. While early research suggested a minimum sample size of two to the power of the number of clustering variables (Formann 1984), more recent rules-of-thumb are as follows:

- In the simplest case where clusters are of equal size, Qiu and Joe (2009) recommend a sample size at least ten times the number of clustering variables multiplied by the number of clusters.
- Dolnicar et al. (2014) recommend using a sample size of 70 times the number of clustering variables.
- Dolnicar et al. (2016) find that increasing the sample size from 10 to 30 times the number of clustering variables substantially improves the clustering solution. This improvement levels off subsequently, but is still noticeable up to a sample size of approximately 100 times the number of clustering variables.

These rules-of-thumb provide only rough guidance as the required sample size depends on many factors, such as the survey data characteristics (e.g., nonresponse, sampling error, response styles), relative cluster sizes, and the degree to which the clusters overlap (Dolnicar et al. 2016). However, these rules also jointly suggest that a minimum of 10 times the number of clustering variables should be considered the bare minimum. Keep in mind that no matter how many variables are used and

[1]Tonks (2009) provides a discussion of segment design and the choice of clustering variables in consumer markets.

no matter how small the sample size, cluster analysis will almost always provide a result. At the same time, however, the quality of results shows decreasing marginal returns as the sample size increases. Since cluster analysis is an exploratory technique whose results should be interpreted by taking practical considerations into account, it is not necessary to increase the sample size massively.

Are the clustering variables highly correlated?

If there is strong correlation between the variables, they are not sufficiently unique to identify distinct market segments. If highly correlated variables are used for cluster analysis, the specific aspects that these variables cover will be overrepresented in the clustering solution. In this regard, absolute correlations above 0.90 are always problematic. For example, if we were to add another variable called *brand preference* to our analysis, it would almost cover the same aspect as *brand loyalty*. The concept of being attached to a brand would therefore be overrepresented in the analysis, because the clustering procedure does not conceptually differentiate between the clustering variables. Researchers frequently handle such correlation problems by applying cluster analysis to the observations' factor scores derived from a previously carried out principal component or factor analysis. However, this **factor-cluster segmentation** approach is subject to several limitations, which we discuss in Box 9.1.

Box 9.1 Issues with Factor-Cluster Segmentation

Dolnicar and Grün (2009) identify several problems of the factor-cluster segmentation approach (see Chap. 8 for a discussion of principal component and factor analysis and related terminology):

1. The data are pre-processed and the clusters are identified on the basis of transformed values, not on the original information, which leads to different results.
2. In factor analysis, the factor solution does not explain all the variance; information is thus discarded before the clusters have been identified or constructed.
3. Eliminating variables with low loadings on all the extracted factors means that, potentially, the most important pieces of information for the identification of niche clusters are discarded, making it impossible to ever identify such groups.
4. The interpretations of clusters based on the original variables become questionable, given that these clusters were constructed by using factor scores.

(continued)

Box 9.1 (continued)

Several studies have shown that the factor-cluster segmentation reduces the success of finding useable clusters significantly.[2] Consequently, you should reduce the number of items in the questionnaire's pre-testing phase, retaining a reasonable number of relevant, non-overlapping questions that you believe differentiate the clusters well. However, if you have doubts about the data structure, factor-clustering segmentation may still be a better option than discarding items.

Are the data underlying the clustering variables of high quality?

Ultimately, the choice of clustering variables always depends on contextual influences, such as the data availability or the resources to acquire additional data. Market researchers often overlook that the choice of clustering variables is closely connected to data quality. Only those variables that ensure that high quality data can be used should be included in the analysis (Dolnicar and Lazarevski 2009). Following our discussions in Chaps. 3, 4 and 5, data are of high quality if the questions. . .

- . . . have a strong theoretical basis,
- . . . are not contaminated by respondent fatigue or response styles, and
- . . . reflect the current market situation (i.e., they are recent).

The requirements of other functions in the organization often play a major role in the choice of clustering variables. Consequently, you have to be aware that the choice of clustering variables should lead to segments acceptable to the different functions in the organization.

9.3.2 Select the Clustering Procedure

By choosing a specific clustering procedure, we determine how clusters should be formed. This forming of clusters always involves optimizing some kind of criterion, such as minimizing the within-cluster variance (i.e., the clustering variables' overall variance of the objects in a specific cluster), or maximizing the distance between the clusters. The procedure could also address the question of how to determine the (dis)similarity between objects in a newly formed cluster and the remaining objects in the dataset.

There are many different clustering procedures and also many ways of classifying these (e.g., overlapping versus non-overlapping, unimodal versus

[2]See Arabie and Hubert (1994), Sheppard (1996), and Dolnicar and Grün (2009).

multimodal, exhaustive versus non-exhaustive). Wedel and Kamakura (2000), Dolnicar (2003), and Kaufman and Rousseeuw (2005) offer reviews of clustering techniques. A practical distinction is the differentiation between hierarchical and partitioning methods (especially *k*-means), which we will discuss in the next sections.

9.3.2.1 Hierarchical Clustering Methods

Understanding Hierarchical Clustering Methods

Hierarchical clustering methods are characterized by the tree-like structure established in the course of the analysis. Most hierarchical methods fall into a category called **agglomerative clustering**. In this category, clusters are consecutively formed from objects. Agglomerative clustering starts with each object representing an individual cluster. The objects are then sequentially merged to form clusters of multiple objects, starting with the two most similar objects. Similarity is typically defined in terms of the distance between objects. That is, objects with smaller distances between one another are considered more similar, whereas objects with larger distances are considered more dissimilar. After the merger of the first two most similar (i.e., closest) objects, the agglomerative clustering procedure continues by merging another pair of objects or adding another object to an already existing cluster. This procedure continues until all the objects have been merged into one big cluster. As such, agglomerative clustering establishes a hierarchy of objects from the bottom (where each object represents a distinct cluster) to the top (where all objects form one big cluster). The left-hand side of Fig. 9.4 shows how agglomerative clustering merges objects (represented by circles) step-by-step with other objects or clusters (represented by ovals).

Hierarchical clustering can also be interpreted as a top-down process, where all objects are initially merged into a single cluster, which the algorithm then gradually splits up. This approach to hierarchical clustering is called **divisive clustering**. The right-hand side of Fig. 9.4 illustrates the divisive clustering concept. As we can see, in both agglomerative and divisive clustering, a cluster on a higher level of the hierarchy always encompasses all clusters from a lower level. This means that if an object is assigned to a certain cluster, there is no possibility of reassigning this object to another cluster (hence, hierarchical clustering). This is an important distinction between hierarchical and partitioning methods, such as *k*-means, which we will explore later in this chapter.

Divisive procedures are rarely used in market research and not implemented in statistical software programs such as Stata as they are computationally very intensive for all but small datasets.[3] We therefore focus on (agglomerative) hierarchical clustering.

[3]Whereas agglomerative methods have the large task of checking $N \cdot (N-1)/2$ possible first combinations of observations (note that N represents the number of observations in the dataset), divisive methods have the almost impossible task of checking $2^{(N-1)}-1$ combinations.

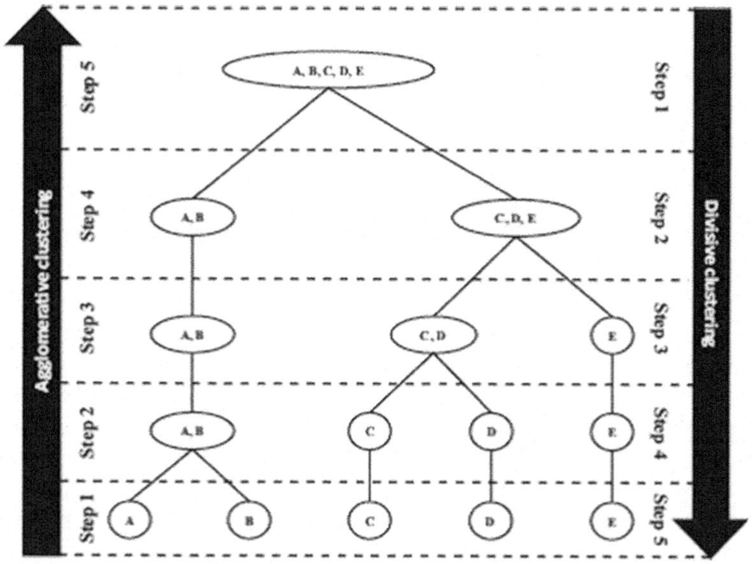

Fig. 9.4 Agglomerative and divisive clustering

Linkage algorithms

When using agglomerative hierarchical clustering, you need to specify a **linkage algorithm.** Linkage algorithms define the distance from a newly formed cluster to a certain object, or to other clusters in the solution. The most popular linkage algorithms include the following:

- **Single linkage** (nearest neighbor): The distance between two clusters corresponds to the shortest distance between any two members in the two clusters.
- **Complete linkage** (furthest neighbor): The oppositional approach to single linkage assumes that the distance between two clusters is based on the longest distance between any two members in the two clusters.
- **Average linkage**: The distance between two clusters is defined as the average distance between all pairs of the two clusters' members. **Weighted average linkage** performs the same calculation, but weights distances based on the number of objects in the cluster. Thus, the latter method is preferred when clusters are not of approximately equal size.
- **Centroid linkage**: In this approach, the geometric center (centroid) of each cluster is computed first. This is done by computing the clustering variables' average values of all the objects in a certain cluster. The distance between the two clusters equals the distance between the two centroids.
- **Ward's linkage**: This approach differs from the previous ones in that it does not combine the two closest or most similar objects successively. Instead, Ward's linkage combines those objects whose merger increases the overall within-

cluster variance (i.e., the homogeneity of clusters) to the smallest possible degree. The approach is generally used in combination with (squared) Euclidean distances, but can be used in combination with any other (dis)similarity measure.

Figures 9.5, 9.6, 9.7, 9.8 and 9.9 illustrate these linkage algorithms for two clusters, which are represented by white circles surrounding a set of objects. Each of these linkage algorithms can yield totally different results when used on the same dataset, as each has specific properties:

- The single linkage algorithm is based on minimum distances; it tends to form one large cluster with the other clusters containing only one or a few objects each. We can make use of this **chaining effect** to detect outliers, as these will be merged with the remaining objects—usually at very large distances—in the last steps of the analysis. Single linkage is considered the most versatile algorithm.
- The complete linkage method is strongly affected by outliers, as it is based on maximum distances. Clusters produced by this method are likely to be compact and tightly clustered.
- The average linkage and centroid linkage algorithms tend to produce clusters with low within-cluster variance and with similar sizes. The average linkage is affected by outliers, but less than the complete linkage method.
- Ward's linkage yields clusters of similar size with a similar degree of tightness. Prior research has shown that the approach generally performs very well. However, outliers and highly correlated variables have a strong bearing on the algorithm.

To better understand how the linkage algorithms work, let's manually examine some calculation steps using single linkage as an example. Let's start by looking at the distance matrix in Table 9.2, which shows the distances between objects A-G from our initial example. In this distance matrix, the non-diagonal elements express the distances between pairs of objects based on the Euclidean distance—we will discuss this distance measure in the following section. The diagonal elements of the matrix represent the distance from each object to itself, which is, of course, 0. In our example, the distance matrix is an 8×8 table with the lines and rows representing the objects under consideration (see Table 9.1). As the distance between objects B and C (in this case, 21.260 units) is the same as between C and B, the distance matrix is symmetrical. Furthermore, since the distance between an object and itself is 0, you only need to look at either the lower or upper non-diagonal elements.

In the very first step, the two objects exhibiting the smallest distance in the matrix are merged. Since the smallest distance occurs between B and C (d(B,C) = 21.260; printed in bold in Table 9.2), we merge these two objects in the first step of the analysis.

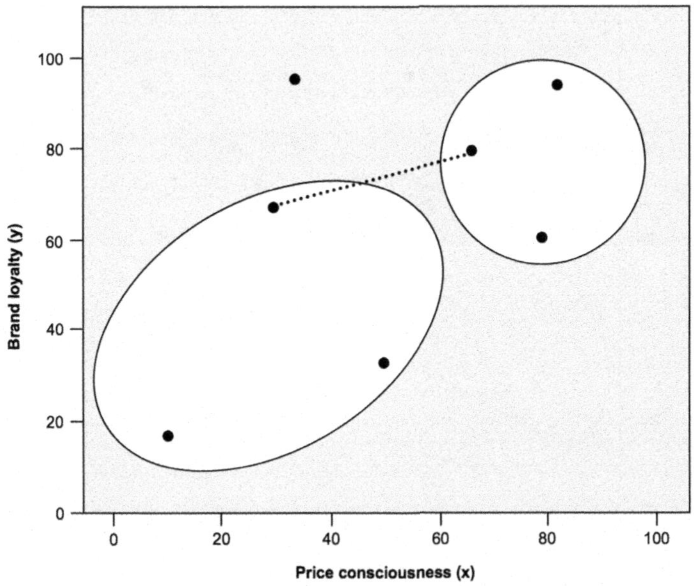

Fig. 9.5 Single linkage

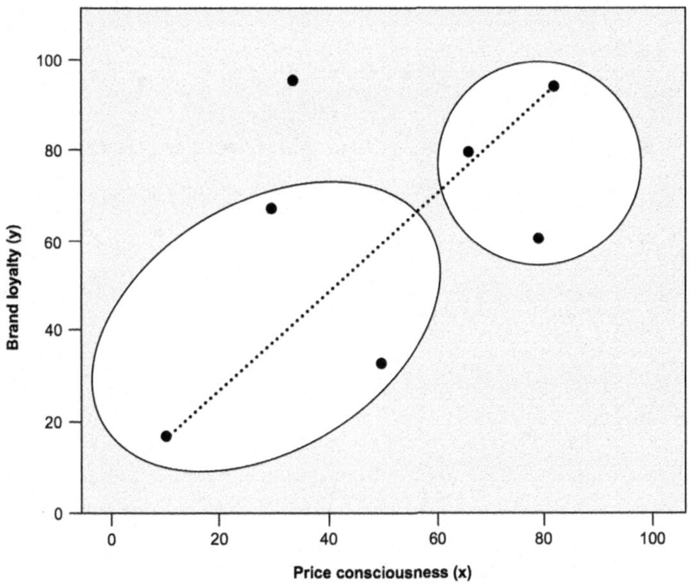

Fig. 9.6 Complete linkage

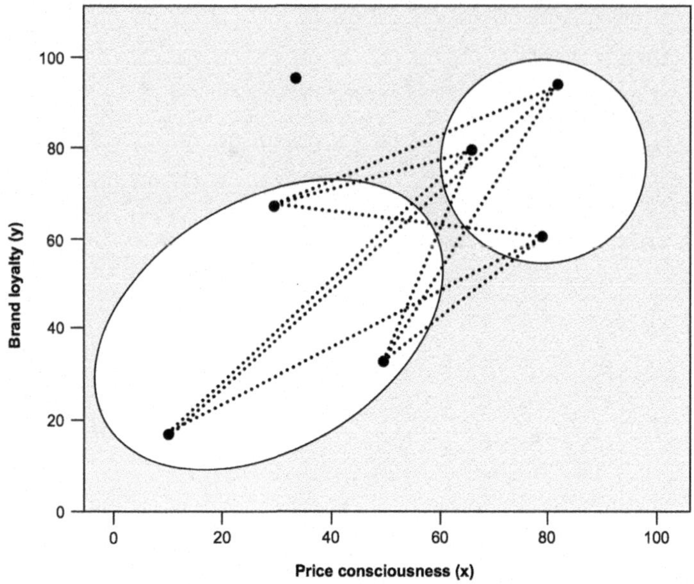

Fig. 9.7 Average linkage

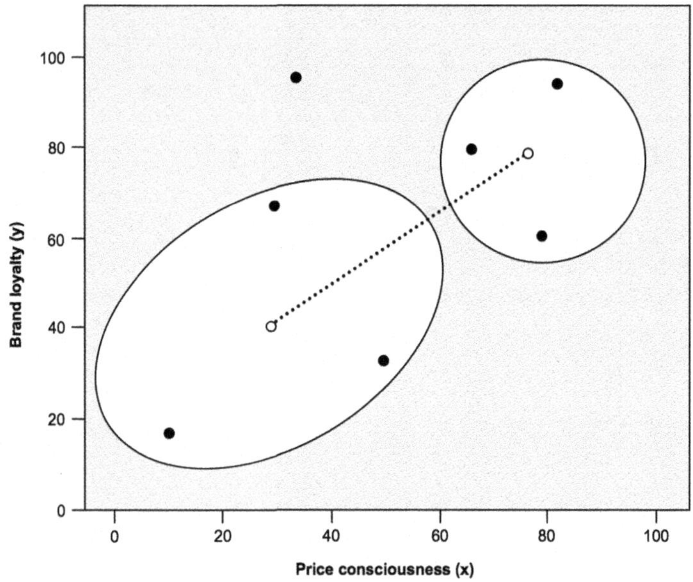

Fig. 9.8 Centroid linkage

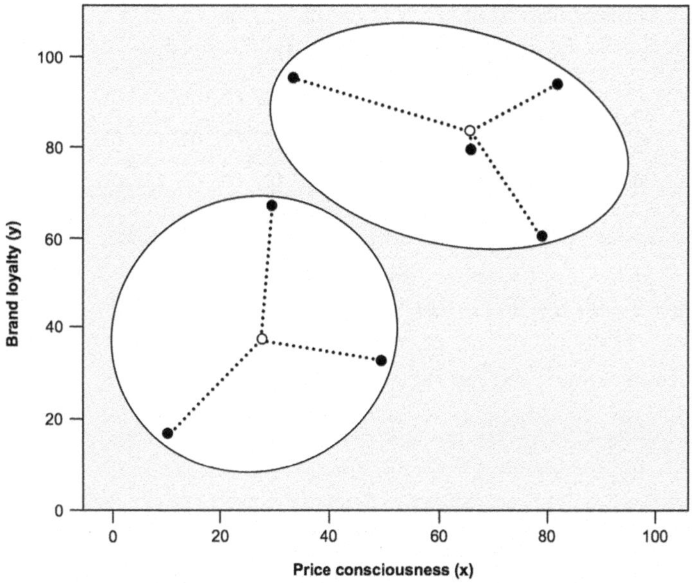

Fig. 9.9 Ward's linkage

Agglomerative clustering procedures always merge those objects with the smallest distance, regardless of the linkage algorithm used (e.g., single or complete linkage).

In the next step, we form a new distance matrix by considering the single linkage decision rule as discussed above. Using this linkage algorithm, we need to compute the distance from the newly formed cluster [B,C] (clusters are indicated by squared brackets) to all the other objects. For example, with regard to the distance from the cluster [B,C] to object A, we need to check whether A is closer to object B or to object C. That is, we look for the minimum value in d(A,B) and d(A,C) from Table 9.2. As d(A,C) = 36.249 is smaller than d(A,B) = 49.010, the distance from A to the newly formed cluster is equal to d(A,C); that is, 36.249. We also compute the distances from cluster [B,C] to all the other objects (i.e., D, E, F, G). For example, the distance between [B,C] and D is the minimum of d(B,D) = 58.592 and d(C,D) = 38.275 (Table 9.2). Finally, there are several distances, such as d(D,E) and d(E,F), which are not affected by the merger of B and C. These distances are simply copied into the new distance matrix. This yields the new distance matrix shown in Table 9.3.

Continuing the clustering procedure, we simply repeat the last step by merging the objects in the new distance matrix that exhibit the smallest distance and calculate the distance from this new cluster to all the other objects. In our case,

Table 9.2 Euclidean distance matrix

Objects	A	B	C	D	E	F	G
A	0						
B	49.010	0					
C	36.249	**21.260**	0				
D	28.160	58.592	38.275	0			
E	57.801	34.132	23.854	40.497	0		
F	64.288	68.884	49.649	39.446	39.623	0	
G	81.320	105.418	84.291	53.852	81.302	43.081	0

Note: Smallest distance is printed in bold

Table 9.3 Distance matrix after first clustering step (single linkage)

Objects	A	B, C	D	E	F	G
A	0					
B, C	36.249	0				
D	28.160	38.275	0			
E	57.801	**23.854**	40.497	0		
F	64.288	49.649	39.446	39.623	0	
G	81.320	84.291	53.852	81.302	43.081	0

Note: Smallest distance is printed in bold

Table 9.4 Distance matrix after second clustering step (single linkage)

Objects	A	B, C, E	D	F	G
A	0				
B, C, E	36.249	0			
D	**28.160**	38.275	0		
F	64.288	39.623	39.446	0	
G	81.320	81.302	53.852	43.081	0

Note: Smallest distance is printed in bold

Table 9.5 Distance matrix after third clustering step (single linkage)

Objects	A, D	B, C, E	F	G
A, D	0			
B, C, E	**36.249**	0		
F	39.446	39.623	0	
G	53.852	81.302	43.081	0

Note: Smallest distance is printed in bold

the smallest distance (23.854, printed in bold in Table 9.3) occurs between the newly formed cluster [B, C] and object E. The result of this step is described in Table 9.4.

Try to calculate the remaining steps yourself and compare your solution with the distance matrices in the following Tables 9.5, 9.6 and 9.7.

Table 9.6 Distance matrix after fourth clustering step (single linkage)

Objects	A, B, C, D, E	F	G
A, B, C, D, E	0		
F	**39.446**	0	
G	53.852	43.081	0

Note: Smallest distance is printed in bold

Table 9.7 Distance matrix after fifth clustering step (single linkage)

Objects	A, B, C, D, E, F	G
A, B, C, D, E, F	0	
G	43.081	0

By following the single linkage procedure, the last steps involve the merger of cluster [A,B,C,D,E,F] and object G at a distance of 43.081. Do you get the same results? As you can see, conducting a basic cluster analysis manually is not that hard at all—not if there are only a few objects.

9.3.2.2 Partitioning Methods: *k*-means

Partitioning clustering methods are another important group of procedures. As with hierarchical clustering, there is a wide array of different algorithms; of these, *k*-means is the most popular for market research.

Understanding *k*-means Clustering

The ***k*-means** method follows an entirely different concept than the hierarchical methods discussed above. The initialization of the analysis is one crucial difference. Unlike with hierarchical clustering, we need to specify the number of clusters to extract from the data prior to the analysis. Using this information as input, *k*-means then assigns all the objects to the number of clusters that the researcher specifies. This starting partition comes in different forms. Examples of these forms include:

- randomly select *k* objects as starting centers for the *k* clusters (*K unique random observations* in Stata),
- use the first or last *k* objects as starting centers for the *k* clusters (*First K observations* and *Last K observations* in Stata),
- randomly allocate all the objects into *k* groups and compute the means (or medians) of each group. These means (or medians) then serve as starting centers (*Group means from K random partitions of the data* in Stata), and
- provide an initial grouping variable that defines the groups among the objects to be clustered. The group means (or medians) of these groups are used as the starting centers (*Group means from partitions defined by initial grouping variables* in Stata).

After the initialization, *k*-means successively reassigns the objects to other clusters with the aim of minimizing the within-cluster variation. This within-cluster variation is equal to the squared distance of each observation to the center of the associated cluster (i.e., the centroid). If the reallocation of an object to another cluster decreases the within-cluster variation, this object is reassigned to that cluster.

Since cluster affiliations can change in the course of the clustering process (i.e., an object can move to another cluster in the course of the analysis), *k*-means does not build a hierarchy, which hierarchical clustering does (Fig. 9.4). Therefore, *k*-means belongs to the group of **non-hierarchical clustering methods**.

For a better understanding of the approach, let's take a look at how it works in practice. Figures 9.10, 9.11, 9.12 and 9.13 illustrate the four steps of the *k*-means clustering process—research has produced several variants of the original algorithm, which we briefly discuss in Box 9.2.

- **Step 1:** The researcher needs to specify the number of clusters that *k*-means should retain from the data. Using this number as the input, the algorithm selects a center for each cluster. In our example, two cluster centers are randomly initiated, which CC1 (first cluster) and CC2 (second cluster) represent in Fig. 9.10.
- **Step 2:** Euclidean distances are computed from the cluster centers to every object. Each object is then assigned to the cluster center with the shortest distance to it. In our example (Fig. 9.11), objects A, B, and C are assigned to the first cluster, whereas objects D, E, F, and G are assigned to the second. We now have our initial partitioning of the objects into two clusters.
- **Step 3:** Based on the initial partition in step 2, each cluster's geometric center (i.e., its centroid) is computed. This is done by computing the mean values of the objects contained in the cluster (e.g., A, B, C in the first cluster) in terms of each of the variables (price consciousness and brand loyalty). As we can see in Fig. 9.12, both clusters' centers now shift to new positions (CC1' in the first and CC2' in the second cluster; the inverted comma indicates that the cluster center has changed).
- **Step 4:** The distances are computed from each object to the newly located cluster centers and the objects are again assigned to a certain cluster on the basis of their minimum distance to other cluster centers (CC1' and CC2'). Since the cluster centers' position changed with respect to the initial situation in the first step, this could lead to a different cluster solution. This is also true of our example, because object E is now—unlike in the initial partition—closer to the first cluster center (CC1') than to the second (CC2'). Consequently, this object is now assigned to the first cluster (Fig. 9.13).

The *k*-means procedure is now repeated until a predetermined number of iterations are reached, or convergence is achieved (i.e., there is no change in the cluster affiliations).

Three aspects are worth noting in terms of using *k*-means:

- *k*-means is implicitly based on pairwise Euclidean distances, because the sum of the squared distances from the centroid is equal to the sum of the pairwise squared Euclidean distances divided by the number of objects. Therefore, the method should only be used with metric and, in case of equidistant scales, ordinal variables. Similarly, you should only use (squared) Euclidean distances with *k*-means.

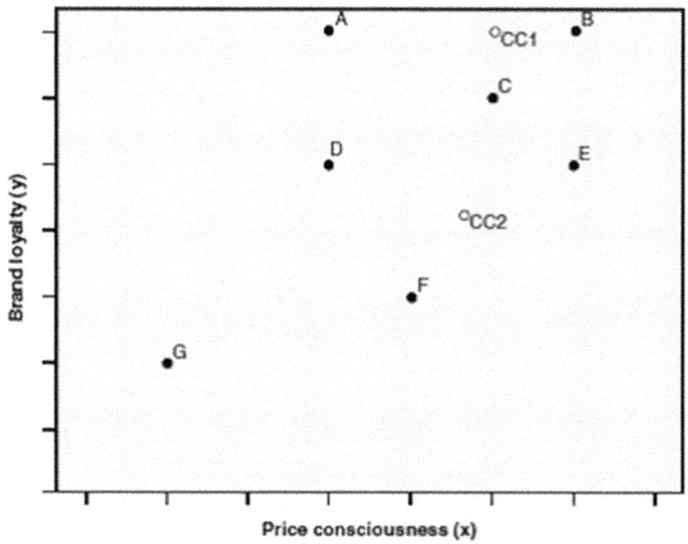

Fig. 9.10 *k*-means procedure (step 1: placing random cluster centers)

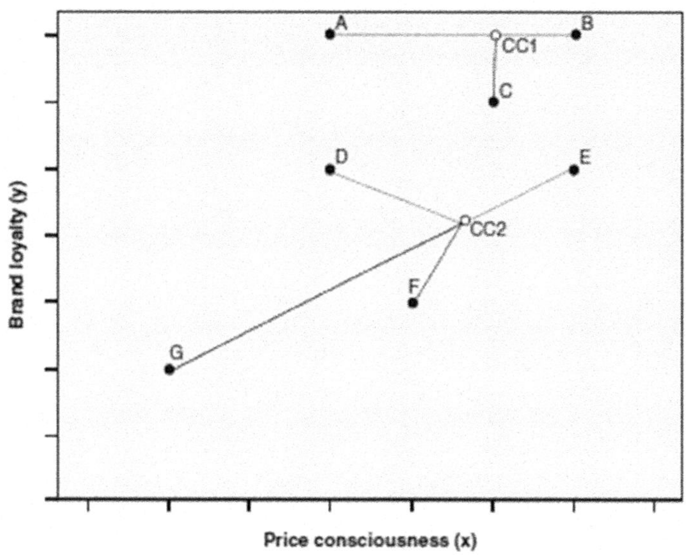

Fig. 9.11 *k*-means procedure (step 2: assigning objects to the closest cluster center)

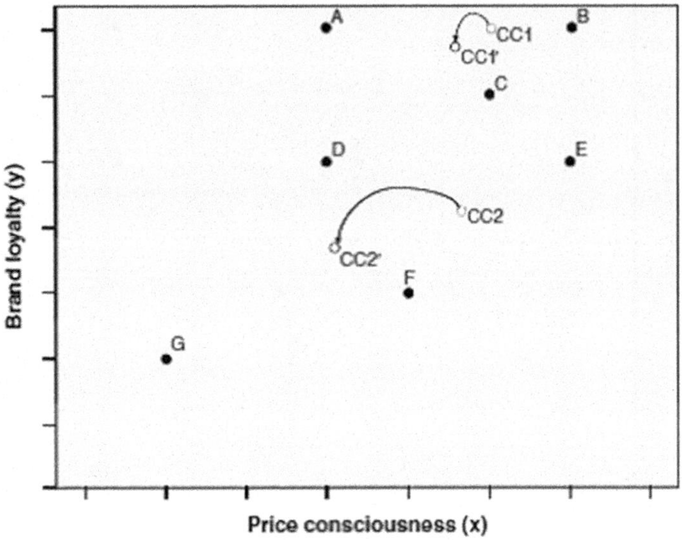

Fig. 9.12 *k*-means procedure (step 3: recomputing cluster centers)

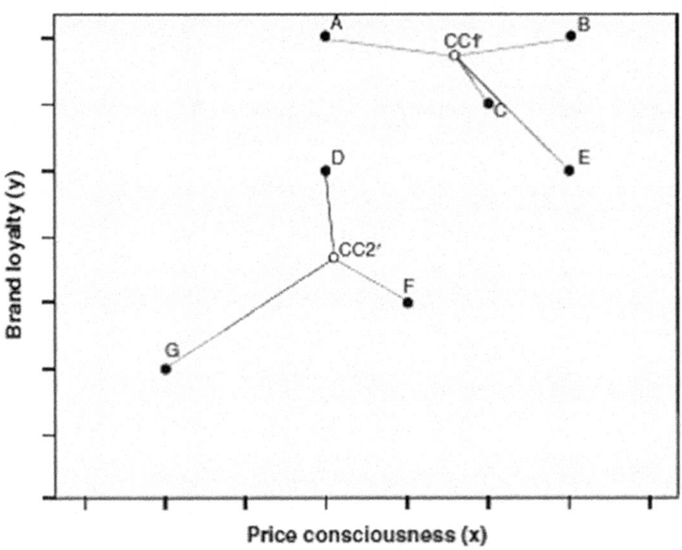

Fig. 9.13 *k*-means procedure (step 4: reassigning objects to the closest cluster center)

– Results produced by *k*-means depend on the starting partition. That is, *k*-means produce different results, depending on the starting partition chosen by the researcher or randomly initiated by the software. As a result, *k*-means may converge in a **local optimum**, which means that the solution is only optimal

compared to similar solutions, but not globally. Therefore, you should run *k*-means multiple times using different options for generating a starting partition.
- *k*-means is less computationally demanding than hierarchical clustering techniques. The method is therefore generally preferred for sample sizes above 500, and particularly for *big data* applications.
- Running *k*-means requires specifying the number of clusters to retain prior to running the analysis. We discuss this issue in the next section.

Box 9.2 Variants of the Original *k*-means Method
k*-medians** is a popular variant of *k*-means and has also been implemented in Stata. This procedure essentially follows the same logic and procedure as *k*-means. However, instead of using the cluster mean as a reference point for the calculation of the within cluster variance, *k*-medians minimizes the absolute deviations from the cluster medians, which equals the city-block distance. Thus, *k*-medians does *not* optimize the squared deviations from the mean as in *k*-means, but absolute distances. In this way, *k*-medians avoids the possible effect of extreme values on the cluster solution. Further variants, which are not menu-accessible in Stata, use other cluster centers (e.g., ***k-medoids; Kaufman and Rousseeuw 2005; Park and Jun 2009), or optimize the initialization process (e.g., ***k-means++***; Arthur and Vassilvitskii 2007).

9.3.3 Select a Measure of Similarity or Dissimilarity

In the previous section, we discussed different linkage algorithms used in agglomerative hierarchical clustering as well as the *k*-means procedure. All these clustering procedures rely on measures that express the (dis)similarity between pairs of objects. In the following section, we introduce different measures for metric, ordinal, nominal, and binary variables.

9.3.3.1 Metric and Ordinal Variables

Distance Measures

A straightforward way to assess two objects' proximity is by drawing a straight line between them. For example, on examining the scatter plot in Fig. 9.1, we can easily see that the length of the line connecting observations B and C is much shorter than the line connecting B and G. This type of distance is called **Euclidean distance** or **straight line distance**; it is the most commonly used type for analyzing metric variables and, if the scales are equidistant (Chap. 3), ordinal variables. Statistical software programs such as Stata simply refer to the Euclidean distance as *L2*, as it is a specific type of the more general Minkowski distance metric with argument 2 (Anderberg 1973). Researchers also often use the **squared Euclidean distance**, referred to as *L2 squared* in Stata. For *k*-means, using the squared Euclidean distance is more appropriate because of the way the method computes the distances from the objects to the centroids (see Section 9.3.2.2).

In order to use a hierarchical clustering procedure, we need to express these distances mathematically. Using the data from Table 9.1, we can compute the Euclidean distance between customer B and customer C (generally referred to as d(B,C)) by using variables x and y with the following formula:

$$d_{Euclidean}(B,C) = \sqrt{(x_B - x_C)^2 + (y_B - y_C)^2}$$

As can be seen, the Euclidean distance is the square root of the sum of the squared differences in the variables' values. Using the data from Table 9.1, we obtain the following:

$$d_{Euclidean}(B,C) = \sqrt{(82 - 66)^2 + (94 - 80)^2} = \sqrt{452} \approx 21.260$$

This distance corresponds to the length of the line that connects objects B and C. In this case, we only used two variables, but we can easily add more under the root sign in the formula. However, each additional variable will add a dimension to our research problem (e.g., with six clustering variables, we have to deal with six dimensions), making it impossible to represent the solution graphically. Similarly, we can compute the distance between customer B and G, which yields the following:

$$d_{Euclidean}(B,G) = \sqrt{(82 - 10)^2 + (94 - 17)^2} = \sqrt{11,113} \approx 105.418$$

Likewise, we can compute the distance between all other pairs of objects and summarize them in a distance matrix. Table 9.2, which we used as input to illustrate the single linkage algorithm, shows the Euclidean distance matrix for objects A-G.

There are also alternative distance measures: The **city-block distance** (called *L1* in Stata) uses the sum of the variables' absolute differences. This distance measure is referred to as the **Manhattan metric** as it is akin to the walking distance between two points in a city like New York's Manhattan district, where the distance equals the number of blocks in the directions North-South and East-West. Using the city-block distance to compute the distance between customers B and C (or C and B) yields the following:

$$d_{City-block}(B,C) = |x_B - x_C| + |y_B - y_C| = |82 - 66| + |94 - 80| = 30$$

The resulting distance matrix is shown in Table 9.8.

Lastly, when working with metric (or ordinal) data, researchers frequently use the **Chebychev distance** (called *Linfinity* in Stata), which is the maximum of the

Table 9.8 City-block distance matrix

Objects	A	B	C	D	E	F	G
A	0						
B	50	0					
C	48	30	0				
D	31	79	49	0			
E	81	37	33	56	0		
F	79	93	63	54	56	0	
G	101	149	119	70	112	56	0

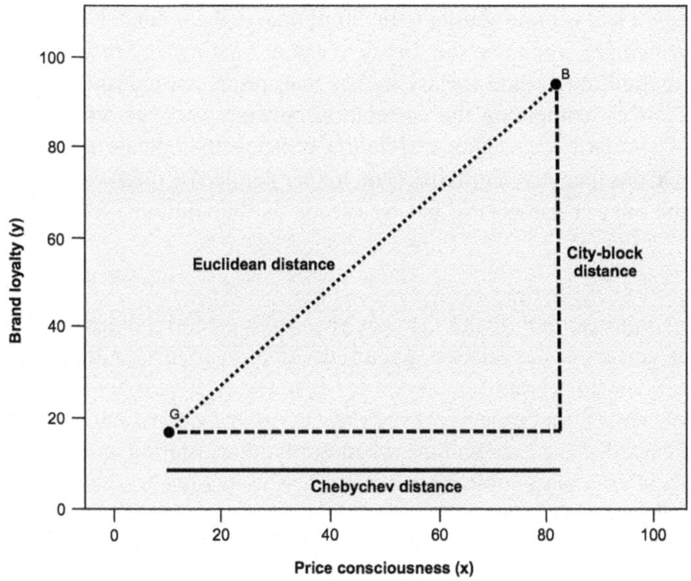

Fig. 9.14 Distance measures

absolute difference in the clustering variables' values. In respect of customers B and C, this result is:

$$d_{Chebychev}(B, C) = \max(|x_B - x_C|, |y_B - y_C|) = \max(|82 - 66|, |94 - 80|) = 16$$

Figure 9.14 illustrates the interrelation between these three distance measures regarding two objects (here: B and G) from our example.

Research has brought forward a range of other distance measures suitable for specific research settings. For example, the Stata menu offers the **Canberra distance**, a weighted version of the city-block distance, which is typically used for clustering data scattered widely around an origin. Other distance measures, such as the **Mahalanobis distance**, which compensates for collinearity between the clustering variables, are accessible via Stata syntax.

Different distance measures typically lead to different cluster solutions. Thus, it is advisable to use several measures, check for the stability of results, and compare them with theoretical or known patterns.

Association Measures

The (dis)similarity between objects can also be expressed by means of *association measures* (e.g., correlations). For example, suppose a respondent rated price consciousness 2 and brand loyalty 3, a second respondent indicated 5 and 6, whereas a third rated these variables 3 and 3. Euclidean and city-block, distances indicate that the first respondent is more similar to the third than to the second. Nevertheless, one could convincingly argue that the first respondent's ratings are more similar to the second's, as both rate brand loyalty higher than price consciousness. This can be accounted for by computing the correlation between two vectors of values as a measure of similarity (i.e., high correlation coefficients indicate a high degree of similarity). Consequently, similarity is no longer defined by means of the difference between the answer categories, but by means of the similarity of the answering profiles.

Whether you use one of the distance measures or correlations depends on whether you think the relative magnitude of the variables within an object (which favors correlation) matters more than the relative magnitude of each variable across the objects (which favors distance). Some researchers recommended using correlations when applying clustering procedures that are particularly susceptible to outliers, such as complete linkage, average linkage or centroid linkage. Furthermore, correlations implicitly standardize the data, as differences in the scale categories do not have a strong bearing on the interpretation of the response patterns. Nevertheless, distance measures are most commonly used for their intuitive interpretation. Distance measures best represent the concept of proximity, which is fundamental to cluster analysis. Correlations, although having widespread application in other techniques, represent patterns rather than proximity.

Standardizing the Data

In many analysis tasks, the variables under consideration are measured in different units with hugely different variance. This would be the case if we extended our set of clustering variables by adding another metric variable representing the customers' gross annual income. Since the absolute variation of the income variable would be much higher than the variation of the remaining two variables (remember, x and y are measured on a scale from 0 to 100), this would clearly distort our analysis results. We can resolve this problem by standardizing the data prior to the analysis (Chap. 5).

Different standardization methods are available, such as z-standardization, which rescales each variable to a mean of 0 and a standard deviation of 1 (Chap. 5). In cluster analysis, however, *range standardization* (e.g., to a range of 0 to 1) typically works better (Milligan and Cooper 1988).

9.3.3.2 Binary and Nominal Variables

Whereas the distance measures presented thus far can be used for variables measured on a metric and, in general, on an ordinal scale, applying them to binary and nominal variables is problematic. When nominal variables are involved, you should rather select a similarity measure expressing the degree to which the variables' values share the same category. These **matching coefficients** can take different forms, but rely on the same allocation scheme as shown in Table 9.9. In this crosstab, cell a is the number of characteristics present in both objects, whereas cell d describes the number of characteristics absent in both objects. Cells b and c describe the number of characteristics present in one, but not the other, object (see Table 9.10 for an example).

The allocation scheme in Table 9.9 applies to binary variables (i.e., nominal variables with two categories). For nominal variables with more than two categories, you need to convert the categorical variable into a set of binary variables in order to use matching coefficients. For example, a variable with three categories needs to be transformed into three binary variables, one for each category (see the following example).

Based on the allocation scheme in Table 9.9, we can compute different matching coefficients, such as the **simple matching (SM) coefficient** (called *Matching* in Stata):

Table 9.9 Allocation scheme for matching coefficients

		Second object	
		Presence of a characteristics (1)	Absence of a characteristic (0)
First object	Presence of a characteristic (1)	a	b
	Absence of a characteristic (0)	c	d

Table 9.10 Recoded measurement data

Object	*Gender* (binary)		*Customer* (binary)		*Country of residence* (binary)		
	Male	Female	Yes	No	GER	UK	USA
A	1	0	1	0	1	0	0
B	1	0	0	1	0	0	1
C	0	1	0	1	0	0	1

$$SM = \frac{a + d}{a + b + c + d}$$

This coefficient takes both the joint presence and the joint absence of a characteristic (as indicated by cells *a* and *d* in Table 9.9) into account. This feature makes the simple matching coefficient particularly useful for symmetric variables where the joint presence and absence of a characteristic carry an equal degree of information. For example, the binary variable *gender* has the possible states "male" and "female." Both are equally valuable and carry the same weight when the simple matching coefficient is computed. However, when the outcomes of a binary variable are not equally important (i.e., the variable is asymmetric), the simple matching coefficient proves problematic. An example of an asymmetric variable is the presence, or absence, of a relatively rare attribute, such as customer complaints. While you say that two customers who complained have something in common, you cannot say that customers who did not complain have something in common. The most important outcome is usually coded as 1 (present) and the other is coded as 0 (absent). The agreement of two 1s (i.e., a positive match) is more significant than the agreement of two 0s (i.e., a negative match). Similarly, the simple matching coefficient proves problematic when used on nominal variables with many categories. In this case, objects may appear very similar, because they have many negative matches rather than positive matches.

Given this issue, researchers have proposed several other matching coefficients, such as the **Jaccard coefficient (JC)** and the **Russell and Rao coefficient** (**RR**, called *Russell* in Stata), which (partially) omit the *d* cell from the calculation. Like the simple matching coefficient, these coefficients range from 0 to 1 with higher values indicating a greater degree of similarity.[4] They are defined as follows:

$$JC = \frac{a}{a + b + c}$$

$$RR = \frac{a}{a + b + c + d}$$

To provide an example that compares the three coefficients, consider the following three variables:

- *gender*: male, female
- *customer*: yes, no
- *country of residence*: GER, UK, USA

[4]There are many other matching coefficients, such as Yule's Q, Kulczynski, or Ochiai, which are also menu-accessible in Stata. However, since most applications of cluster analysis rely on metric or ordinal data, we will not discuss these. See Wedel and Kamakura (2000) for more information on alternative matching coefficients.

We first transform the measurement data into binary data by recoding the original three variables into seven binary variables (i.e., two for *gender* and *customer*; three for *country of residence*). Table 9.10 shows a binary data matrix for three objects A, B, and C. Object A is a male customer from Germany; object B is a male non-customer from the United States; object C is a female non-customer, also from the United States.

Using the allocation scheme from Table 9.9 to compare objects A and B yields the following results for the cells: $a = 1$, $b = 2$, $c=2$, and $d = 2$.

This means that the two objects have only one shared characteristic ($a = 1$), but two characteristics, which are absent from both objects ($d = 2$). Using this information, we can now compute the three coefficients described earlier:

$$SM(A, B) = \frac{1+2}{1+2+2+2} = 0.571,$$

$$JC(A, B) = \frac{1}{1+2+2} = 0.2, \text{ and}$$

$$RR(A, B) = \frac{1}{1+2+2+2} = 0.143$$

As can be seen, the simple matching coefficient suggests that objects A and B are reasonably similar. Conversely, the Jaccard coefficient, and particularly the Russel Rao coefficient, suggests that they are not.

Try computing the distances between the other object pairs. Your computation should yield the following: *SM(A,C)* = 0.143, *SM(B,C)* = 0.714, *JC(A,C)* = 0, *JC (B,C)* = 0.5, *RR(A,C)* = 0, and *RR(B,C)* = 0.286.

9.3.3.3 Mixed Variables

Most datasets contain variables that are measured on multiple scales. For example, a market research questionnaire may require the respondent's gender, income category, and age. We therefore have to consider variables measured on a nominal, ordinal, and metric scale. How can we simultaneously incorporate these variables into an analysis?

A common approach is to dichotomize all the variables and apply the matching coefficients discussed above. For metric variables, this involves specifying categories (e.g., low, medium, and high age) and converting these into sets of binary variables. In most cases, the specification of categories is somewhat arbitrary. Furthermore, this procedure leads to a severe loss in precision, as we disregard more detailed information on each object. For example, we lose precise information on each respondent's age when scaling this variable down into age categories.

Gower (1971) introduced a dissimilarity coefficient that works with a mix of binary and continuous variablesa. **Gower's dissimilarity coefficient** is a composite measure that combines several measures into one, depending on each variable's scale level. If binary variables are used, the coefficient takes the value 1 when two

Table 9.11 Recoded measurement data

	Gender (binary)		*Customer* (binary)			
Object	Male	Female	Yes	No	*Income category* (ordinal)	*Age* (metric)
A	1	0	1	0	2	21
B	1	0	0	1	3	37
C	0	1	0	1	1	29

objects do not share a certain characteristic (cells *b* and *c* in Table 9.9), and 0 else (cells *a* and *d* in Table 9.9). Thus, when all the variables are binary and symmetric, Gower's dissimilarity coefficient reduces to the simple matching coefficient when expressed as a distance measure instead of a similarity measure (i.e., 1 – *SM*). If binary and asymmetric variables are used, Gower's dissimilarity coefficient equals the Jaccard coefficient when expressed as a distance measure instead of a similarity measure (i.e., 1 – *JC*). If continuous variables are used, the coefficient is equal to the city-block distance divided by each variable's range. Ordinal variables are treated as if they were continuous, which is fine when the scale is equidistant (see Chap. 3). Gower's dissimilarity coefficient welds the measures used for binary and continuous variables into one value that is an overall measure of dissimilarity.

To illustrate Gower's dissimilarity coefficient, consider the following example with the two binary variables *gender* and *customer*, the ordinal variable *income category* (1 = "low", 2 = "medium", 3 = "high"), and the metric variable *age*. Table 9.11 shows the data for three objects A, B, and C.

To compute Gower's dissimilarity coefficient for objects A and B, we first consider the variable *gender*. Since both objects A and B are male, they share two characteristics (male = "yes", female = "no"), which entails a distance of 0 for both variable levels. With regard to the *customer* variable, the two objects have different characteristics, hence a distance of 1 for each variable level. The ordinal variable income category is treated as continuous, using the city-block distance (here: |2–3|) divided by the variable's range (here: 3 – 1). Finally, the distance with regard to the *age* variable is $|21-37|/(37-21)=1$. Hence, the resulting Gower distance is:

$$d_{Gower}(A,B) = \frac{1}{6}(0+0+1+1+0.5+1) \approx 0.583$$

Computing the Gower distance between the other two object pairs yields $d_{Gower}(A,C) \approx 0.833$, and $d_{Gower}(B,C) \approx 0.583$.

9.3.4 Decide on the Number of Clusters

An important question we haven't yet addressed is how to decide on the number of clusters. A misspecified number of clusters results in under- or oversegmentation, which easily leads to inaccurate management decisions on, for example, customer targeting, product positioning, or determining the optimal marketing mix (Becker et al. 2015).

We can select the number of clusters pragmatically, choosing a grouping that "works" for our analysis, but sometimes we want to select the "best" solution that the data suggest. However, different clustering methods require different approaches to decide on the number of clusters. Hence, we discuss hierarchical and portioning methods separately.

9.3.4.1 Hierarchical Methods: Deciding on the Number of Clusters

To guide this decision, we can draw on the distances at which the objects were combined. More precisely, we can seek a solution in which an additional combination of clusters or objects would occur at a greatly increased distance. This raises the issue of what a great distance is.

We can seek an answer by plotting the distance level at which the mergers of objects and clusters occur by using a **dendrogram**. Figure 9.15 shows the dendrogram for our example as produced by Stata. We read the dendrogram from the bottom to the top. The horizontal lines indicate the distances at which the objects were merged. For example, according to our calculations above, objects B and C were merged at a distance of 21.260. In the dendrogram, the horizontal line linking the two vertical lines that go from B and C indicates this merger. To decide on the number of clusters, we cut the dendrogram horizontally in the area where no merger has occurred for a long distance. In our example, this is done when moving from a four-cluster solution, which occurs at a distance of 28.160 (Table 9.4), to a three-cluster solution, which occurs at a distance of 36.249 (Table 9.5). This result suggests a four-cluster solution [A,D], [B,C,E], [F], and [G], but this conclusion is not clear-cut. In fact, the dendrogram often does not provide a clear indication, because it is generally difficult to identify where the cut should be made. This is particularly true of large sample sizes when the dendrogram becomes unwieldy.

Research has produced several other criteria for determining the number of clusters in a dataset (referred to as *stopping rules* in Stata).[5] One of the most prominent criteria is Calinski and Harabasz's (1974) **variance ratio criterion** (**VRC**; also called *Calinski-Harabasz pseudo-F* in Stata). For a solution with n objects and k clusters, the VRC is defined as:

$$VRC_k = (SS_B/(k-1))/(SS_W/(n-k)),$$

where SS_B is the sum of the squares between the clusters and SS_W is the sum of the squares within the clusters. The criterion should seem familiar, as it is similar to the F-value of a one-way ANOVA (see Chap. 6). To determine the appropriate number of clusters, you should choose the number that maximizes the VRC. However, as the VRC usually decreases with a greater number of clusters, you should compute

[5]For details on the implementation of these stopping rules in Stata, see Halpin (2016).

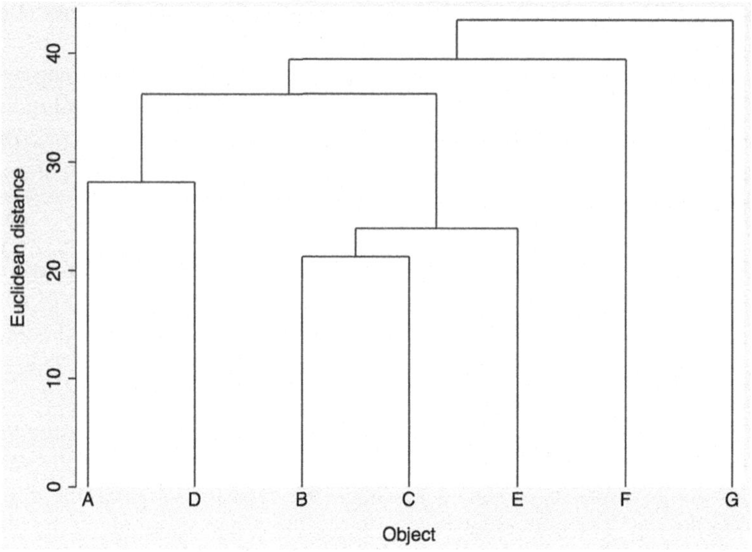

Fig. 9.15 Dendrogram

the difference in the VRC values ω_k of each cluster solution, using the following formula:[6]

$$\omega_k = (VRC_{k+1} - VRC_k) - (VRC_k - VRC_{k-1}).$$

The number of clusters k that minimizes the value in ω_k indicates the best cluster solution. Prior research has shown that the VRC reliably identifies the correct number of clusters across a broad range of constellations (Miligan and Cooper 1985). However, owing to the term VRC_{k-1}, which is not defined for a one-cluster solution, the minimum number of clusters that can be selected is three, which is a disadvantage when using the ω_k statistic.

Another criterion, which works well for determining the number of clusters (see Miligan and Cooper 1985) is the **Duda-Hart index** (Duda and Hart 1973). This index essentially performs the same calculation as the VRC, but compares the SS_W values in a pair of clusters to be split both before and after this split. More precisely, the Duda-Hart index is the SS_W in the two clusters (*Je(2)*) divided by the SS_W in one cluster (*Je(1)*); that is:

$$Duda - Hart = \frac{Je(2)}{Je(1)}$$

[6]In the ⭳ Web Appendix (→Downloads), we offer a Stata.ado file to calculate the ω_k called chomega.ado. We also offer an Excel sheet (*VRC.xlsx*) to calculate the ω_k manually.

To determine the number of clusters, you should choose the solution, which *maximizes* the Je(2)/Je(1) index value.

Duda et al. (2001) have also proposed a modified version of the index, which is called the *pseudo T-squared*. This index takes the number of observations in both groups into account. Contrary to the Duda-Hart index, you should choose the number of clusters that *minimizes* the pseudo T-squared.

Two aspects are important when using the Duda-Hart indices:

- The indices are not appropriate in combination with single linkage clustering, as chaining effects may occur. In this case, both indices will produce ambiguous results, as evidenced in highly similar values for different cluster solutions (Everitt and Rabe-Hesketh 2006).
- The indices are considered "local" in that they do not consider the entire data structure in their computation, but only the SS_W in the group being split. With regard to our example above, the Je(2)/Je(1) index for a two-cluster solution would only consider the variation in objects A to F, but not G. This characteristic makes the Duda-Hart indices somewhat inferior to the VRC, which takes the entire variation into account (i.e., the criterion is "global").

In practice, you should combine the VRC and the Duda-Hart indices by selecting the number of clusters that yields a large VRC, a large Je(2)/Je (1) index, and a small pseudo T-squared value. These values do not necessarily have to be the maximum or minimum values. Note that the VRC and Duda-Hart indices become less informative as the number of objects in the clusters becomes smaller.

Overall, the above criteria can often only provide rough guidance regarding the number of clusters that should be selected; consequently, you should instead take practical considerations into account. Occasionally, you might have a priori knowledge, or a theory on which you can base your choice. However, first and foremost, you should ensure that your results are interpretable and meaningful. Not only must the number of clusters be small enough to ensure manageability, but each segment should also be large enough to warrant strategic attention.

9.3.4.2 Partitioning Methods: Deciding on the Number of Clusters

When running partitioning methods, such as *k*-means, you have to pre-specify the number of clusters to retain from the data. There are varying ways of guiding this decision:

- Compute the VRC (see discussion in the context of hierarchical clustering) for an alternating number of clusters and select the solution that maximizes the VRC or minimizes ω_k. For example, compute the VRC for a three- to five-cluster

solution and select the number of clusters that minimizes ω_k. Note that the Duda-Hart indices are not applicable as they require a hierarchy of objects and mergers, which partitioning methods do not produce.
- Run a hierarchical procedure to determine the number of clusters by using the dendrogram and run k-means afterwards.[7] This approach also enables you to find starting values for the initial cluster centers to handle a second problem, which relates to the procedure's sensitivity to the initial classification (we will follow this approach in the example application).
- Rely on prior information, such as earlier research findings.

9.3.5 Validate and Interpret the Clustering Solution

Before interpreting the cluster solution, we need to assess the stability of the results. Stability means that the cluster membership of individuals does not change, or only changes a little when different clustering methods are used to cluster the objects. Thus, when different methods produce similar results, we claim stability.

The aim of any cluster analysis is to differentiate well between the objects. The identified clusters should therefore differ substantially from each other and the members of different clusters should respond differently to different marketing-mix elements and programs.

Lastly, we need to profile the cluster solution by using observable variables. **Profiling** ensures that we can easily assign new objects to clusters based on observable traits. For example, we could identify clusters based on loyalty to a product, but in order to use these different clusters, their membership should be identifiable according to tangible variables, such as income, location, or family size, in order to be actionable.

The key to successful segmentation is to critically revisit the results of different cluster analysis set-ups (e.g., by using different algorithms on the same data) in terms of managerial relevance. The following criteria help identify a clustering solution (Kotler and Keller 2015; Tonks 2009).

- *Substantial*: The clusters are large and sufficiently profitable to serve.
- *Reliable*: Only clusters that are stable over time can provide the necessary basis for a successful marketing strategy. If clusters change their composition quickly, or their members' behavior, targeting strategies are not likely to succeed. Therefore, a certain degree of stability is necessary to ensure that marketing strategies can be implemented and produce adequate results. Reliability can be evaluated by critically revisiting and replicating the clustering results at a later date.
- *Accessible*: The clusters can be effectively reached and served.

[7]See Punj and Stewart (1983) for additional information on this sequential approach.

- *Actionable*: Effective programs can be formulated to attract and serve the clusters.
- *Parsimonious*: To be managerially meaningful, only a small set of substantial clusters should be identified.
- *Familiar*: To ensure management acceptance, the cluster composition should be easy to relate to.
- *Relevant*: Clusters should be relevant in respect of the company's competencies and objectives.

9.3.5.1 Stability

Stability is evaluated by using different clustering procedures on the same data and considering the differences that occur. For example, you may first run a hierarchical clustering procedure, followed by *k*-means clustering to check whether the cluster affiliations of the objects change. Alternatively, running a hierarchical clustering procedure, you can use different distance measures and evaluate their effect on the stability of the results. However, note that it is common for results to change even when your solution is adequate. As a rule of thumb, if more than 20% of the cluster affiliations change from one technique to the other, you should reconsider the analysis and use, for example, a different set of clustering variables, or reconsider the number of clusters. Note, however, that this percentage is likely to increase with the number of clusters used.

When the data matrix exhibits identical values (referred to as **ties**), the ordering of the objects in the dataset can influence the results of the hierarchical clustering procedure. For example, the distance matrix based on the city-block distance in Table 9.8 shows the distance of 56 for object pairs (D,E), (E,F), and (F,G). Ties can prove problematic when they occur for the minimum distance in a distance matrix, as the decision about which objects to merge then becomes ambiguous (i.e., should we merge objects D and E, E and F, or F and G if 56 was the smallest distance in the matrix?). To handle this problem, van der Kloot et al. (2005) recommend re-running the analysis with a different input order of the data. The downside of this approach is that the labels of a cluster may change from one analysis to the next. This issue is referred to as *label switching*. For example, in the first analysis, cluster 1 may correspond to cluster 2 in the second analysis. Ties are, however, more the exception than the rule in practical applications—especially when using (squared) Euclidean distances—and generally don't have a pronounced impact on the results. However, if changing the order of the objects also drastically changes the cluster compositions (e.g., in terms of cluster sizes), you should reconsider the set-up of the analysis and, for example, re-run it with different clustering variables.

9.3.5.2 Differentiation of the Data

To examine whether the final partition differentiates the data well, we need to examine the cluster centroids. This step is highly important, as the analysis sheds light on whether the clusters are truly distinct. Only if objects across two (or more) clusters exhibit significantly different means in the clustering variables (or any other relevant variable) can they be distinguished from each other. This can be easily ascertained by comparing the means of the clustering variables across the clusters with independent *t*-tests or ANOVA (see Chap. 6).

Furthermore, we need to assess the solution's criterion validity. We do this by focusing on the criterion variables that have a theoretical relationship with the clustering variables, but were not included in the analysis. In market research, criterion variables are usually managerial outcomes, such as the sales per person, or willingness-to-pay. If these criterion variables differ significantly, we can conclude that the clusters are distinct groups with criterion validity.

9.3.5.3 Profiling

As indicated at the beginning of the chapter, cluster analysis usually builds on unobservable clustering variables. This creates an important problem when working with the final solution: How can we decide to which cluster a new object should be assigned if its unobservable characteristics, such as personality traits, personal values, or lifestyles, are unknown? We could survey these attributes and make a decision based on the clustering variables. However, this is costly and researchers therefore usually try to identify observable variables (e.g., demographics) that best mirror the partition of the objects. More precisely, these observable variables should partition the data into similar groups as the clustering variables do. Using these observable variables, it is then easy to assign a new object (whose cluster membership is unknown) to a certain cluster. For example, assume that we used a set of questions to assess the respondents' values and learned that a certain cluster comprises respondents who appreciate self-fulfillment, enjoyment of life, and a sense of accomplishment, whereas this is not the case in another cluster. If we were able to identify explanatory variables, such as gender or age, which distinguish these clusters adequately, then we could assign a new person to a specific cluster on the basis of these observable variables whose value traits may still be unknown.

9.3.5.4 Interpret the Clustering Solution

The interpretation of the solution requires characterizing each cluster by using the criterion or other variables (in most cases, demographics). This characterization should focus on criterion variables that convey why the cluster solution is relevant. For example, you could highlight that customers in one cluster have a lower willingness to pay and are satisfied with lower service levels, whereas customers in another cluster are willing to pay more for a superior service. By using this information, we can also try to find a meaningful name or label for each cluster; that is, one that adequately reflects the objects in the cluster. This is usually a challenging task, especially when unobservable variables are involved.

> While companies develop their own market segments, they frequently use standardized segments, based on established buying trends, habits, and customers' needs to position their products in different markets. The

(continued)

PRIZM lifestyle by Nielsen is one of the most popular segmentation databases. It combines demographic, consumer behavior, and geographic data to help marketers identify, understand, and reach their customers and prospective customers. PRIZM defines every US household in terms of more than 60 distinct segments to help marketers discern these consumers' likes, dislikes, lifestyles, and purchase behaviors.

An example is the segment labeled "Connected Bohemians," which Nielsen characterizes as a "collection of mobile urbanites, Connected Bohemians represent the nation's most liberal lifestyles. Its residents are a progressive mix of tech savvy, young singles, couples, and families ranging from students to professionals. In their funky row houses and apartments, Bohemian Mixers are the early adopters who are quick to check out the latest movie, nightclub, laptop, and microbrew." Members of this segment are between 25 and 44 years old, have a midscale income, own a hybrid vehicle, eat at Starbucks, and go skiing/snowboarding. (http://www.MyBestSegments.com).

Table 9.12 summarizes the steps involved in a hierarchical and *k*-means clustering when using Stata. The syntax code shown in the cells comes from the case study, which we introduce in the following section.

Table 9.12 Steps involved in carrying out a cluster analysis in Stata

Theory	Action
Research problem	
Identification of homogenous groups of objects in a population	
Select clustering variables to form segments	Select relevant variables that potentially exhibit high degrees of criterion validity with regard to a specific managerial objective.
Requirements	
Sufficient sample size	Make sure that the relationship between the objects and the clustering variables is reasonable. Ten times the number of clustering variables is the bare minimum, but 30 to 70 times is recommended. Ensure that the sample size is large enough to guarantee substantial segments.
Low levels of collinearity among the variables	► Statistics ► Summaries, tables and tests ► Summary and descriptive statistics ► Pairwise correlations
	`pwcorr e1 e5 e9 e21 e22`
	In case of highly correlated variables (correlation coefficients > 0.90), delete one variable of the offending pair.
Specification	
Choose the clustering procedure	If there is a limited number of objects in your dataset, rather use hierarchical clustering:
	► Statistics ► Multivariate analysis ► Cluster analysis ► Cluster Data ► Choose a linkage algorithm
	`cluster wardslinkage e1 e5 e9 e21 e22, measure (L2squared) name(wards_linkage)`

(continued)

Table 9.12 (continued)

Theory	Action
	If there are many observations (> 500) in your dataset, rather use *k*-means clustering: ▶ Statistics Multivariate analysis ▶ Cluster analysis ▶ Cluster Data ▶ kmeans
	`cluster kmeans e1 e5 e9 e21 e22, k(2) measure (L2squared) start(krandom) name(kmeans)`
Select a measure of (dis) similarity	*Hierarchical methods*:
	Select from the (dis)similarity measure menu, depending on the clustering variables' scale level.
	Depending on the scale level, select the measure; convert variables with multiple categories into a set of binary variables and use matching coefficients; Choose Gower's dissimilarity coefficient for mixed variables.
	When the variables are measured on different units, standardize the variables to a range from 0 to 1 prior to the analysis, using the following commands:
	`summarize e1`
	`return list`
	`gen e1_rsdt =.`
	`replace e1_rsdt = (e1- r(min)) / (r(max)-r (min))`
	Partitioning methods:
	Use the squared Euclidean distance from the (dis) similarity menu.
Deciding on the number of clusters	*Hierarchical clustering:*
	Examine the dendrogram:
	▶ Statistics ▶ Multivariate analysis ▶ Cluster analysis ▶ Postclustering ▶ Dendrogram
	`cluster dendrogram wards_linkage, cutnumber (10) showcount`
	Examine the VRC and Duda-Hart indices:
	▶ Statistics Multivariate analysis ▶ Cluster analysis ▶ Postclustering ▶ Cluster analysis stopping rules.
	For VRC: `cluster stop wards_linkage, rule (calinski) groups(2/11)`
	For Duda-Hart: `cluster stop wards_linkage, rule (duda) groups(1/10)`
	Include practical considerations in your decision.
	Partitioning methods:
	Run a hierarchical cluster analysis and decide on the number of segments based on a dendrogram, the VRC, and the Duda-Hart indices; use the resulting partition as starting partition.
	▶ Statistics Multivariate analysis ▶ Cluster analysis ▶ Postclustering ▶ Cluster analysis stopping rules.
	`cluster kmeans e1 e5 e9 e21 e22, k(3) measure (L2squared) name(kmeans) start(group (cluster_wl))`
	Include practical considerations in your decision.

(continued)

Table 9.12 (continued)

Theory	Action
Validating and interpreting the cluster solution	
Stability	Re-run the analysis using different clustering procedures, linkage algorithms or distance measures. For example, generate a cluster membership variable and use this grouping as starting partition for *k*-means clustering.
	`cluster generate cluster_wl = groups(3), name (wards_linkage) ties(error)`
	`cluster kmeans e1 e5 e9 e21 e22, k(3) measure (L2squared) name (kmeans) start(group (cluster_wl))`
	Examine the overlap in the clustering solutions. If more than 20% of the cluster affiliations change from one technique to the other, you should reconsider the set-up.
	`tabulate cluster_wl kmeans`
	Change the order of objects in the dataset (hierarchical clustering only).
Differentiation of the data	Compare the cluster centroids across the different clusters for significant differences.
	`mean e1 e5 e9 e21 e22, over(cluster_wl)`
	If possible, assess the solution's criterion validity.
Profiling	Identify observable variables (e.g., demographics) that best mirror the partition of the objects based on the clustering variables.
	`tabulate cluster_wl flight_purpose, chi2 V`
Interpretating of the cluster solution	Identify names or labels for each cluster and characterize each cluster by means of observable variables.

9.4 Example

Let's go back to the Oddjob Airways case study and run a cluster analysis on the data. Our aim is to identify a manageable number of segments that differentiates the customer base well. To do so, we first select a set of clustering variables, taking the sample size and potential collinearity issues into account. Next, we apply hierarchical clustering based on the squared Euclidean distances, using the Ward's linkage algorithm. This analysis will help us determine a suitable number of segments and a starting partition, which we will then use as the input for *k*-means clustering.

9.4.1 Select the Clustering Variables

The Oddjob Airways dataset (⤓ Web Appendix → Downloads) offers several variables for segmenting its customer base. Our analysis draws on the following set of variables, which we consider promising for identifying distinct segments based on customers' expectations regarding the airline's service quality (variable names in parentheses):

- ... with Oddjob Airways you will arrive on time (*e1*),
- ... Oddjob Airways provides you with a very pleasant travel experience (*e5*),
- ... Oddjob Airways gives you a sense of safety (*e9*),
- ... Oddjob Airways makes traveling uncomplicated (*e21*), and
- ... Oddjob Airways provides you with interesting on-board entertainment, service, and information sources (*e22*).

With five clustering variables, our analysis meets even the most conservative rule-of-thumb regarding minimum sample size requirements. Specifically, according to Dolnicar et al. (2016), the cluster analysis should draw on 100 times the number of clustering variables to optimize cluster recovery. As our sample size of 1,065 is clearly higher than $5 \cdot 100 = 500$, we can proceed with the analysis. Note, however, that the actual sample size used in the analysis may be substantially lower when using casewise deletion. This also applies to our analysis, which ultimately draws on 969 observations (i.e., after casewise deletion).

To begin with, it is good practice to examine a graphical display of the data. With multivariate data such as ours, the best way to visualize the data is by means of a scatterplot matrix (see Chaps. 5 and 7). To generate a scatterplot matrix, go to ▶ Graphics ▶ Scatterplot matrix and enter the variables *e1*, *e5*, *e9*, *e21*, and *e22* into the **Variables** box (Fig. 9.16). To ensure that the variable labels fit the diagonal boxes of the scatterplot, enter **0.9** next to **Scale text**. Because there are so many observations in the dataset, we choose a different marker symbol. To do so, click on **Marker properties** and select **Point** next to **Symbol**. Confirm by clicking on **Accept**, followed by **OK**. Stata will generate a scatterplot similar to the one shown in Fig. 9.17.

The resulting scatterplots do not suggest a clear pattern except that most observations are in the moderate to high range. But the scatterplots also assure us that all observations fall into the 0 to 100 range. Even though some observations with low values in (combinations of) expectation variables can be considered as extreme, we do not delete them, as they occur naturally in the dataset (see Chap. 5).

In a further check, we examine the variable correlations by clicking on ▶ Statistics ▶ Summaries, tables and tests ▶ Summary and descriptive statistics ▶ Pairwise correlations. Next, enter all variables into the **Variables** box (Fig. 9.18). Click on **OK** and Stata will display the results (Table 9.13).

The results show that collinearity is not at a critical level. The variables *e1* and *e21* show the highest correlation of 0.6132, which is clearly lower than the 0.90 threshold. We can therefore proceed with the analysis, using all five clustering variables.

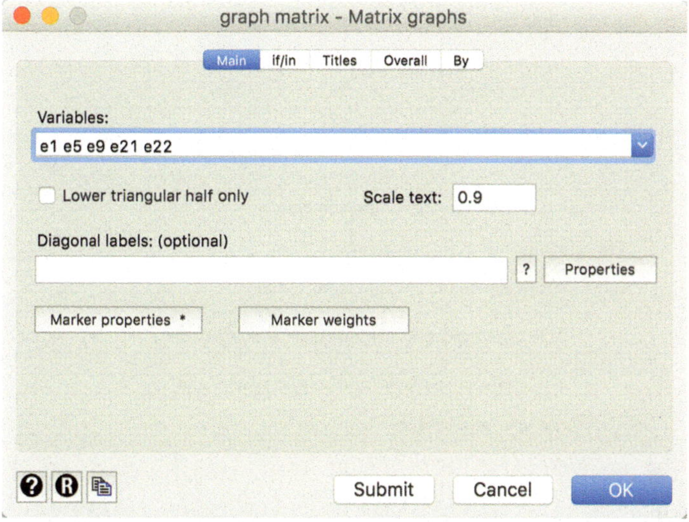

Fig. 9.16 Scatterplot matrix dialog box

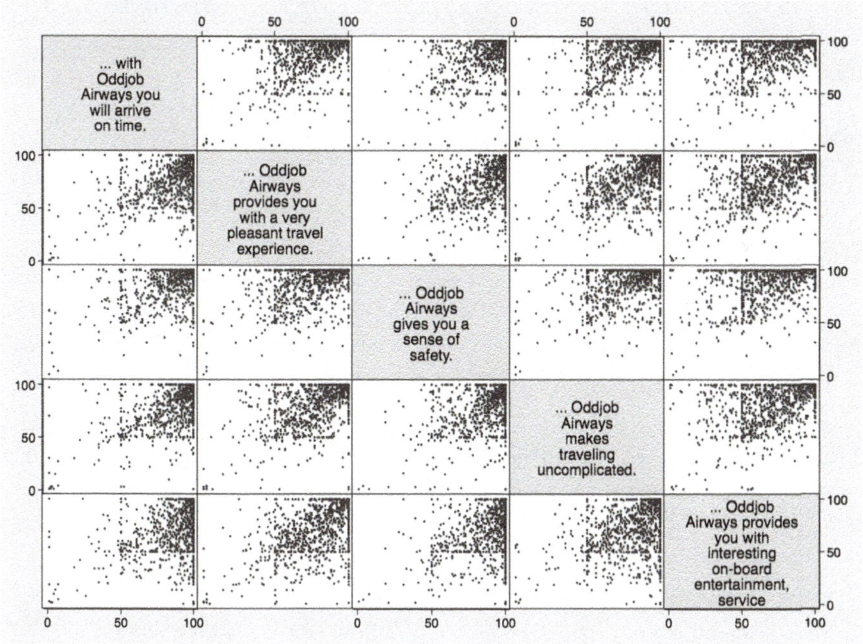

Fig. 9.17 Scatterplot matrix

Note that all the variables used in our analysis are metric and are measured on a scale from 0 to 100. However, if the variables were measured in different units with

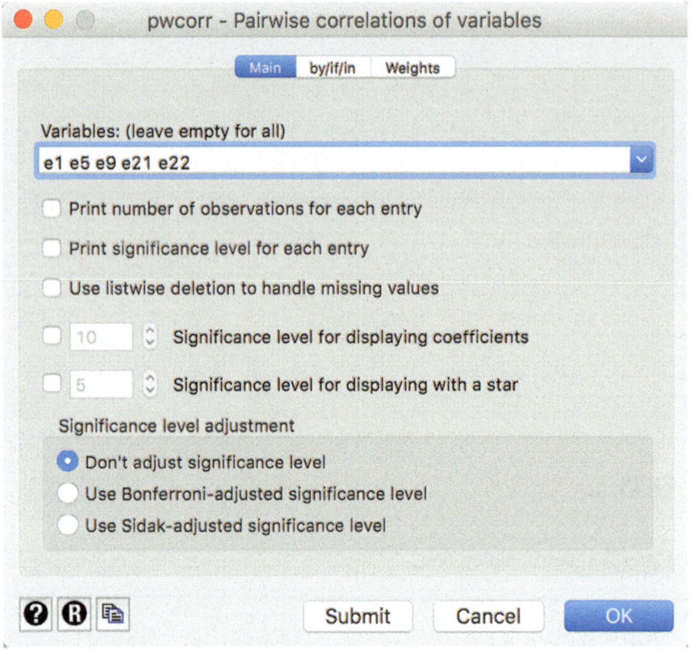

Fig. 9.18 Pairwise correlations dialog box

Table 9.13 Pairwise correlations

```
pwcorr e1 e5 e9 e21 e22

             |       e1        e5        e9       e21       e22
-------------+--------------------------------------------------
          e1 |   1.0000
          e5 |   0.5151    1.0000
          e9 |   0.5330    0.5255    1.0000
         e21 |   0.6132    0.5742    0.5221    1.0000
         e22 |   0.3700    0.5303    0.4167    0.4246    1.0000
```

different variances, we would need to standardize them in order to avoid the variables with the highest variances dominating the analysis. In Box 9.3, we explain how to standardize the data in Stata.

Box 9.3 Standardization in Stata

Stata's menu-based dialog boxes only allow for *z*-standardization (see Chap. 5), which you can access via ► Data ► Create or change data ► Create new variable (extended). In cluster analysis, however, the clustering variables should be standardized to a scale of 0 to 1. There is no menu option

(continued)

Box 9.3 (continued)
or command to do this directly in Stata, but we can improvise by using the `summarize` command. When using this command, Stata saves the minimum and maximum values of a certain variable as scalars. Stata refers to these scalars as *r(max)* and *r(min)*, which we can use to calculate new versions of the variables, standardized to a scale from 0 to 1. To run this procedure for the variable *e1* type in the following:

```
summarize e1
  Variable |        Obs        Mean    Std. Dev.     Min     Max
-----------+----------------------------------------------------
        e1 |      1,038    86.08189     19.3953     1       100
```

We can let Stata display the results of the `summarize` command by typing `return list` in the command window.

```
scalars:
                  r(N)  =  1038
              r(sum_w)  =  1038
               r(mean)  =  86.08188824662813
                r(Var)  =  376.1774729981067
                 r(sd)  =  19.395295125316
                r(min)  =  1
                r(max)  =  100
                r(sum)  =  89353
```

Next, we compute a new variable called *e1_rstd*, which uses the minimum and maximum values as input to compute a standardized version of *e1* (see Chap. 5 for the formula).

```
gen e1_rsdt =.
replace e1_rsdt = (e1- r(min)) / (r(max)-r(min))
```

Similar commands create standardized versions of the other clustering variables.

9.4.2 Select the Clustering Procedure and Measure of Similarity or Dissimilarity

To initiate hierarchical clustering, go to ► Statistics ► Multivariate analysis ► Cluster analysis ► Cluster data. The resulting menu offers a range of hierarchical and partitioning methods from which to choose. Because of its versatility and

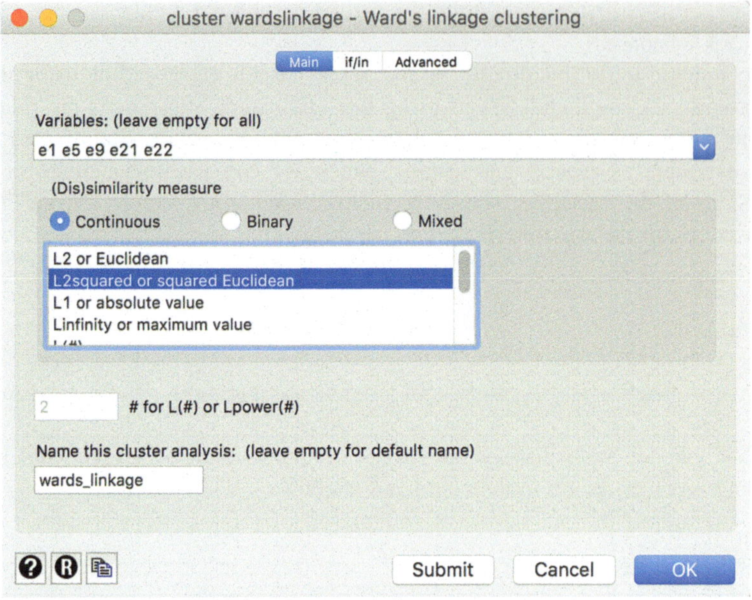

Fig. 9.19 Hierarchical clustering with Ward's linkage dialog box

general performance, we choose **Ward's linkage**. Clicking on the corresponding menu option opens a dialog box similar to Fig. 9.19.

Enter the variables *e1*, *e5*, *e9*, *e21*, and *e22* into the **Variables** box and select the squared Euclidean distance (**L2squared or squared Euclidean**) as the (dis)similarity measure. Finally, specify a name, such as *wards_linkage*, in the **Name this cluster analysis** box. Next, click on **OK**.

Nothing seems to happen (aside from the following command, which gets issued: `cluster wardslinkage e1 e5 e9 e21 e22, measure (L2squared) name(wards_linkage)`), although you might notice that our dataset now contains three additional variables called *wards_linkage_id*, *wards_linkage_ord*, and *wards_linkage_hgt*. While these new variables are not directly of interest, Stata uses them as input to draw the dendrogram.

9.4.3 Decide on the Number of Clusters

To decide on the number of clusters, we start by examining the dendrogram. To display the dendrogram, go to ► Statistics ► Multivariate analysis ► Cluster analysis ► Postclustering ► Dendrogram. Given the great number of observations, we need to limit the display of the dendrogram (see Fig. 9.20). To do so, select **Plot top branches only** in the **Branches** menu. By specifying **10** next to **Number of branches**, we limit the view of the top 10 branches of the dendrogram, which Stata labels *G1* to *G10*. When selecting **Display number of observations for each branch**, Stata will display the number of observations in each of the ten groups.

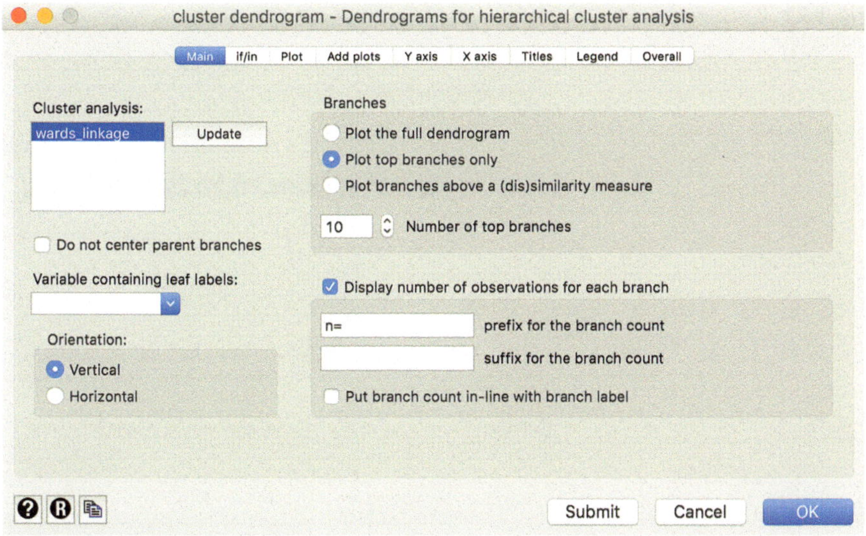

Fig. 9.20 Dendrogram dialog box

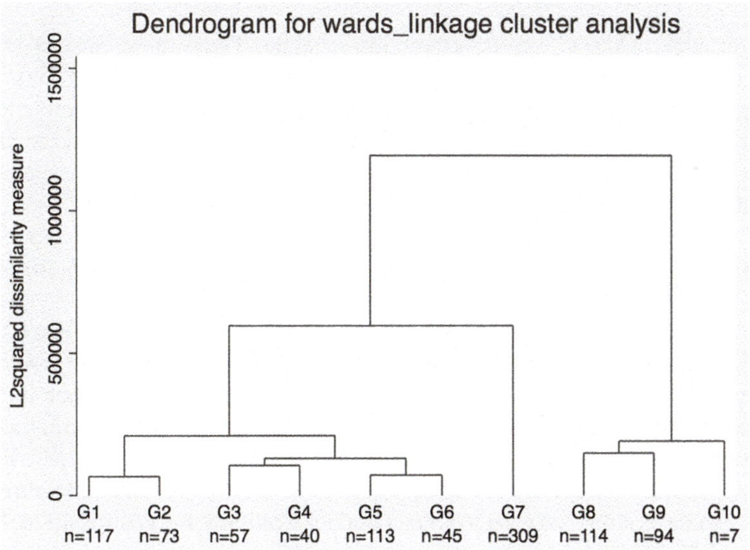

Fig. 9.21 Dendrogram

After clicking on **OK**, Stata will open a new window with the dendrogram (Fig. 9.21).

Reading the dendrogram from the bottom to the top, we see that clusters *G1* to *G6* are merged in quick succession. Clusters *G8* to *G10* are merged at about the

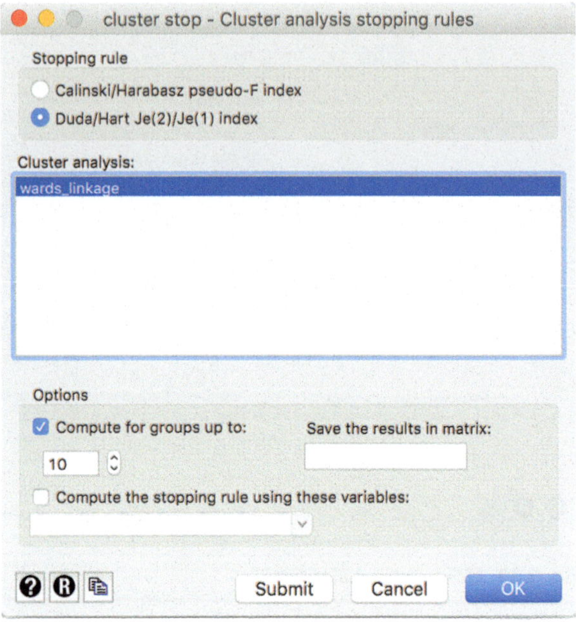

Fig. 9.22 Postclustering dialog box

same distance, while *G7* initially remains separate. These three clusters remain stable until, at a much higher distance, *G7* merges with the first cluster. This result clearly suggests a three-cluster solution, because reducing the cluster number to two requires merging the first cluster with *G7*, which is quite dissimilar to it. Increasing the number of clusters appears unreasonable, as many mergers take place at about the same distance.

The VRC and Duda-Hart indices allow us to further explore the number of clusters to extract. To request these measures in Stata, go to ► Statistics ► Multivariate analysis ► Cluster analysis ► Postclustering ► Cluster analysis stopping rules. In the dialog box that follows (Fig. 9.22), select **Duda/Hart Je(2)/J2(1) index** and tick the box next to **Compute for groups up to**. As we would like to consider a maximum number of ten clusters, enter **10** into the corresponding box and click on **OK**. Before interpreting the output, continue this procedure, but, this time, choose the **Calinski/Harabasz pseudo F-index** to request the VRC. Recall that we can also compute the VRC-based ω_k statistic. As this statistic requires the VRC_{k+1} value as input, we need to enter **11** under **Compute for groups up to**. Next, click on **OK**. Tables 9.14 and 9.15 show the postclustering outputs.

Looking at Table 9.14, we see that the Je(2)/Je(1) index yields the highest value for three clusters (**0.8146**), followed by a six-cluster solution (**0.8111**). Conversely, the lowest pseudo T-squared value (**37.31**) occurs for ten clusters. Looking at the VRC values in Table 9.15, we see that the index decreases with a greater number of clusters.

To calculate the ω_k criterion, we can use a file that has been specially programmed for this, called chomega.ado ⭳ Web Appendix (→ Downloads). This

Table 9.14 Duda-Hart indices

```
cluster stop wards_linkage, rule(duda) groups(1/10)
```

Number of clusters	Duda/Hart Je(2)/Je(1)	Duda/Hart pseudo T-squared
1	0.6955	423.29
2	0.6783	356.69
3	0.8146	100.83
4	0.7808	59.79
5	0.7785	58.59
6	0.8111	58.94
7	0.6652	47.82
8	0.7080	64.34
9	0.7127	75.79
10	0.7501	37.31

Table 9.15 VRC

```
cluster stop wards_linkage, rule(calinski) groups(2/10)
```

Number of clusters	Calinski/ Harabasz pseudo-F
2	423.29
3	406.02
4	335.05
5	305.39
6	285.26
7	273.24
8	263.12
9	249.61
10	239.73
11	233.17

Table 9.16 ω_k statistic

```
chomega
omega_3  is -53.691
omega_4  is  41.300
omega_5  is   9.534
omega_6  is   8.110
omega_7  is   1.899
omega_8  is  -3.394
omega_9  is   3.636
Minimum value of omega: -53.691 at 3 clusters
```

file should first be run before we can use it, just like the add-on modules discussed in Chap. 5, Section 5.8.2. To do this, download the chomega.ado file and drag it into the Stata command box, and add `do "` before and `"` after the text that appears in the

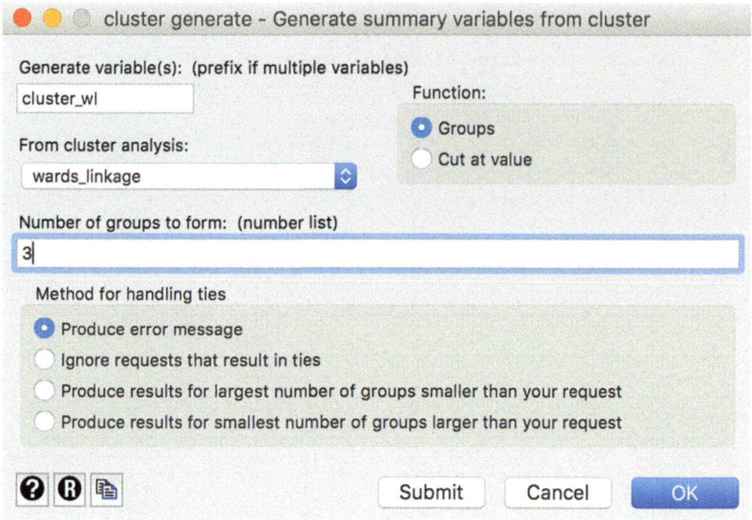

Fig. 9.23 Summary variables dialog box

Stata **Command** window (see Chap. 5). Then click on *enter*. Then you should type `chomega`. Note that this only works if you have first performed a cluster analysis.

The output is included in Table 9.16. We find that the smallest ω_k value of −**53.691** occurs for a three-cluster solution. The smallest value is shown at the top and bottom of Table 9.16. Note again that since ω_k requires $VRC_{k\text{-}1}$ as input, the statistic is only defined for three or more clusters. Taken jointly, our analyses of the dendrogram, the Duda-Hart indices, and the VRC clearly suggest a three-cluster solution.

9.4.4 Validate and Interpret the Clustering Solution

In the next step, we create a cluster membership variable, which indicates the cluster to which each observation belongs. To do so, go to ► Statistics ► Multivariate analysis ► Cluster analysis ► Postclustering ► Generate summary variables from cluster. In the dialog box that opens (Fig. 9.23), enter a name, such as *cluster_wl*, for the variable to be created in the **Generate variable(s)** box. In the dropdown list **From cluster analysis**, we can choose on which previously run cluster analysis the cluster membership variable should be based. As this is our first analysis, we can only select *wards_linkage*. Finally, specify the number of clusters to extract (**3**) under **Number of groups to form** and proceed by clicking **OK**.

Stata generates a new variable *cluster_wl*, which indicates the group to which each observation belongs. We can now use this variable to describe and profile the clusters. In a first step, we would like to tabulate the number of observations in each

Table 9.17 Cluster sizes

```
tabulate cluster_wl, missing

 cluster_wl |      Freq.     Percent        Cum.
------------+-----------------------------------
          1 |        445       41.78       41.78
          2 |        309       29.01       70.80
          3 |        215       20.19       90.99
          . |         96        9.01      100.00
------------+-----------------------------------
      Total |      1,065      100.00
```

Table 9.18 Comparison of means

```
tabstat e1 e5 e9 e21 e22, statistics( mean ) by(cluster_wl)

Summary statistics: mean
  by categories of: cluster_wl

cluster_wl |        e1        e5        e9       e21       e22
-----------+--------------------------------------------------
         1 |  92.39326  75.50562  89.74607   81.7191  62.33933
         2 |   97.1068  95.54693  97.50809  96.63754  92.84466
         3 |   59.4186  58.28372  71.62791  56.72558  58.03256
-----------+--------------------------------------------------
     Total |  86.57998  78.07534  88.20124  80.93086  71.11146
--------------------------------------------------------------
```

cluster by going to ► Statistics ► Summary, tables, and tests ► Frequency tables ► One-way table. Simply select *cluster_wl* in the drop-down menu under **Categorical variable**, tick the box next to **Treat missing values like other values** and click on **OK**. The output in Table 9.17 shows that the cluster analysis assigned **969** observations to the three segments; **96** observations are not assigned to any segment due to missing values. The first cluster comprises **445** observations, the second cluster **309** observations, and the third cluster **215** observations.

Next, we would like to compute the centroids of our clustering variables. To do so, go to ► Statistics ► Summaries, tables, and tests ► Other tables ► Compact table of summary statistics and enter *e1 e5 e9 e21 e22* into the **Variables** box. Next, click on **Group statistics by variable** and select *cluster_wl* from the list. Under **Statistics to display**, tick the first box and select **Mean**, followed by **OK**. Table 9.18 shows the resulting output.

Comparing the variable means across the three clusters, we find that respondents in the first cluster strongly emphasize punctuality (e_1), while comfort (e_5) and, particularly, entertainment aspects (*e22*) are less important. Respondents in the second cluster have extremely high expectations regarding all five performance features, as evidenced in average values well above 90. Finally, respondents in the third cluster do not express high expectations in general, except in terms of security (e_9). Based on these results, we could label the first cluster "on-time is enough," the

Table 9.19 Crosstab

```
tabulate cluster_wl flight_purpose, chi2 V

           |  Do you normaly fly
           |   for business or
           |  leisure purposes?
cluster_wl |  Business    Leisure |     Total
-----------+----------------------+----------
         1 |       239        206 |       445
         2 |       130        179 |       309
         3 |       114        101 |       215
-----------+----------------------+----------
     Total |       483        486 |       969

          Pearson chi2(2) =  10.9943   Pr = 0.004
               Cramér's V =   0.1065
```

second cluster "the demanding traveler," and the third cluster "no thrills." We could further check whether these differences in means are significant by using a one-way ANOVA as described in Chap. 6.

In a further step, we can try to profile the clusters using sociodemographic variables. Specifically, we use crosstabs (see Chap. 5) to contrast our clustering with the variable *flight_purpose*, which indicates whether the respondents primarily fly for business purposes (*flight_purpose=1*) or private purposes (*flight_purpose=2*). To do so, click on ► Statistics ► Summaries, tables, and tests ► Frequency tables ► Two-way table with measures of association. In the dialog box that opens, enter *cluster_wl* into the **Row variable** box and *flight_purpose* into the **Column variable** box. Select **Pearson's chi-squared** and **Cramer's V** under **Test statistics** and click on **OK**. The results in Table 9.19 show that the majority of respondents in the first and third cluster are business travelers, whereas the second cluster primarily comprises private travelers. The χ^2-test statistic (**Pr** = **0.004**) indicates a significant relationship between these two variables. However, the strength of the variables' association is rather small, as indicated by the Cramer's V of **0.1065**.

The Oddjob Airways dataset offers various other variables such as *age*, *gender*, or *status*, which could be used to further profile the cluster solution. However, instead of testing these variables' efficacy step-by-step, we proceed and assess the solution's stability by running an alternative clustering procedure on the data. Specifically, we apply the *k*-means method, using the grouping from the Ward's linkage analysis as input for the starting partition. To do so, go to:

► Statistics ► Multivariate statistics ► Cluster analysis ► Cluster data ► Kmeans. In the dialog box that opens, enter *e1*, *e5*, *e9*, *e21*, and *e22* into the **Variables** box, choose **3** clusters, and select **L2squared or squared Euclidean** under **(Dis)similarity measure** (Fig. 9.24). Under **Name this cluster analysis**, make sure that you specify an intuitive name, such as *kmeans*. When clicking on the **Options** tab, we can choose between different options of how *k*-means should derive a starting partition for the analysis. Since we want to use the clustering from

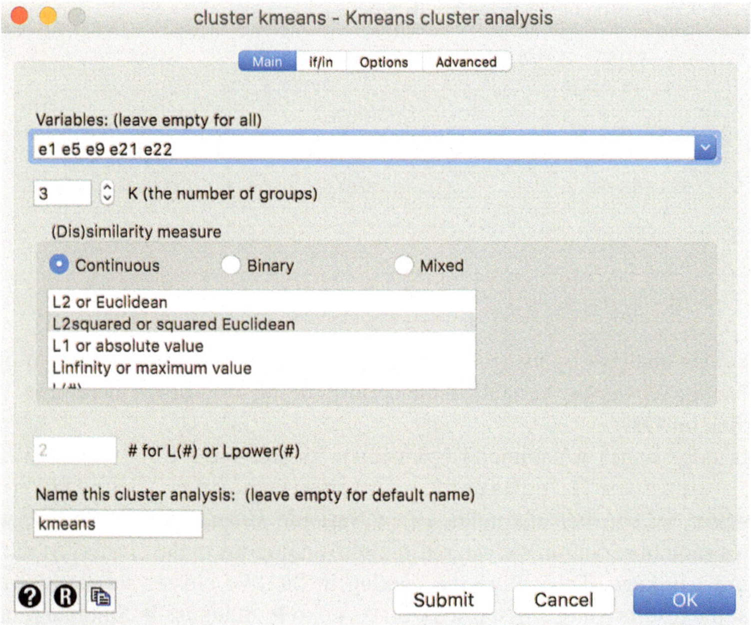

Fig. 9.24 *k*-means dialog box

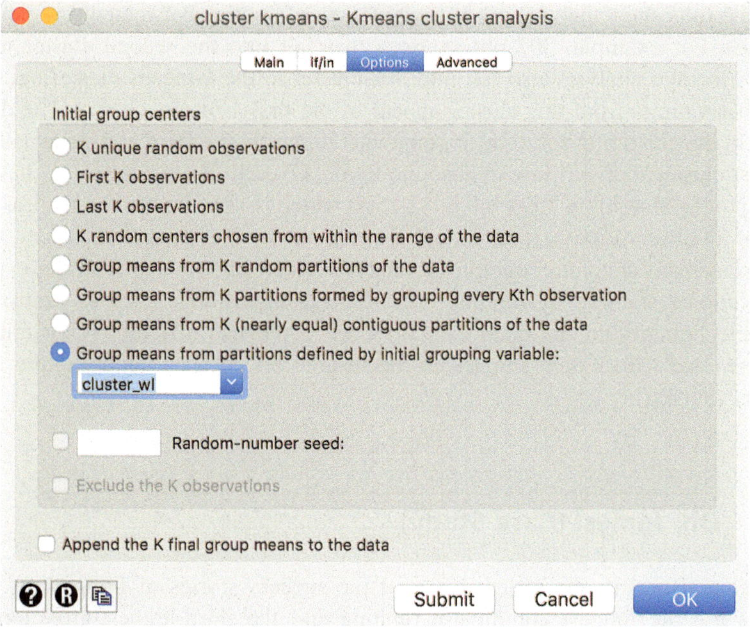

Fig. 9.25 Options in the *k*-means dialog box

Table 9.20 Comparison of clustering results

```
tabulate cluster_wl kmeans

           |              kmeans
cluster_wl |         1          2          3 |     Total
-----------+---------------------------------+----------
         1 |       320        107         18 |       445
         2 |         2        307          0 |       309
         3 |        36         10        169 |       215
-----------+---------------------------------+----------
     Total |       358        424        187 |       969
```

our previous analysis by using Ward's linkage, we need to choose the last option and select the *cluster_wl* variable in the corresponding drop-down menu (Fig. 9.25). Now click on **OK**.

Stata only issues a command (`cluster kmeans e1 e5 e9 e21 e22, k (3) measure(L2squared) name(kmeans) start(group (cluster_wl))`) but also adds a new variable *kmeans* to the dataset, which indicates each observation's cluster affiliation, analogous to the *cluster_wl* variable for Ward's linkage. To explore the overlap in the two cluster solutions, we can contrast the results using crosstabs. To do so, go to ► Statistics ► Summary, tables, and tests ► Frequency tables ► Two-way table with measures of association and select *cluster_wl* under **Row variable** and *kmeans* under **Column variable**. After clicking on **OK**, Stata will produce an output similar to Table 9.20.

The results show that there is a strong degree of overlap between the two cluster analyses. For example, **307** observations that fall into the second cluster in the Ward's linkage analysis also fall into this cluster in the *k*-means clustering. Only two observations from this cluster appear in the first *k*-means cluster. The divergence in the clustering solutions is somewhat higher in the third and, especially, in the first cluster, but still low in absolute terms. Overall, the two analyses have an overlap of (320 + 307 + 169)/969 = 82.15%, which is very satisfactory as less than 20% of all observations appear in a different cluster when using *k*-means.

This analysis concludes our cluster analysis. However, we could further explore the solution's stability by running other linkage algorithms, such as centroid or complete linkage, on the data. Similarly, we could use different (dis)similarity measures and assess their impact on the results. So go ahead and explore these options yourself!

9.5 Oh, James! (Case Study)

The James Bond movie series is one of the success stories of filmmaking. The movies are the longest continually running and the third-highest-grossing film series to date, which started in 1962 with Dr. No, starring Sean Connery as James Bond. As of 2016, there have been 24 movies with six actors having played James

Bond. Interested in the factors that contributed to this running success, you decide to investigate the different James Bond movies' characteristics. Specifically, you want to find out whether the movies can be grouped into clusters, which differ in their box-office revenues. To do so, you draw on Internet Movie Database (www.imdb.com) and collect data on all 24 movies based on the following variables (variable names in parentheses):

- Title. (*title*)
- Actor playing James Bond. (*actor*)
- Year of publication. (*year*)
- Budget in USD, adjusted for inflation. (*budget*)
- Box-office revenues in the USA, adjusted for inflation. (*gross_usa*)
- Box-office revenues worldwide, adjusted for inflation. (*gross_worldwide*)
- Runtime in minutes. (*runtime*)
- Native country of the villain actor. (*villain_country*)
- Native country of the bondgirl. (*bondgirl_country*)
- Haircolor of the bondgirl. (*bondgirl_hair*)

Use the dataset *jamesbond.dta* (⭳ Web Appendix → Downloads) to run a cluster analysis—despite potential objections regarding the sample size. Answer the following questions:

1. Which clustering variables would you choose in light of the study objective, their levels of measurement, and correlations?
2. Given the levels of measurement, which clustering method would you prefer? Carry out a cluster analysis using this procedure.
3. Interpret and profile the obtained clusters by examining cluster centroids. Compare the differences across clusters on the box-office revenue variables.
4. Use a different clustering method to test the stability of your results.

9.6 Review Questions

1. In your own words, explain the objective and basic concept of cluster analysis.
2. What are the differences between hierarchical and partitioning methods? When do we use hierarchical or partitioning methods?
3. Repeat the manual calculations of the hierarchical clustering procedure from the beginning of the chapter, but use complete linkage as the clustering method. Compare the results with those of the single linkage method.
4. Explain the different options to decide on the number of clusters to extract from the data? Should you rather on statistical measures or rather on practical reasoning?
5. Run the *k*-means analysis on the Oddjob Airways data again (*oddjob.dta*, ⭳ Web Appendix → Downloads). Assume a three-cluster solution and try the different

options for obtaining a starting partition that Stata offers. Compare the results with those obtained by the hierarchical clustering.
6. Which clustering variables could be used to segment:
   - The market for smartphones?
   - The market for chocolate?
   - The market for car insurances?

9.7 Further Readings

Bottomley, P., & Nairn, A. (2004). Blinded by science: The managerial consequences of inadequately validated cluster analysis solutions. *International Journal of Market Research, 46*(2), 171–187.

In this article, the authors investigate if managers could distinguish between cluster analysis outputs derived from real-world and random data. They show that some managers feel able to assign meaning to random data devoid of a meaningful structure, and even feel confident formulating entire marketing strategies from cluster analysis solutions generated from such data. As such, the authors provide a reminder of the importance of validating clustering solutions with caution.

Dolnicar, S., Grün, B., & Leisch, F. (2016). Increasing sample size compensates for data problems in segmentation studies. *Journal of Business Research, 69*(2), 992–999.

Using artificial datasets of known structure, the authors examine the effects of data problems such as respondent fatigue, sampling error, and redundant items on segment recovery. The study nicely shows how insufficient sample size of the segmentation base can have serious negative consequences on segment recovery and that increasing the sample size represents a simple measure to compensate for the detrimental effects caused by poor data quality.

Punj, G., & Stewart, D. W. (1983). Cluster analysis in marketing research: Review and suggestions for application. *Journal of Marketing Research, 20*(2),134–148.

In this seminal article, the authors discuss several issues in applications of cluster analysis and provide further theoretical discussion of the concepts and rules of thumb that we included in this chapter.

Romesburg, C. (2004). *Cluster analysis for researchers.* Morrisville: Lulu Press.

Charles Romesburg nicely illustrates the most frequently used methods of hierarchical cluster analysis for readers with limited backgrounds in mathematics and statistics.

Wedel, M., & Kamakura, W. A. (2000). *Market segmentation: Conceptual and methodological foundations* (2nd ed.). Boston: Kluwer Academic.

This book is a clear, readable, and interesting presentation of applied market segmentation techniques. The authors explain the theoretical concepts of recent analysis techniques and provide sample applications. Probably the most comprehensive text in the market.

References

Anderberg, M. R. (1973). *Cluster analysis for applications*. New York: Academic.

Arabie, P., & Hubert, L. (1994). Cluster analysis in marketing research. In R. P. Bagozzi (Ed.), *Advanced methods in marketing research* (pp. 160–189). Cambridge: Basil Blackwell & Mott, Ltd..

Arthur, D., & Vassilvitskii, S. (2007). k-means++: The advantages of careful seeding. In *Proceedings of the eighteenth annual ACM-SIAM symposium on discrete algorithms* (pp. 1027–1035). Philadelphia: Society for Industrial and Applied Mathematics.

Becker, J.-M., Ringle, C. M., Sarstedt, M., & Völckner, F. (2015). How collinearity affects mixture regression results. *Marketing Letters, 26*(4), 643–659.

Caliński, T., & Harabasz, J. (1974). A dendrite method for cluster analysis. *Communications in Statistics—Theory and Methods, 3*(1), 1–27.

Dolnicar, S. (2003). Using cluster analysis for market segmentation—typical misconceptions, established methodological weaknesses and some recommendations for improvement. *Australasian Journal of Market Research, 11*(2), 5–12.

Dolnicar, S., & Grün, B. (2009). Challenging "factor-cluster segmentation". *Journal of Travel Research, 47*(1), 63–71.

Dolnicar, S., & Lazarevski, K. (2009). Methodological reasons for the theory/practice divide in market segmentation. *Journal of Marketing Management, 25*(3–4), 357–373.

Dolnicar, S., Grün, B., Leisch, F., & Schmidt, F. (2014). Required sample sizes for data-driven market segmentation analyses in tourism. *Journal of Travel Research, 53*(3), 296–306.

Dolnicar, S., Grün, B., & Leisch, F. (2016). Increasing sample size compensates for data problems in segmentation studies. *Journal of Business Research, 69*(2), 992–999.

Duda, R. O., & Hart, P. E. (1973). *Pattern classification*. Hoboken: Wiley.

Duda, R. O., Hart, P. E., & Stork, D. G. (2001). *Pattern classification* (2nd ed.). Hoboken: Wiley.

Everitt, B. S., & Rabe-Hesketh, S. (2006). *Handbook of statistical analyses using Stata* (4th ed.). Boca Raton: Chapman & Hall/CRC.

Formann, A. K. (1984). *Die Latent-Class-Analyse: Einführung in die Theorie und Anwendung*. Beltz: Weinheim.

Gower, J. C. (1971). A general coefficient of similarity and some of its properties. *Biometrics, 27*(4), 857–871.

Halpin, B. (2016). *Cluster analysis stopping rules in Stata*. University of Limerick. Department of Sociology Working Paper Series, WP2016-01. http://ulsites.ul.ie/sociology/sites/default/files/wp2016-01.pdf

Kaufman, L., & Rousseeuw, P. J. (2005). *Finding groups in data. An introduction to cluster analysis*. Hoboken: Wiley.

Kotler, P., & Keller, K. L. (2015). *Marketing management* (15th ed.). Upper Saddle River: Prentice Hall.

Milligan, G. W., & Cooper, M. (1985). An examination of procedures for determining the number of clusters in a data set. *Psychometrika, 50*(2), 159–179.

Milligan, G. W., & Cooper, M. (1988). A study of variable standardization. *Journal of Classification, 5*(2), 181–204.

Park, H.-S., & Jun, C.-H. (2009). A simple and fast algorithm for K-medoids clustering. *Expert Systems with Applications, 36*(2), 3336–3341.

Punj, G., & Stewart, D. W. (1983). Cluster analysis in marketing research: Review and suggestions for application. *Journal of Marketing Research, 20*(2), 134–148.

Qiu, W., & Joe, H. (2009). Cluster generation: Random cluster generation (with specified degree of separation). R package version 1.2.7.

Sheppard, A. (1996). The sequence of factor analysis and cluster analysis: Differences in segmentation and dimensionality through the use of raw and factor scores. *Tourism Analysis, 1*(1), 49–57.

Tonks, D. G. (2009). Validity and the design of market segments. *Journal of Marketing Management, 25*(3/4), 341–356.

Wedel, M., & Kamakura, W. A. (2000). *Market segmentation: Conceptual and methodological foundations* (2nd ed.). Boston: Kluwer Academic.

van der Kloot, W. A., Spaans, A. M. J., & Heinser, W. J. (2005). Instability of hierarchical cluster analysis due to input order of the data: The PermuCLUSTER solution. *Psychological Methods, 10*(4), 468–476.

Lilien, G. L., & Rangaswamy, A. (2004). *Marketing engineering. Computer-assisted marketing analysis and planning* (2nd ed.). Bloomington: Trafford Publishing.

John H. R., Kayande, U., & Stremersch, S. (2014). From academic research to marketing practice: Exploring the marketing science value chain. *International Journal of Research in Marketing, 31*(2), 127–140

Communicating the Results 10

Keywords

Ethics • KISS principle • Minto principle • Pyramid structure for presentations • Self-contained figure • Self-contained table • Visual aids

Learning Objectives
After reading this chapter, you should understand:

- Why communicating the results is a crucial element of every market research study.
- The elements that should be included in a written research report and how to structure these elements.
- How to communicate the findings in an oral presentation.
- The ethical issues concerning communicating the report findings to the client.

10.1 Introduction

Communicating results is key in any market research project. This includes giving clear answers to the investigated research questions and recommending a course of action, where appropriate. The importance of communicating marketing research results should not be underestimated. Even if the research has been carefully conducted, the recipients will find it difficult to understand the implications of the results and to appreciate the study's quality if you spend too little time and energy on communicating these. Clear communication may also set the stage for follow-up research. If you communicate the findings effectively, the clients, who may know little about market research and may even be unfamiliar with the specific market

E. Mooi et al., *Market Research*, Springer Texts in Business and Economics,
DOI 10.1007/978-981-10-5218-7_10

research project, will understand them. Hence, the communication must be relevant for the addressed audience and provide a clear picture of the project.

Market researchers usually present their findings in the form of an oral presentation and written report. This report is the written evidence of the research effort and includes the details. Identifying the addressed audience is critical for both these points, as this determines how you can best communicate the findings. In this chapter, we discuss guidelines on how to effectively communicate research findings orally and in writing. We first discuss written reports before listing the basics of oral presentations. We also provide hints on how to acquire research follow-up. At the end of the chapter, we briefly review the ethical issues related to market research.

10.2 Identify the Audience

When providing reports (and presentations), you should keep the audience's characteristics and needs in mind and should tailor the report to their objectives. Imagine you are dealing with the marketing department of a company planning to launch a new product and needing to learn more about the potential customers' buying behavior. The knowledge and level of interest in the study might differ greatly within the department. While the managers, who commissioned the study, are generally familiar with its objective and design, others, who might be unaware of the background (e.g., the marketing director or the sales staff), must be informed about the research to allow them to understand the research findings. When preparing the report, you should consider the following questions:

- Who will read the report?
- Why will they read the report?
- Which parts of the report are of specific interest to them?
- What do they already know about the study?
- What information will be new to them?
- What is the most important point for them to know after they have read the report?
- What can be done with the research findings?

These questions help you determine the level of detail that should be included in your report. Furthermore, they reveal information that requires specific focus during the project. Remember, a successful report meets its audience's needs! However, not everything that you consider appropriate for your audience is appropriate. Nothing is worse than suggesting an idea that the audience finds unpalatable (e.g., saying that a specific senior management behavior or attitude is a major obstacle to success), or proposing a course of action that has been attempted before. Informal talks with the client are therefore vital before you present the results—never present findings formally without discussing them with the client first!

Further, you need to ask clients about their expectations and the recommendations they think will be made early in the project. Why would clients spend $100,000 if you merely give them the answers they expect to get? Such discussions may help exceed clients' expectations in a way that is useful to them.

10.3 Guidelines for Written Reports

You should always keep the people addressed in written report in mind. Decision makers are generally unfamiliar with statistical details, but would like to know how the findings can help them make decisions. You should therefore avoid research jargon and state the key insights clearly without omitting important facts. There are several major rules to consider when writing a report (Armstrong 2010; Churchill and Iacobucci 2009):

1. The report must be *complete*; that is, it must contain all information that the reader needs in order to understand the research. Technical or ambiguous terms, as well as abbreviations and symbols, should be clearly defined and illustrated. Although you know what terms like heteroscedasticity or eigenvalue mean, the report reader probably won't! In addition, the report must provide enough detail to enable the reader to verify and replicate the findings if necessary. Bear in mind that the staff turnover in many organizations is high and that reports should therefore be stand-alone to allow those with little knowledge of the background to read and understand them.
2. The report must be *accurate*. The readers will base their assessment of the entire research project's quality on the presented report. Consequently, the report must be well written. For example, grammar and spelling must be correct, no slang should be used, tables must be labeled correctly, and page numbers should be included. If there are small errors, the reader may believe they are due to your lack of care and generalize about your analysis! Therefore, proofread (a proofreader should preferably do this) to eliminate obvious errors. Lastly, objectivity is an important attribute of any report. This means that any subjective conclusions should be clearly stated as such.
3. The report must be *clear* and language simple and concise:
   - Use short sentences.
   - Use simple and unambiguous words.
   - Use concrete examples to illustrate aspects of the research (e.g., unexpected findings). These can also be helpful if the audience has strong beliefs that are not consistent with your recommendation, which are often not implemented, because the client does not believe them.
   - Use the active voice to make the report easy to read and to help understanding.
   - Avoid negative words.
   - Use business language.
   - Avoid exclamation marks and do not use caps unnecessarily. Avoid the use of bold or italics for more than just a few words.

4. Follow the **KISS principle**: Keep it short and simple! This principle requires the report to be *concise*. And since it needs to be action-driven, the reader must immediately understand its purpose and the results, so start off with these. You should present the results clearly and simply. Important details can be shown in the appendix or appendices of the report, which should also not be overloaded with irrelevant material. In addition, keep in mind that each section's first sentences are the most important ones: They should summarize the main idea you want to convey in this section.
5. The report must be *structured* logically. This applies to the general structure of the report (see Table 10.1) and to the line of argumentation in each section. Make sure you avoid style elements that may distract the reader:
   - Avoid cross-references. Having to search elsewhere for important results is disruptive. For example, do not put important tables in the appendix.
   - Use footnotes instead of endnotes and as few as possible.
   - The structure should not be too detailed. As a rule of thumb, you should avoid using more than four levels.
   - A new level must include at least two sections. For example, if there is a Sect. 3.1.1, there must also be a Sect. 3.1.2.

10.4 Structure the Written Report

When preparing a written report, a clear structure helps readers navigate it to quickly and easily find those elements that interest them. Although all reports differ, we include a suggested structure for a research report in Table 10.1.

Table 10.1 Suggested structure for a written research report

Title Page
Executive Summary
Table of Contents
1. Introduction
1.1 Problem definition
1.2 Research objectives
1.3 Research questions and/or hypotheses[a]
2. Methodology
2.1 Population, sampling method, and sample description
2.2 Quantitative and qualitative methods used for data analysis
3. Results
4. Conclusions and Recommendations
5. Limitations
6. Appendix

[a]In practice, the word hypotheses may be replaced by research question(s) or proposition(s)

10.4.1 Title Page

The title page should state the title of the report, the name of the client who commissioned the report, the organization or researcher submitting it, and the date of release. The heading should clearly state the nature and scope of the report. It may simply describe the research (e.g., "Survey of Mobile Phone Usage") or may outline the objectives of the study in the form of an action title (e.g., "How to Increase the Adoption of Wearable Technologies").

10.4.2 Executive Summary

The executive summary should appear first and is essential, because it is often the only section that executives read. This summary helps set the expectations of those who read more. Hence, this section must be short to allow busy executives to read it and should give them the essence (findings and recommendations) of the research. As a rule of thumb, the executive summary should not exceed 150 words. It should contain key findings and recommendations, and help the reader understand the full study. The executive summary also requires more structure. A common way of giving structure is to tell a story. Begin with a description of the problem, thereafter introducing the issues that make this difficult or complicated and describing how these give rise to a number of questions. Finally, lead the reader through your line of reasoning to the answer:

- *Situation*: Background information.
- *Difficulty or complication*: A short window of opportunity; a change from the previously stable situation; lack of performance due to unknown causes (i.e., the reason for your research study).
- *Question*: The scope and goal of your research study.
- *Answer*: Your findings and conclusions (and if the client requires this, also your recommendations).

10.4.3 Table of Contents

The table of contents helps the reader locate specific aspects of the report. The table of contents should correspond to the main report headings. It should also include lists of tables and figures with page references.

10.4.4 Introduction

This section should explain the project context to the reader. Questions to be answered include:

- Why was the study undertaken?
- What were the objectives and which key questions are answered?
- Is the study related to other studies and, if so, which findings did they produce?
- How is the report structured?

Besides introducing the background and purpose of the research, the introduction should briefly explain how the objectives and key questions are addressed. You should briefly mention the hypotheses or propositions tested during the research and how the research was approached (e.g., cluster analysis). You should ensure that Critical terms are defined. For example, aviation terms such as CASM (cost per available seat mile) require explanation. As a rule, the following three questions on the research should be answered in the introduction, but should be neither too detailed nor too technical:

- What was done?
- How was it done?
- Why was it done?

Keep in mind that the introduction should set the stage for the body of the report and the presentation of the results, but no more than this. You should only provide a detailed description of how you collected and analyzed the data in the next section of the report. Lastly, you should provide a brief summary of how the report is organized at the end of the introduction.

10.4.5 Methodology

In this section, you should describe the research procedure and the different (statistical) methods used to analyze the data. These must be presented precisely and coherently, allowing the reader to understand the analyses' process and basic principles. Always keep your audience in mind! If the audience is familiar with research methodology, you can describe the procedures in detail and skip the basics. If the client has little knowledge of research methodology, you should introduce these briefly. If you have an audience of whom some have a little and others more knowledge, you might want to move the basics to an appendix.

If not already stated in the previous section, you should define whether the study is exploratory, descriptive, or causal by nature and whether the results are based on primary or secondary data. If primary data are used, their source should be specified (e.g., observation or questionnaire). If a questionnaire was used, you should state whether it was administered by means of face-to-face interviews, telephone interviews, or through web or mail surveys. Also explain why you chose this specific method.

The reader should also know the demographic or other relevant characteristics of the targeted survey population. This includes the geographical area, age group, and gender. While it is usually sufficient to describe the population in a few sentences, the sampling method needs more explanation: How was the sample selected? Which sampling frames were chosen (e.g., random, systematic, stratified)? In addition, information on the sample size, response rate, and sample characteristics is essential, as this indicates the results' reliability and validity.

You should include a copy of the actual instruments used, such as the questionnaire or the interview guide, the data collection protocol, and the detailed statistical analyses of the results, in the appendix, or present them separately. Although these are important to fully understand the characteristics of the research project, including them in the main text would make reading the report more difficult.

10.4.6 Results

In this section, you need to present the findings and describe how they relate to a possible solution to the research problem and how they influence the recommendations. There are several ways of presenting the results logically. You could, for instance, use the different research objectives as a guideline to structure this section and then analyze them one by one.

Another way is to first summarize the overall findings and then analyze them in relevant subgroups, such as the type of customer or geographical regions. Alternatively, you can classify the findings according to the data type or the research method if several were used. For example, you could first present the conclusions of the secondary data collection and then those derived from an analysis of the questionnaire.

Use tables and graphs when presenting statistical data, as they make the report and the data more interesting. Tables and graphs also structure information, thus facilitating understanding. Graphs often allow the researcher to visually present complex data, which might not be possible when only using tables. However, graphs can also be misleading, as they may be adjusted to favor a specific viewpoint (see the next section for examples).

Results are often presented in Excel or Word format. Fortunately, Stata has built-in capabilities to export its results to Excel by means of export `excel` and `putexcel`. You can also output results to Word, using a package called `outreg2`, which is a Stata add-on. The following videos offer step-by-step

(continued)

introductions to exporting results to Excel (first mobile tag) and Word (second mobile tag):

https://www.youtube.com/watch?v=MUQ3E8hIQZE

https://www.youtube.com/watch?v=UemL7uYM6Lc

10.4.6.1 Window Dressing with Graphs

While graphs have the advantage that they can present complex information in a way that is easily understandable, they can be used to mislead the reader. Experience with generating and interpreting graphs will help you spot this. In this section, we show examples of how graphs can mislead. By shortening the x-axis in Fig. 10.1 (i.e., removing the years 2003–2007), it suggests a growth in the units sold (Fig. 10.2).

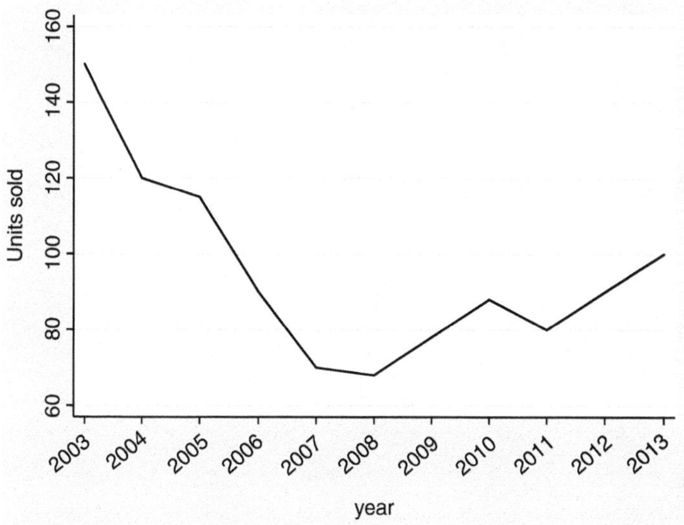

Fig. 10.1 What year does the curve start? (I)

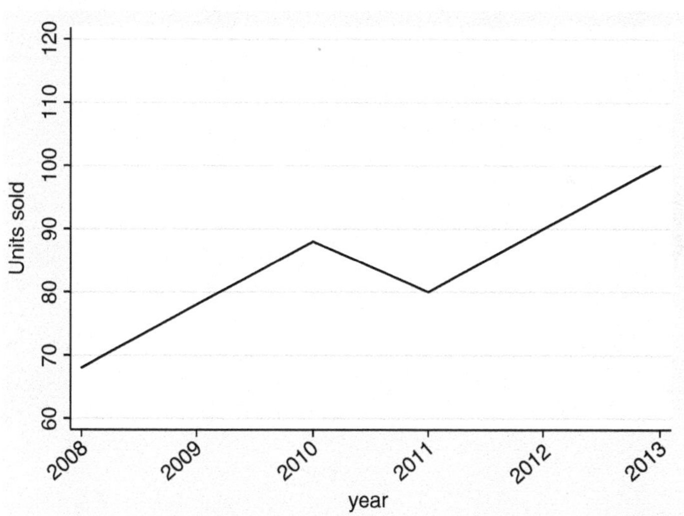

Fig. 10.2 What year does the curve start? (II)

Likewise, we can modify the scale range (Fig. 10.2 vs. Fig. 10.3). Specifically, reducing the *y*-axis to a range from 68 to 100 units with 4 unit increments, suggests faster growth (Fig. 10.3). Another example is the "floating" *y*-axis (Fig. 10.4 vs. Fig. 10.5), which increases the scale range along the *y*-axis from 0 to 190 with 30-unit increments, thus making the drop in the number of units sold over the period 2005 to 2008 less visually pronounced.

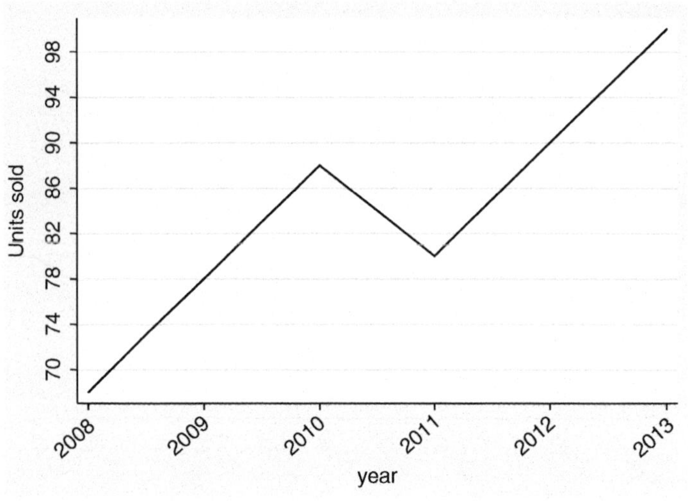

Fig. 10.3 Shortening the *y*-axis

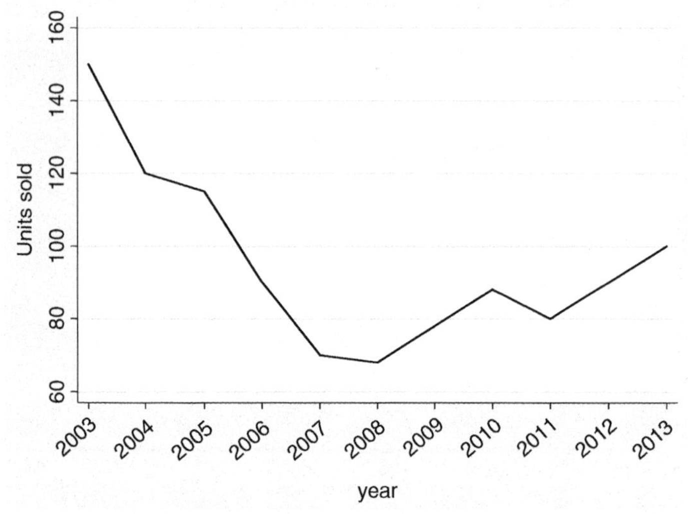

Fig. 10.4 The "floating" *y*-axis (I)

Data are often presented by means of three-dimensional figures, such as in Fig. 10.6. While these can be visually appealing, they are also subject to window-dressing. In this example, the lengths of all the edges were doubled to correspond to the 100% increase in turnover.

However, the resulting area is not twice but four times as large as the original image, thus presenting a false picture of the increase. These are just some common examples; Huff's (1993) classical text offers more on this topic.

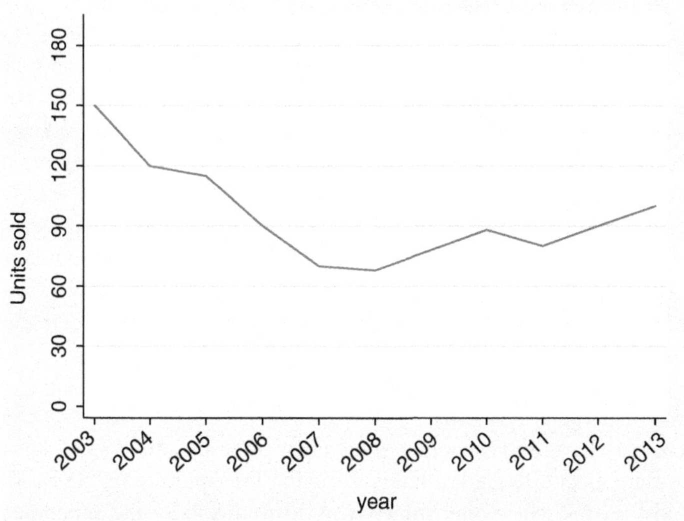

Fig. 10.5 The “floating” y-axis (II)

Fig. 10.6 Doubling the edge length quadruples the area

Tables are generally less susceptible to manipulation, as they contain data in numbers, which the reader can check and understand. As a rule of thumb, each table or graph in the report should be numbered sequentially and have a meaningful title so that it can be understood without reading the text. This is called a **self-contained table** or **self-contained figure**. Some rules of thumb:

- Put data to be compared in columns, not rows.
- Round data off to whole percentages, thousands or millions for sales, and two or three digits for academic purposes.
- Highlight data to reinforce conclusions (e.g., making the key numbers bold).
- Clearly state the units of measurement.

10.4.6.2 Presenting Statistical Data

In this section, we describe various ideas that you can use to convey statistical results in a reader-friendly manner. In the results section it is common to start presenting the descriptive statistics first. This comprises text that is visually supported by graphs, to offer information and context to those readers with the required background. Graphs offer two major advantages: first, they organize and simplify complex and dense information, which is especially useful with large data samples (Tufte 2001); second, graphs can summarize dense information efficiently. There are, of course, many kinds of graphs and each type has its advantages and disadvantages. The sample size and the nature of your data may constrain the choice of graphs. Sometimes your client may even have specific requirements. Here are some tips to help you with the presentation of your data:

Summarize your results efficiently

Graphs, like bar charts and especially dot charts, offer a useful way of summarizing descriptive data most efficiently; that is, by using less space (Cox 2008). Bar charts are generally useful where the objective is to depict how the outcome variable varies across two or more grouping variables. Figure 10.7 uses the *Oddjob.dta* dataset to illustrate this point by plotting the average overall satisfaction level with the price of the airline over the (1) respondents' gender, (2) flight frequency, and (3) country of residence. Note that, in a complete presentation, it is important to include a title and subtitle, to label the *y*-axis and the *x*-axis, and to list the source of the data below the figure. Details of the syntax used are shown in the ⭳ Web Appendix (→ Downloads).

Combine several plots

Stata offers many more graphical options, many of which we discussed in Chap. 5. This chapter will therefore not review the many graphical options in Stata. However, it is worth mentioning `coefplot`, a practical user-written program for plotting model estimates with Stata (Jann 2014). To install the program, type `help coefplot` in the command window and follow the instructions. The `coefplot` program offers many useful options for plotting statistical results innovatively. These options range from plotting descriptive data to more complex statistical outputs. It can also combine coefficients from different models by saving each estimated model separately before matching the coefficients and equations in a combined plot. Figure 10.8 offers an example and depicts the proportion of male and female travelers with different levels of disagreement or agreement with Oddjob Airways' price along the seven different points of the overall satisfaction scale. The inclusion of confidence intervals in the graph—indicated by the vertical lines at the end of each bar—show that the proportion of differences between males and females are not statistically significant. This is because the vertical lines of each of the seven pairs of bars overlap. This matter could be unclear if the confidence intervals are not included and may lead to misinterpretation by knowledgeable readers, who assume that such differences are statistically significant. Details of the syntax used are included in the ⭳ Web Appendix (→ Downloads).

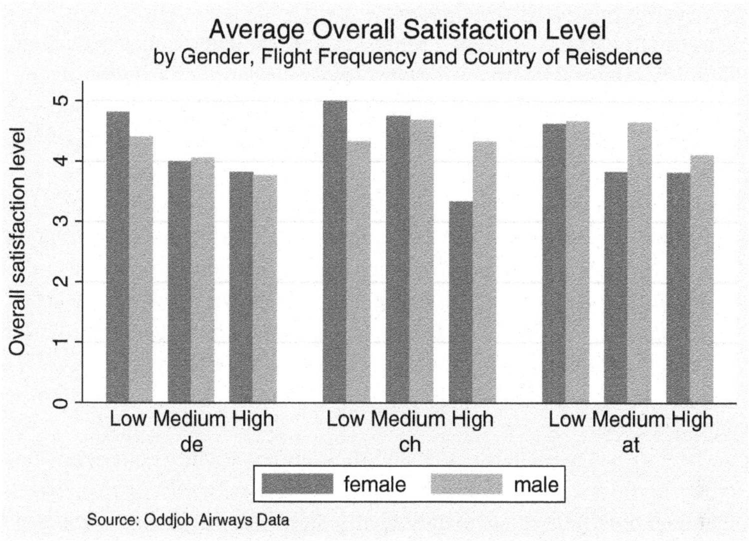

Fig. 10.7 Bar chart presentation

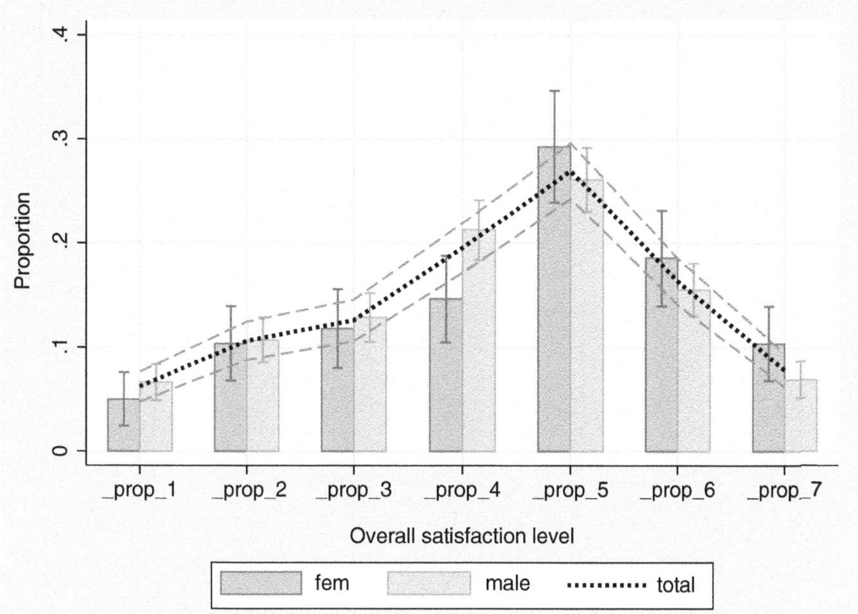

Fig. 10.8 Combination of different estimations and charts using `coefplot`

Consider plotting regression coefficients
It can be useful to plot the estimated regression coefficients, rather than show them in a table. This is not a substitute for presenting your careful data analysis in tables, but an aid and a complement. A graphical presentation of the regression coefficient estimates can be an efficient way of depicting both the significance and direction of the predicted effects.

In Stata, regression coefficient estimates are plotted in three steps. First, a regression model is estimated. Then the regression coefficients are predicted in a second step. Third, the estimated coefficients with their corresponding confidence intervals are plotted in a profile plot. Figure 10.9 shows the result of these steps using the *Oddjob.dta* dataset. The figure depicts how *commitment* to fly with the airline relates to *nflightsx* (flight frequency), *age*, and the travelers' *gender*. In the ⭳ Web Appendix (→ Downloads), you find details on how to plot regression coefficients yourself.

Figure 10.9 shows that coefficients and their corresponding confidence intervals do not cross the **0** line on the *x*-axis (the vertical line on the right-hand side). This means that each coefficient has a statistically significant association (at $p < 0.05$) with *commitment*. The figure also shows the direction of the predicted effects, with negative effects left of the **0** line and positive effects right of the line. This means that *nflightsx* and *gender* are negatively related to customers' commitment, while *age* is positively related.

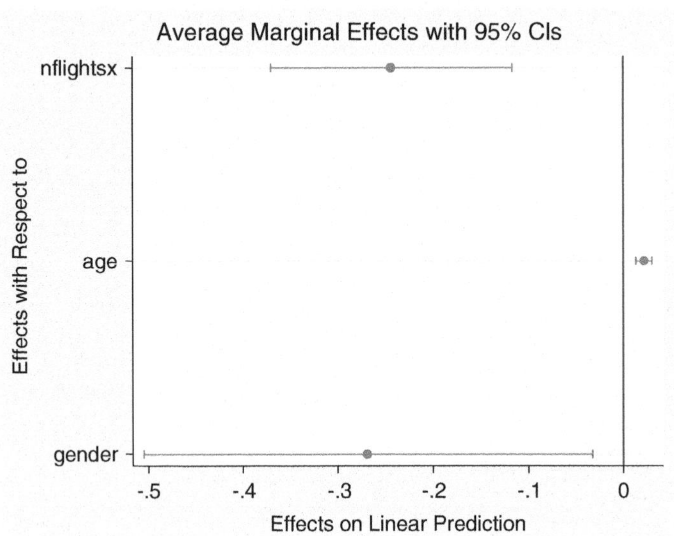

Fig. 10.9 Graphical presentation of regression coefficient estimates

Finally, `coefplot` is not limited to regression, but can be extended to any other estimation method, varying from logistic regression models, in which the outcome variable is binary by nature, to multinomial and poisson regression models with categorical and count type outcome variables. See Jann (2014) for further details on the program.

Make concise and clear (regression) tables with Stata

Research projects often require running several analyses on the same set of data by using slightly different settings in the data analysis technique. For example, researchers generally run a regression analysis on multiple models, which differ with regard to the number of independent variables or the type of dependent variable. Typing the output tables for such multiple analyses by hand is time-consuming and often leads to errors. Fortunately, Stata offers different ways of creating tables that avoid such mistakes. Before showing how Stata can be helpful in this respect, it is important to understand what the different rows and columns of a table mean. When multiple models are estimated, these are usually presented in adjacent columns to make comparisons easier. In regression tables, each column represents the results of one regression analysis. The rows indicate each independent variable's influence on the dependent variable by means of the (standardized or unstandardized) regression coefficient estimates. To create a clear table, include the:

1. *Model Title*: A first step in any table is to label each model that you are presenting. The label should be self-explanatory ("Commitment to Oddjob Airlines"). For academic purposes, a model number (e.g., Model 1, Model 2, etc.) or a title that best represents the model (e.g., Baseline model; Extended model, etc.) is sometimes used. This is particularly useful when writing up the results as you can then refer to and compare the estimates of the different models in the text. The choice of model name depends on the audience and the formatting guidelines.
2. *Independent variables*: In a (regression) table, the rows refer to the independent variables in the model. Give the variables a straightforward name to make it easier for the reader to understand. Make sure that these variable names are identical to those used in other tables and graphs (if any).
3. *Coefficient estimates*: Depending on your audience, the (regression) table needs to specify whether standardized or unstandardized coefficient estimates are being presented. This can be included as a subtitle and explained above the table. In Table 10.2, the type of coefficient estimates is indicated as **b/se.** Sometimes this is listed directly below the table.
4. *Significance level*: Putting asterisks (*) behind the estimated regression coefficients is a common way of presenting the coefficients' significance levels. Usually, one asterisk indicates a significance level of 0.10, two asterisks a significance level of 0.05, and three asterisks a significance level of 0.01. Depending on the audience, researchers sometimes present only effects with a

Table 10.2 A regression table made using `estout`

	Model 1	Model 2	Model 3
	b/se	b/se	b/se
nflightsx	-0.316***	-0.271***	-0.243***
	(0.06)	(0.06)	(0.06)
Age		0.022***	0.022***
		(0.00)	(0.00)
Gender			-0.269*
			(0.12)
constant	4.810***	3.614***	4.022***
	(0.14)	(0.27)	(0.33)
R-squared	0.023	0.046	0.051
N	1065	1065	1065
df_m	1063.0	1062.0	1061.0
BIC	4189.6	4170.8	4172.7

* p<0.05, ** p<0.01, *** p<0.001

significance of 0.05. Whatever strategy you choose, make sure you add a note below your table indicating the level of significance that the asterisks represent.

5. *Standard error* or *t-values:* In addition to the significance levels, you need to present the corresponding *t*-value or standard error of the coefficient estimate, which is usually placed in brackets below the coefficient estimates. Both presentation methods are accepted and the choice depends on your audience and the formatting criteria.
6. *Sample size and degrees of freedom:* For a complete presentation, you should also include the sample size and the models' degrees of freedom after the model estimation. These statistics are respectively indicated as **N** and **df_m** in Table 10.2. Reporting these statistics can reveal differences in the sample sizes of the different estimated models, which can be due to missing values in specific independent variables. If this happens, a comparison of the different models may make little sense, given that the models are based on different sample sizes with different characteristics. It is therefore important to understand what causes the large sample size differences between the models before taking any further action or drawing conclusions.
7. *Model fit*: Finally, depending on the type of estimated model, statistics indicating the model significance, such as the R^2 or η^2 (eta-squared), and the relative model fit, such as the AIC or BIC statistics, should be part of the table. In Table 10.2 we have included the **R-squared** and **BIC** to indicate the model fit.

You can produce tables with Stata by using the user-written package `estout`. As with any user-written program, you can install this Stata package immediately by typing `help estout` in the command window and then following the instructions. Table 10.2 shows an example of a regression table produced by using the `estout` command. It comprises three models that add several variables, at a time, containing some of the key elements of the Table that we mentioned earlier. These are: the model title, indicated as **Model 1**, **Model 2** and **Model 3**, with a caption indicating that the (unstandardized) type of coefficient estimates and their pertaining standard errors **b/se** are presented. Next, the independent variables, with the corresponding coefficients, standard errors (in brackets), and significance level are listed in the first column. For example, in **Model 1**, the first row presents the unstandardized coefficient for *nfligtsx* (**−0.316**), together with the pertaining standard errors (e.g., **(0.06)**), and significance level (*******) for which a caption is included at the bottom of the table. Next, the sample size (**N**) and the models' degrees of freedom (**df_m**) in terms of each estimated model are shown in Table 10.2. Finally, both the **R-squared** and the **BIC** values of each estimated model are included to indicate the model fit.

Further information that enables you to produce similar tables yourself can be found in the ⭳ Web Appendix (→ Downloads).

10.4.7 Conclusion and Recommendations

Having presented the findings, the next step is to summarize the most relevant points and interpret them in the light of the research objectives. You should write the conclusions in such a way that they present information that is relevant for managerial decision-making. Keep in mind that, for the client, the quality of the marketing research depends heavily on how well decision makers can use the information! The research must provide the client with clear benefits, which could lead to further research assignments.

Researchers are increasingly asked to go beyond stating facts and to provide recommendations or to advise. Whereas conclusions based solely on the research should be unbiased and impersonal, specific recommendations are based on a personal and (at least partially) subjective opinion on how the results can be most favorably used in the client's interest. You should therefore make sure that recommendations are recognizable as such. During the negotiations prior to the start of a project, the client needs to determine the extent to which the research report should include recommendations. This will also depend on the researcher's expertise in the area. Researchers may provide logical recommendations based up the their findings, but these might be unrealistic or impossible for the client to implement due to issues such as insufficient budgets, predetermined methods, or specific policies, regulations, and politics. Make sure that you or another member of your research team is familiar with the overall context, including the regulatory and legal issues, to avoid such issues. Furthermore, before making recommendations, review them with the client to determine whether these are acceptable and actionable (see Box 10.1 for an example).

Box 10.1 Bad Recommendations
A candy company wishes to know how it can increase its sales and has commissioned a research organization to gain insights into its different customer segments. The researchers find that teenagers are the most important target for the given brand and suggest that vending machines in schools would increase the company's revenue. Although this might boost sales, the recommendation does not help the company if vending machines are not allowed in schools. And even if they were allowed, they might lead to negative media reports.

10.4.8 Limitations

Finally, you should explain the extent to which the findings can be generalized. All research studies have limitations due to time, budget, and other constraints. Furthermore, errors might have occurred during the data collection. Not mentioning potential weaknesses (e.g., the use of a convenience sample, or a small sample size) for whatever reason reduces the research's credibility. Not disclosing important facts also violates common codes of industry conduct, such as those drafted by ESOMAR. Taking all these factors into regard, the research results should always be discussed objectively and in a balanced way. You should neither overly belittle the importance and validity of the research, nor try to conceal sources of errors and, hence, potentially mislead managers. Finally, some modesty is in order as, in hindsight, many reports have been proved inaccurate or even wrong. Few, for example, predicted the global financial crisis, the Trump presidency, or Brexit.

10.4.9 Appendix

All material not directly required for an understanding of the project, but still related to the study, should be included in the appendix or appendices. This includes questionnaires, interview guides, detailed data analyses, and other types of data or material.

10.5 Guidelines for Oral Presentations

Most clients want an oral presentation to accompany the written report. One could deliver such a presentation in the form of an interim report during the research, or at the end to explain the findings to the management and other staff. Members of the client staff often present the research findings to the management and do not ask the market research company to do so. Satisfaction with the delivered report may increase if a member of the client staff, such as an internal market researcher or business analyst, delivers the presentation, because the client feels they know and accept the content.

If asked to deliver an oral presentation, you should keep the principles of a written report in mind. It is especially important to identify and understand your audience, and to prepare the presentation thoroughly. A professional and interesting presentation might increase interest in the written report! Furthermore, since the oral presentation allows for interaction, interesting points can be highlighted and discussed in more detail. However, if you are not well prepared for the presentation, nor understand your audience's expectations, needs, and wants, you could face an unpleasant situation. You should always keep the following golden rule in mind: *Never deliver a presentation you wouldn't want to sit through*!

10.6 Visual Aids in Oral Presentations

It is useful to provide the audience with a written summary or a handout so that they do not have to take notes, but can focus on the presentation. If focus group interviews were conducted, for example, you could show excerpts from the recordings to provide concrete examples in support of a finding. The saying "a picture says more than a thousand words" is also true of the oral presentation. **Visual aids**, such as overhead transparencies, flip charts, or computer slide shows (e.g., PowerPoint or Prezi at http://www.prezi.com) not only help emphasize important points, but also facilitate the communication of difficult ideas. In the following, we summarize some suggestions (Armstrong 2010).

Use of Visual Aids:

- Use a simple master slide and avoid fancy animations.
- Use a sufficiently large font size (as a rule of thumb, 16pt. or higher and never less than 12pt.) so that everyone attending the presentation can read the slides.
- Use high contrasts for text. Use black and white. Do not write on illustrations or wallpapers.
- Use contrasting colors to emphasize specific points, but not too many.
- Use simple graphs, diagrams or short sentences rather than tables.

Arranging Visual Aids:

- Do not have too much information on one slide (generally, one key issue per slide). Never put a block of text on a page.
- Organize the material so that the different modes reinforce one another. For example, you do not want people running ahead of you, so either explain each point as you discuss it on a slide, or use many simple slides.
- Use a small number of slides compared to the time available for the presentation. The focus should be on the presenter and not on the slides. Having more slides than minutes available is not a good idea. Good presenters often use between 3 and 5 min to discuss a slide.
- Prepare (color) handouts for all members of the audience.

– If you intend to use media elements in your presentation, make sure that the equipment supports them (e.g., that the sound equipment is working, or that your video formats are supported).

10.7 Structure the Oral Presentation

Be aware that an oral presentation cannot cover the same amount of information as a written report. You must be selective and structure the presentation content clearly and logically. There are two ways of creating a presentation:

1. A common way of starting your presentation is by structuring the introduction in the classic narrative pattern of story-telling (situation → difficulty or complication → question → answer) introduced earlier in the context of written reports. Limit the introduction to what the audience can accept. Nothing could be worse than triggering resistance of what you are presenting right from the start of your oral presentation. Next, move on to the main part of your presentation. Based on a brief description of your major findings, capture the audience's attention by presenting answers to the logical questions that arise from the project, such as: "How were these results achieved?" or "How did we reach this conclusion?"
2. An alternative is to follow the **Minto principle,** according to which presentations have a **pyramid structure**, starting with the conclusion. This raises question in the audience's mind that has to be subsequently answered. Figure 10.10 illustrates this concept by using the example of a mobile phone study, which found that a novel smartphone should be introduced in white.

You begin by introducing the result of the study (i.e., the smartphone should be introduced in white) and then work your way down. Begin by explaining that a comprehensive market analysis was carried out, after which you discuss the elements of the analysis (i.e., focus group interviews, lead user interviews, and a customer survey). Finally, present the results of each element of the analysis (e.g., that lead users perceived black as too conservative, silver as too cheap, while white was perceived as modern). Once at the bottom of the pyramid, it is time to pause and to provide a summary, before moving from the first key line, which you have just presented, to the next key line, and so on. This process forces you to only provide the information relevant to the question under consideration. Moving from top to bottom and then bottom to top, helps you answer the questions: "Why so?" and "So what?," while being both exhaustive and mutually exclusive regarding the results and the concepts you have presented. Ensure you never provide findings that do not lead to specific conclusions and do not offer conclusions not based on findings. Ultimately, this pyramid approach helps the audience grasp the line of reasoning better. This technique is also frequently called the Minto principle or Minto pyramid after its creator Barbara Minto (2009).

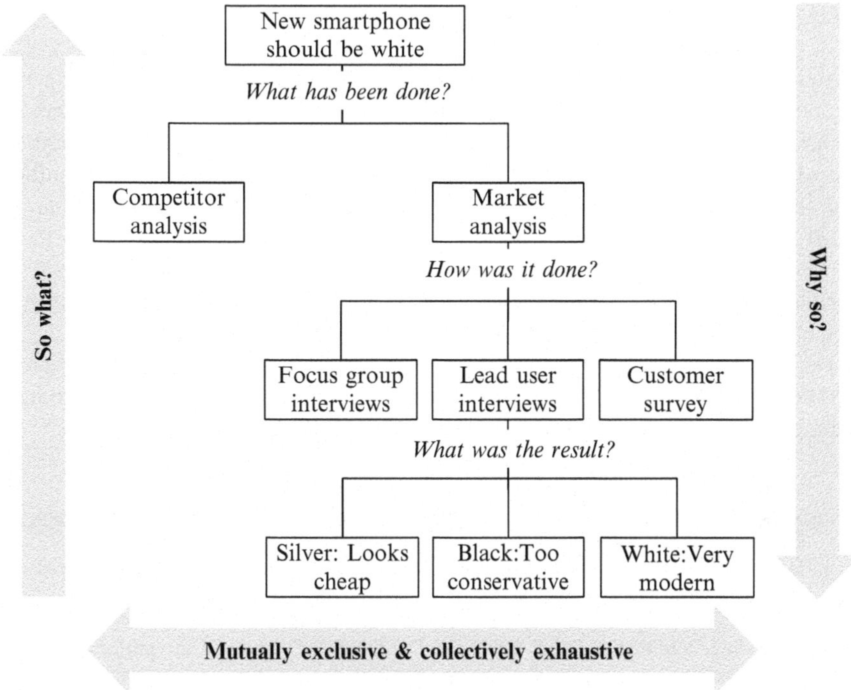

Fig. 10.10 Pyramid structure for presentations

10.8 Follow-Up

Having delivered the written report and oral presentation, two tasks remain: First, you may need to help the client implement the findings. This includes answering questions that may arise from the written report and oral presentation, helping select a product, advertising agency, marketing actions, etc., or incorporate information from the report into the firm's marketing information system or decision support system (see Chap. 3). This provides an opportunity to discuss other research projects. For example, you might agree on repeating the study after 1 year to see whether the marketing actions were effective. Second, you need to evaluate the market research project internally and with the client. Only (critical) feedback can disclose potential problems that may have occurred and, thus, provide the necessary grounds for improving your work. Using uniform questionnaires for the evaluation of different projects helps compare the feedback from different projects conducted simultaneously or at different points in time. However, some market research companies do not want to be involved in implementation.

10.9 Ethics in Research Reports

Ethics is an important topic in marketing research, because research interacts with human beings at several stages (e.g., during data collection and the communication of the findings). There are two "problematic" relations that can ultimately lead to ethical dilemmas. First, ethical issues arise when the researcher's interests conflict with those of the participants. For instance, the researcher's interest is to gather as much information as possible from the respondents, but they often require their answers to be treated confidentially and to remain anonymous. Second, in addition to researchers' legal and professional responsibilities towards their respondents, they also have reporting responsibilities.

For example, the European Society for Opinion and Marketing Research (ESOMAR) has established a code which sets minimum standards of ethical conduct to be followed by all researchers (ESOMAR 2007, p. 4):

1. Market researchers shall conform to all relevant national and international laws.
2. Market researchers shall behave ethically and shall not do anything which might damage the reputation of market research.
3. Market researchers shall take special care when carrying out research among children and young people.
4. Respondents' cooperation is voluntary and must be based on adequate, and not misleading, information about the general purpose and nature of the project when their agreement to participate is being obtained and all such statements shall be honoured.
5. The rights of respondents as private individuals shall be respected by market researchers and they shall not be harmed or adversely affected as the direct result of cooperating in a market research project.
6. Market researchers shall never allow personal data they collect in a market research project to be used for any purpose other than market research.
7. Market researchers shall ensure that projects and activities are designed, carried out, reported and documented accurately, transparently and objectively.
8. Market researchers shall conform to the accepted principles of fair competition.

In practice, researchers face an ethical dilemma. They are paid by the client and may feel forced to deliver "good" results. In this sense, they might be tempted to interpret results in a way that fits the client's perspective or the client's presumed interests. For instance, researchers might ignore data because they would reveal an inconvenient truth (e.g., the client's brand has low awareness, or customers do not like the product design).

Remember that researchers should never mislead the audience! For instance, it would be ethically questionable to modify the scales of a graph so that the results look more impressive, as shown in Figs. 10.1, 10.2, 10.3, and 10.4. Furthermore, researchers have a duty to treat information and research results confidentially, to store data securely, and to only use data for the research purpose agreed upon. Above all, you should keep in mind that marketing research is based on trust. Thus, when writing the report, you should respect the profession's ethical standards in order to maintain this trust.

10.10 Review Questions

1. What are the basic elements of any written research report?
2. Revisit the case study on Oddjob Airlines in Chap. 7 and prepare an outline for a written research report.
3. Consider the following situations. Do you think they confront the market researcher with ethical issues?
   (a) The client asks the researcher for a list of respondents to allow him/her to target selling activities at them.
   (b) The client asks the researcher not to disclose part of the research to his organization.
   (c) The client asks the researcher to present other recommendations.
   (d) The client asks the researcher to re-consider the analysis, because the findings seem implausible to him/her.
   (e) The client wishes to know the name of a particular customer who was very negative about the quality of service provided.

10.11 Further Readings

Huff D. (1993). *How to lie with statistics*. New York: Norton & Company.

First published in 1954, this book remains relevant as a wake-up call for people unaccustomed to the slippery world of means, correlations, and graphs. Although many of the examples used in the book are dated, the conclusions are timeless.

Durate N. (2008). *Slideology. The art and science of crafting great presentations*. Sebastopol: O'Reilly Media.

In this book, the author presents a rich source for effective visual expression in presentations. It is full of practical approaches to visual story development that can be used to connect with your audience. The text provides good hints to fulfill the golden rule to never deliver a presentation you wouldn't want to sit through.

Market Research Society at http://www.mrs.org.uk/standards/guidelines.htm

Under this link you find the (ethical) guidelines of the Market Research Society. The guidelines discuss, for example, the ethical issues surrounding research using children or the elderly as participants.

References

Armstrong, J. S. (2010). *Persuasive advertising: Evidence-based principles*. New York: Palgrave Macmillan.

Churchill, G. A., Jr., & Iacobucci, D. (2009). *Marketing research: Methodological foundations* (10th ed.). Mason: South-Western College Publishers.

Cox, N. J. (2008). Speaking Stata: Between tables and graphs. *Stata Journal, 8*(2), 269–289.

European Society for Opinion and Marketing Research (ESOMAR). (2007). ICC/ESOMAR International Code On Market And Social Research. http://www.esomar.org/uploads/public/knowledge-and-standards/codes-and-guidelines/ICCESOMAR_Code_English_.pdf

Huff, D. (1993). *How to lie with statistics*. New York: W. W. Norton & Company.

Jann, B. (2014). Plotting regression coefficients and other estimates. *Stata Journal, 14*(4), 708–737.

Minto, B. (2009). *The pyramid principle: Logic in writing and thinking* (3rd ed.). Harlow: Pearson.

Tufte, E. R. (2001). *The visual display of quantitative information* (2nd ed.). Cheshire: Graphics Press.

Glossary

α-Error occurs when erroneously rejecting a true null hypothesis. Also referred to as *type I error*.

α-Inflation results when multiple tests are conducted simultaneously on the same data. The result is that you are more likely to claim a significant result when this is not so (i.e., an increase or inflation in the type I error).

Acquiescence describes the tendency of respondents from different cultures to agree with statements (e.g., as formulated in a Likert scale item) regardless of their content.

Adjusted coefficient of determination is a modified measure of goodness-of-fit that takes the number of independent variables and the sample size into account. The statistic is useful for comparing regression models with different numbers of independent variables, sample sizes, or both.

Adjusted R^2 See *Adjusted coefficient of determination.*

Agglomerative clustering is a type of hierarchical clustering method in which clusters are consecutively formed from objects. It starts with each object representing an individual cluster. The objects are then sequentially merged to form clusters of multiple objects, starting with the two most similar.

Aggregation is a type of scale transformation in which variables measured at a lower level are taken to a higher level.

Akaike information criterion (AIC) is a relative measure of goodness-of-fit, which can be used to assess the various statistical models such as regression or factor analysis. Compared to an alternative solution with a different number of variables or factors, smaller AIC values indicate a better fit.

American Marketing Association (AMA) is the world's leading association for marketing professionals.

Analysis of variance (ANOVA) is a multivariate data analysis technique that allows testing whether the means of (typically) three or more groups differ significantly on one (one-way ANOVA) or two (two-way ANOVA) metric dependent variable(s). There are numerous extensions to more dependent variables and to differently scaled independent variables.

Anti-image is a measure used in principal component and factor analysis to determine whether the items correlate sufficiently. The anti-image describes

E. Mooi et al., *Market Research*, Springer Texts in Business and Economics,
DOI 10.1007/978-981-10-5218-7

the portion of an item's variance that is independent of another item in the analysis.

Arithmetic mean See *mean.*

Armstrong and Overton procedure is used to assess the degree of non-response bias. This procedure calls for comparing the first 50% respondents with the last 50% with regard to key demographic variables. The concept behind this procedure is that later respondents more closely match the characteristics of non-respondents.

Autocorrelation occurs when the residuals from a regression analysis are correlated.

Average see *mean.*

Average linkage is a linkage algorithm in hierarchical clustering methods in which the distance between two clusters is defined as the average distance between all pairs of objects in the two clusters.

Average marginal effects average change (in percentage) of one variable when another variable increases by one unit.

β-Error occurs when erroneously accepting a false null hypothesis. Also referred to as *type II error.*

Back-translation is a translation method used in survey research in which a survey is being translated and then back-translated into the original language by another person.

Balanced scale describes a scale with an equal number of positive and negative scale categories.

Bar chart is a graphical representation of a single categorical variable indicating each category's frequency of occurrence. Bar charts are primarily useful for describing nominal and ordinal variables.

Bartlett method is a procedure to generate factor scores in principal component analysis. The resulting factor scores have a zero mean and a standard deviation larger than one.

Bayes information criterion (BIC) is a relative measure of goodness-of-fit, which is similar to the AIC. Compared to the AIC, the BIC applies a greater penalty to statistical analysis that have a greater number of variables or factors.

Big data refers to very large datasets, generally a mix of quantitative and qualitative data in very large volumes.

Binary logistic regression is a type of regression method used when the dependent variable is binary and only takes two values.

Bivariate statistics describes statistics that express the empirical relationship between two variables. Covariance and correlation are key measures that indicate (linear) associations between two variables.

Bonferroni correction is a post hoc test typically used in an ANOVA that maintains the familywise error rate by calculating a new pairwise alpha that divides the statistical significance level α by the number of comparisons made (see also *familywise error rate* and *α-Inflation*).

Box-and-whisker plot See *box plot.*

Box plot shows the distribution of a variable. A box plot is a graph representing a variable's distribution and consists of elements expressing the dispersion of the data. Also referred to as *box-and-whisker plot.*

Breusch-Pagan test is used to test for heteroskedasticity in regression analysis.

Canberra distance is a distance measure used in cluster analysis. The Canberra distance is a weighted version of the city-block distance, typically used for clustering data scattered widely around an origin.

Case is an object such as a customer, a company, or a country in statistical analysis. Also referred to as *observation.*

Causal research is used to understand the relationships between two or more variables. Causal research explains how variables relate.

Census is a procedure of systematically acquiring and recording information about all the members of a given population.

Centroid linkage is a linkage algorithm in hierarchical clustering methods in which the distance between two clusters is defined as the distance between their geometric centers (centroids).

Chaining effect is a solution pattern typically observed when using a single linkage algorithm in cluster analysis.

Chebychev distance is a distance measure used in cluster analysis that uses the maximum of the absolute difference in the clustering variables' values.

City-block distance is a distance measure used in cluster analysis that uses the sum of the variables' absolute differences. Also referred to *as Manhattan metric.*

Closed-ended questions is a type of question format in which respondents have a certain number of response categories from which to choose.

Cluster analysis is a class of methods that groups a set of objects with the goal of obtaining high similarity within the formed groups and high dissimilarity between groups.

Clustering variables are variables used in cluster analysis.

Clusters are groups of objects with similar characteristics.

Codebook contains essential details of a data file, such as variable names and summary statistics.

Coefficient of determination (R^2) is a measure used in regression analysis to express the dependent variable's amount of variance that the independent variables explain.

Collinearity arises when two variables are highly correlated.

Communality describes the amount of a variable's variance that the extracted factors in a principal component and factor analysis reproduce.

Complete linkage is a linkage algorithm in hierarchical clustering methods in which the distance between two clusters corresponds to the longest distance between any two members in the two clusters.

Components are extracted in the course of a principal component analysis. They are also commonly referred to as factors.

Computer-assisted web interviews (CAWI) See *Web surveys.*

Confidence interval provides the lower and upper limit of values within which a population parameter will fall with a certain probability (e.g., 95%).

Confirmatory factor analysis is a special form of factor analysis used to test whether the measures of a construct are consistent with a researcher's understanding of that construct.

Constant sum scale is a type of scale that requires respondents to allocate a certain total number of points (typically 100) to a number of alternatives.

Constant is a characteristic of an object whose value does not change.

Construct scores are composite scores that calculate a value for each construct of each observation. Construct scores are often computed by taking the mean of all the items associated with the construct.

Construct validity is the degree of correspondence between a measure at the conceptual level and its empirical manifestation. Researchers often use this as an umbrella term for content, criterion, discriminant, face, and nomological validity.

Construct measures a concept that is abstract, complex, and cannot be directly observed by (multiple) items. Also referred to as *latent variable*.

Content validity refers to the extent to which a measure represents all facets of a given construct.

Correlation residuals are the differences between the original item correlations and the reproduced item correlations in a principal component and factor analysis.

Correlation is a measure of how strongly two variables relate to each other. Correlation is a scaled version of the covariance.

Covariance is a measure of how strongly two variables relate to each other.

Covariance-based structural equation modeling (CB-SEM) is an approach to structural equation modeling to test relationships between multiple items and constructs.

Criterion validity measures how well one measure predicts the outcome of another measure when both are measured at the same time.

Cronbach's alpha is a measure of internal consistency reliability. Cronbach's alpha generally varies between 0 and 1 with greater values indicating higher degrees of reliability.

Cross validation entails comparing the results of an analysis with those obtained when using a new dataset.

Crosstabs are tables in a matrix format that show the frequency distribution of nominal or ordinal variables.

Customer relationship management (CRM) refers to a system of databases and software used to track and predict customer behavior.

Data entry errors is a mistake in transcribing data during data entry. Erroneous values that fall outside a variable's standard range can easily be identified by means of descriptive statistics (minimum, maximum, and range).

Degrees of freedom (df) represents the amount of information available to estimate a test statistic. In general terms, an estimate's degrees of freedom are equal to the amount of independent information used (i.e., the number of observations) minus the number of parameters estimated.

Dendrogram visualizes the results of a cluster analysis. Horizontal lines in a dendrogram indicate the distances at which the objects have been merged.

Dependence of observations is the degree to which observations are related.

Dependent variables are the concepts a researcher wants to understand, explain, or predict.

Depth interview is an interview type typically used in exploratory research that allows one-to-one probing to foster interaction between the interviewer and the respondent.

Descriptive research is used to detail certain phenomena, characteristics, or functions. Descriptive research often builds on previous exploratory research.

Discriminant validity ensures that a measure is empirically unique and represents phenomena of interest that other measures in a model do not capture.

Distance matrix expresses the distances between pairs of objects.

Disturbance term see *Residual*.

Divisive clustering is a type of hierarchical clustering method in which all objects are initially merged into a single cluster, which the algorithm then gradually splits up.

Double-barreled questions are survey questions to which respondents can agree with one part but not with the other. Also refers to survey questions that cannot be answered without accepting an assumption.

Duda-Hart index is a statistic used in cluster analysis to determine the number of clusters to extract from the data. The statistic compares the sum of the squares in a pair of clusters to be split both before and after this extraction.

Dummy variables are binary variables that indicate whether a certain trait is present or not.

Durbin-Watson test is a test for autocorrelation used in regression analysis.

Eigenvalue indicates the amount of variance reproduced by a specific component or factor.

Eigenvectors are the results of a principal component analysis and include the factor weights.

Equidistance is indicated when the (psychological) distances between a scale's categories are identical.

Equidistant scale is a scale whose scale categories are equidistant.

Error is the difference between the regression line (which represents the regression prediction) and the actual observation.

Error sum of squares quantifies the difference between the observations and the regression line.

ESOMAR is the world organization for market, consumer, and societal research.

Estimation sample is the sample used to run a statistical analysis.

Eta-squared (η^2) is a statistic used in an ANOVA to describe the ratio of the between-group variation to the total variation, thereby indicating the variance accounted for by the sample data. There are two types of η^2: the model η^2, which is identical to the R^2, and each variable's partial η^2, which describes the percentage of the total variance accounted for by that variable. Eta squared also refers to the percentage of variance explained by a single variable in regression analysis.

Ethics are a system of morals and principles which defines a research organization's obligations, for example, with regard to the findings they release being an accurate portrayal of the survey data.

Ethnography is a type of qualitative research in which the researcher interacts with consumers over a period to observe and question them.

Euclidean distance is a distance measure commonly used in cluster analysis. It is the square root of the sum of the squared differences in the variables' values. Also referred to as *straight-line distance*.

Experimental design describes which treatment variables to administer and how these relate to dependent variables. Prominent experimental designs include the one-shot case study, the before-after design, the before-after design with a control group, and the Solomon four-group design.

Experiments are study designs commonly used in causal research in which a researcher controls for a potential cause and observes corresponding changes in hypothesized effects via treatment variables.

Exploratory factor analysis is a type of factor analysis that derives factors from a set of correlated indicator variables without the researcher having to prespecify a factor structure.

Exploratory research is conducted when the researcher has little or no information about a particular problem or opportunity. It is used to refine research questions, discover new relationships, patterns, themes, and ideas or to inform measurement development.

External secondary data are compiled outside a company for a variety of purposes. Sources of secondary data include, for example, governments, trade associations, market research firms, consulting firms, (literature) databases, and social networks.

External validity is the extent to which the study results can be generalized to real-world settings.

Extreme response styles occur when respondents systematically select the endpoints of a response scale.

Face validity is the extent to which a test is subjectively viewed as covering the concept it purports to measure.

Face-to-face interview See *Personal interview*.

Factor analysis is a statistical procedure that uses the correlation patterns among a set of indicator variables to derive factors that represent most of the original variables' variance. Also referred to as *Principal axis factoring*.

Factor loading is the correlation between a (unit-scaled) factor and a variable.

Factor rotation is a technique used to facilitate the interpretation of solutions in principal component and factor analysis.

Factors are (1) independent variables in an ANOVA and (2) the resulting variables of a principal component and factor analysis that summarize the information from a set of indicator variables.

Factor scores are composite scores that calculate a value for each factor of each observation.

Factor variable is a categorical variable used to define the groups (e.g., three types of promotion campaigns) in an ANOVA.

Factor weights express the relationships between variables and factors.

Familywise error rate is the probability of making one or more false discoveries or type I errors when performing multiple hypotheses tests (see also *α-inflation*).

Field experiments are experiments in which the manipulation of a treatment variable occurs in a natural setting, thereby emphasizing the external validity, but potentially compromising internal validity.

Field service firms are companies that focus on conducting surveys, determining samples and sample sizes, and collecting data. Some of these firms also translate surveys or provide addresses and contact details.

Focus groups is a method of data collection in which four to six participants discuss a defined topic under the leadership of a moderator.

Forced-choice scale is an answer scale that omits a neutral category, thereby forcing the respondents to make a positive or negative assessment.

Formative construct is a type of measurement in which the indicators form the construct.

Free-choice scale is an answer scale that includes a neutral choice category. Respondents are therefore not forced to make a positive or negative assessment.

Frequency table is a table that displays the absolute, relative, and cumulative frequencies of one or more variables.

***F*-test of sample variance** see *Levene's test.*

***F*-test** a test statistic used in an ANOVA and regression analysis to test the overall model's significance.

Full service providers are large market research companies, such as The Nielsen Company, Kantar, or GfK, that offer syndicated and customized services.

Gower's dissimilarity coefficient a dissimilarity coefficient used in cluster analysis that works with a mix of binary and continuous variables.

Heteroskedasticity refers to a situation in regression analysis in which the variance of the residuals is not constant.

Heywood cases negative estimates of variances or correlation estimates greater than one in absolute value.

Hierarchical clustering methods develop a treelike structure of objects in the course of the clustering process, which can be top-down (divisive clustering) or bottom-up (agglomerative clustering).

Histogram is a graph that shows how frequently categories derived from a continuous variable occur.

Hypotheses are claims made about effects or relationships in a population.

Inconsistent answers are a respondent's contradictory answer patterns.

Independent samples *t*-test a test using the *t*-statistic that establishes whether two means collected from independent samples differ significantly.

Independent variables are variables that explain or predict a dependent variable.

In-depth interview is a qualitative conversation with participants on a specific topic.

Index consists of a set of variables that defines the meaning of the resulting composite.

Index construction is the procedure of combining several items to form an index.

Indicators See *items*.

Interaction effect refers to how the effect of one variable on another variable is influenced by a third variable.

Intercept is the expected mean value of the dependent variable in a regression analysis, when the independent variables are zero. Also referred to as a constant.

Internal consistency reliability is a form of reliability used to judge the consistency of results across items in the same test. It determines whether the items measuring a construct are highly correlated. The most prominent measure of internal consistency reliability is Cronbach's alpha.

Internal secondary data are data that companies compile for various reporting and analysis purposes.

Internal validity is the extent to which causal claims can be made in respect of the study results.

Interquartile range is the difference between the third and first quartile.

Inter-rater reliability is the degree of agreement between raters expressed by the amount of consensus in their judgment.

Interviewer fraud is an issue in data collection resulting from interviewers making up data or even falsifying entire surveys.

Item non-response occurs when people do not provide answers to certain questions, for example, because they refuse to answer, or forgot to answer.

Items represent measurable characteristics in conceptual models and statistical analysis. Also referred to as *indicators*.

Kaiser criterion is a statistic used in principal component and factor analysis to determine the number of factor to extract from the data. According to this criterion, researchers should extract all factors with an eigenvalue greater than one. Also referred to as *latent root criterion*.

Kaiser-Meyer-Olkin criterion is an index used to assess the adequacy of the data for a principal component and factor analysis. High values indicate that the data are sufficiently correlated. Also referred to as *measure of sampling adequacy (MSA)*.

KISS principle the abbreviation of "keep it short and simple!" and implies that any research report should be as concise as possible.

***k*-means** is a group of clustering methods that starts with an initial partitioning of all the objects into a prespecified number of clusters and then gradually reallocates objects in order to minimize the overall within-cluster variation.

***k*-means++** is a variant of the *k*-means method that uses an improved initialization process.

***k*-medians** is a popular variant of *k*-means that aims at minimizing the absolute deviations from the cluster medians.

***k*-medoids** is a variant of *k*-means that uses other cluster centers rather than the mean or median.

Lab experiments are performed in controlled environments (usually in a company or academic lab) to isolate the effects of one or more treatment variables on an outcome.

Label switching a situation in which the labels of clusters change from one analysis to the other.

Laddering is an interviewing technique where the interviewer pushes a seemingly simple response to a question in order to find subconscious motives. It is typically used in the means-end approach.

Latent concepts represent broad ideas or thoughts about certain phenomena that researchers have established and want to measure in their research.

Latent root criterion See *Kaiser criterion.*

Latent variable measures a concept that is abstract, complex, and cannot be directly observed by (multiple) items. Also referred to as *construct.*

Levene's test tests the equality of the variances between two or more groups of data. Also referred to as *F-test of sample variance.*

Likert scale is a type of answering scale in which respondents have to indicate their degree of agreement to a statement. The degree of agreement is usually set by the scale endpoints, which range from strongly disagree to strongly agree.

Limited service providers are market research companies that specialize in one or more services.

Line chart is a type of chart in which measurement points are ordered (typically according to their x-axis value) and joined with straight-line segments.

Linkage algorithm defines the distance from a newly formed cluster to a certain object or to other clusters in the solution.

Listwise deletion entails deleting cases with one or more missing value(s) in any of the variables used in an analysis.

Little's MCAR test is used to analyze the patterns of missing data by comparing the observed data with the pattern expected if the data were missing completely at random.

Local optimum is an optimal solution when compared with similar solutions, but not a global optimum.

Log transformation is a type of scale transformation commonly used to handle skewed data.

Mahalanobis distance is a distance measure used in cluster analysis that compensates for the collinearity between the clustering variables.

Mail surveys are paper-based surveys sent to respondents via regular mail.

Main effect is the effect of one independent variable (i.e., factor) on the dependent variable, ignoring the effects of all the other independent variables in a two-way ANOVA.

Manhattan metric See *City-block distance.*

Manipulation checks a type of analysis in experiments to check whether the experimental treatment was effective.

Mann-Whitney U test is the nonparametric equivalent of the independent samples t-test used to assess whether two sample means are equal or not.

Marginal mean represents the mean value of one category in respect of each of the other types of categories.

Market segmentation is the segmenting of markets into groups (segments) of objects (e.g., consumers) with similar characteristics (e.g., needs and wants).

Market segments are groups of objects with similar characteristics.

Matching coefficients are similarity measures that express the degree to which the clustering variables' values fall into the same category.

Mean is the most common method of defining a typical value of a list of numbers. It is equal to the sum of a variable's values divided by the number of observations. Also referred to as *arithmetic mean* or simply *average*.

Means-end approach a method used to identify the ends consumers aim to satisfy and the means (consumption) they use to do so.

Measure of sampling adequacy (MSA) See *Kaiser-Meyer-Olkin criterion*.

Measurement scaling refers to (1) the level at which a variable is measured (nominal, ordinal, interval, or ratio scale) and (2) the general act of using a set of variables to measure a construct.

Measures of centrality are statistical indices of a typical or average value of a list of numbers. There are two main types of measures of centrality, the *median* and the *mean*.

Measures of dispersion provide researchers with information about the variability of the data (i.e., how far the values are spread out). There are four main types of measures of dispersion: the *range*, *interquartile range*, *variance*, and *standard deviation*.

Median is a value that separates the lowest 50% of values from the highest 50% of values.

Middle response styles a systematic way of responding to survey items describing respondents' tendency to choose the midpoints of a response scale.

Minto principle a guideline for presentations that starts with the conclusion, raising questions in the audience's mind about the way this conclusion was reached. The presenter subsequently explains the steps involved in the analysis.

Missing at random (MAR) is a missing values pattern in which the probability that data points are missing varies from respondent to respondent.

Missing completely at random (MCAR) is a missing values pattern in which the probability that data points are missing is unrelated to any other measured variable and to the variable with the missing values.

Missing data occur when entire observations are missing (survey non-response) or respondents have not answered all the items (item non-response).

Mixed mode is the act of combining different ways of administering surveys.

Moderation analysis involves assessing whether the effect of an independent variable on a dependent variable depends on the values of a third variable, referred to as a moderator variable.

(Multi)collinearity is a data issue that arises in regression analysis when two or more independent variables are highly correlated.

Multi-item construct is a measurement of an abstract concept that uses several items.

Multinomial logistic regression is a type of regression analysis used when the dependent variable is nominal and takes more than two values.

Multiple imputation is a simulation-based statistical technique that replaces missing observations with a set of possible values (as opposed to a single value) representing the uncertainty about the missing data's true value.

Multiple regression is a type of regression analysis that includes multiple independent variables.

Mystery shopping is a type of observational study in which a trained researcher visits a store or restaurant and consumes their products/services.

Nested models are simpler versions of a complex model.

Net Promoter Score (NPS) is a measure of customer loyalty that uses the single question: "How likely are you to recommend our company/product/service to a friend or colleague?"

Nomological validity is the degree to which a construct behaves as it should in a system of related constructs.

Nonhierarchical clustering methods see *Partitioning methods.*

Nonparametric tests are statistical tests for hypothesis testing that do not assume a specific distribution of the data (typically a normal distribution).

Non-probability sampling is a sampling technique that does not give every individual in the population an equal chance of being included in the sample. The resulting sample is not representative of the population.

Nonrandom missing is a missing values pattern in which the probability that data points are missing depends on the variable and on other unobserved factors.

Null and alternative hypothesis the null hypothesis (indicated as H_0) is a statement expecting no difference or no effect. The alternative hypothesis (indicated as H_1) is the hypothesis against which the null hypothesis is tested.

Oblimin rotation is a popular oblique rotation method used in principal component and factor analysis and principal component analysis.

Oblique rotation is a technique used to facilitate the interpretation of the factor solution in which the independence of a factor to all other factors is not maintained.

Observation is an object, such as a customer, a company, or a country, in statistical analysis. Also referred to as *case.*

Observational studies are procedures for gathering data in which the researcher observes people's behavior in a certain context. Observational studies are normally used to understand what people are doing rather than why they are doing it.

Omega-squared (ω^2) is a statistic used in an ANOVA to describe the ratio of the between-group variation to the total variation, thereby indicating the variance accounted for by the data. It is commonly used for sample sizes of 50 or less and corresponds to the adjusted R^2 of regression analysis. Omega squared is also used to indicate effect sizes of individual variables in regression analysis.

One-sample *t*-test is a parametric test used to compare one mean with a given value.

One-tailed tests are a class of statistical tests frequently used when the hypothesis is expressed directionally (i.e., $<$ or $>$). The region of rejection is on one side of the sampling distribution.

One-way ANOVA is a type of ANOVA that involves a single metric dependent variable and one factor variable with three (or more) levels.

Open-ended questions are a type of question format that provides little or no structure for respondents' answers. Generally, the researcher asks a question and the respondent writes down his or her answer in a box. Also referred to as *verbatim items*.

Operationalization is the process of defining a set of variables to measure a construct. The process defines latent concepts and allows them to be measured empirically.

Ordinary least squares (OLS) is the estimation approach commonly used in regression analysis and involves minimizing the squared deviations from the observations to the regression line (i.e., the residuals).

Orthogonal rotation is a technique used to facilitate the interpretation of a factor solution in which a factor's independence is maintained from all other factors. The correlation between the factors is determined as zero.

Outliers are observations that differ substantially from other observations in respect of one or more characteristics.

Paired samples *t*-test is a statistical procedure used to determine whether there is a significant mean difference between observations measured at two points in time.

Parallel analysis is a statistic used in principal component and factor analysis to determine the number of factors to extract from the data. According to this criterion, researchers should extract all factors whose eigenvalues are larger than those derived from randomly generated data with the same sample size and number of variables.

Parametric tests are statistical tests that assume a specific data distribution (typically normal).

Partial least squares structural equation modeling (PLS-SEM) is a variance-based method to estimate structural equation models. The goal is to maximize the explained variance of the dependent latent variables.

Partial sums of squares is a statistic in an ANOVA indicating the additional portion of variance explained when another variable is added to the analysis.

Partitioning method is a group of clustering procedures that does not establish a treelike structure of objects and clusters, but exchanges objects between clusters to optimize a certain goal criterion. The most popular type of partitioning method is *k*-means.

Path diagram is a visual representation of expected relationships tested in a structural equation modeling analysis.

Personal interview is an interview technique that involves face-to-face contact between the interviewer and the respondent. Also referred to as *face-to-face interviews*.

Pie chart displays the relative frequencies of a variable's values.

Population is a group of objects (e.g., consumers, companies, or products) that a researcher wants to assess.

Post hoc tests are a group of tests used for paired comparisons in an ANOVA. Post hoc tests maintain the familywise error rate (i.e., they prevent excessive type I error).

Power of a test represents the probability of rejecting a null hypothesis when it is in fact false. In other words, the power of a statistical test is the probability of rendering an effect significant when it is indeed significant (defined by $1 - \beta$, where β is the probability of a type II error).

Practical significance refers to whether differences or effects are large enough to influence decision-making processes.

Predictive validity measures how well one measure predicts the outcome of another measure when both are measured at a later point in time.

Primary data are data gathered for a specific research project.

Principal axis factoring See *Factor analysis.*

Principal component analysis is a statistical procedure that uses correlation patterns among a set of indicator variables to derive factors that represent most of the original variables' variance. Different from factor analysis, the procedure uses all the variance in the variables.

Principal components are linear composites of original variables that reproduce the original variables' variance as well as possible.

Principal factor analysis See *Factor analysis.*

Probability sampling is a sampling technique that gives every individual in the population an equal chance, different from zero, of being included in the sample.

Profiling is a step in market segmentation that identifies observable variables (e.g., demographics) that characterize the segments.

Projective technique is a special type of testing procedure, usually used as part of in-depth interviews. This technique provides the participants with a stimulus (e.g., pictures, words) and then gauges their responses (e.g., through sentence completion).

Promax rotation is a popular oblique rotation method used in principal component and factor analysis and principal component analysis.

***p*-value** is the probability of erroneously rejecting a true null hypothesis in a given statistical test.

Pyramid structure for presentations See *Minto principle.*

Qualitative data are audio, pictorial, or textual information that researchers use to answer research questions.

Qualitative research is primarily used to gain an understanding of *why* certain things happen. It can be used in an exploratory context by defining problems in more detail or by developing hypotheses to be tested in subsequent research.

Quantitative data are data to which numbers are assigned to represent specific characteristics.

R^2 See Coefficient of determination.

Ramsey's RESET test is a test for linearity used in regression analysis.

Range standardization is a type of scale transformation in which the values of a scale are standardized to a specific range that the researcher has set.

Range is the difference between the highest and the lowest value in a variable measured, at least, on an ordinal scale.

Rank order scale is an ordinal scale that asks respondents to rank a set of objects or characteristics in terms of, for example, importance, preference, or similarity.

Reflective constructs is a type of measurement in which the indicators are considered manifestations of the underlying construct.

Regression method is a procedure to generate factor scores in principal component analysis. The resulting factor scores have a zero mean and unit standard deviation.

Regression sum of squares quantifies the difference between the regression line and the line indicating the average. It represents the variation in the data that the regression analysis explains.

Reliability is the degree to which a measure is free from random error.

Reliability analysis is an important element of a confirmatory factor analysis and essential when working with measurement scales. See *Reliability*.

Research design describes the general approach to answer a research question related to a marketing opportunity or problem. There are three broad types of research design: exploratory research, descriptive research, and causal research.

Residual is the unexplained variance in a regression model. Also referred to as *disturbance term*.

Reverse-scaled items are items whose statement (if a Likert scale is used) or word pair (if a semantic differential scale is used) is reversed when compared to the other items in the set.

Robust regression is a variant of regression analysis used when heteroskedasticity is present.

Russell and Rao coefficient is a similarity coefficient used in cluster analysis.

Sample size is the number of observations drawn from a population.

Sampling error occurs when the sample and population structure differ on relevant characteristics.

Sampling is the process through which objects are selected from a population.

Scale development is the process of defining a set of variables to measure a construct and which follows an iterative process with several steps and feedback loops. Also referred to as *operationalization*, or, in the case of an index, *index construction*.

Scale transformation is the act of changing a variable's values to ensure comparability with other variables or to make the data suitable for analysis.

Scanner data are collected at the checkout of a supermarket where details about each product sold are entered into a database.

Scatter plot is a graph that represents the relationship between two variables, thus portraying the joint values of each observation in a two-dimensional graph.

Scree plot is a graph used in principal component and factor analysis that plots the number of factors against the eigenvalues, resulting in a distinct break (elbow) that indicates the number of factors to extract. Following the same principle, the scree plot is also used in hierarchical cluster analysis to plot the number of clusters against the distances at which objects were merged.

Secondary data are data that have already been gathered, often for a different research purpose and some time ago. Secondary data comprise internal secondary data, external secondary data, or a mix of both.

Segment specialists are companies that concentrate on specific market segments, such as a particular industry or type of customer.

Self-contained figure is a graph in a market research report that should be numbered sequentially and have a meaningful title so that it can be understood without reading the text.

Self-contained table is a table in a market research report that should be numbered sequentially and have a meaningful title so that it can be understood without reading the text.

Semantic differential scales is a type of answering scale that comprises opposing pairs of words, normally adjectives (e.g., young/old, masculine/feminine) constituting the endpoints of the scale. Respondents then indicate how well one of the word in each pair describes how he or she feels about the object to be rated (e.g., a company or brand).

Sentence completion is a type of projective technique that provides respondents with beginnings of sentences that they have to complete in ways that are meaningful to them.

Sequential sums of squares is a statistic in an ANOVA that indicates the additional portion of variance explained when a set of variables is added to the analysis.

Shapiro-Wilk test is a test for normality (i.e., whether the data are normally distributed).

Significance level is the probability that an effect is incorrectly assumed when there is in fact none. The researcher sets the significance level prior to the analysis.

Simple matching coefficient is a similarity coefficient used in cluster analysis.

Simple regression is the simplest type of regression analysis with one dependent and one independent variable.

Single-item constructs is a measurement of a concept that uses only one item.

Single linkage is a linkage algorithm in hierarchical clustering methods in which the distance between two clusters corresponds to the shortest distance between any two members in the two clusters.

Skewed data occur if a variable is asymmetrically distributed. A positive skew (also called right skewed) occurs when many observations are concentrated on the left side of the distribution, producing a long right tail (the opposite is called negative skew or left skewed).

Social desirability bias occurs when respondents provide socially desirable answers (e.g., by reporting higher or lower incomes than are actually true) or take a position that they believe society favors (e.g., not smoking or drinking).

Social media analytics are methods for analyzing social networking data and comprise text mining, social network analysis, and trend analysis.

Social networking data reflect how people would like others to perceive them and, thus, indicate consumers' intentions. Product or company-related social networking data are of specific interest to market researchers.

Specialized service firms are market research companies that focus on particular products, markets, or market research techniques.

Split-half reliability is a type of reliability assessment in which scale items are divided into halves, and the scores of the halves are correlated.

Split-sample validation involves splitting the dataset into two samples, running the analysis on both samples, and comparing the results.

Stability of the measurement See *Test-retest reliability*.

Standard deviation describes the sample distribution values' variability from the mean. It is the square root of the variance and, therefore, a variant.

Standard error is the sampling distribution of a statistic's standard deviation, mostly from the mean.

Standardized effects express the relative effects of differently measured independent variables in a regression analysis by expressing them in terms of standard deviation changes from the mean.

Standardizing variables have been rescaled (typically to a zero mean and unit standard deviation) to facilitate comparisons between differently scaled variables.

Stata computer package specializing in quantitative data analysis.

Statistical inference is the process of drawing conclusions about populations from data.

Statistical significance occurs when an effect is so large that it is unlikely to have occurred by chance. Statistical significance depends on several factors, including the size of the effect, the variation in the sample data, and the number of observations.

Straight-line distance See *Euclidean distance*.

Straight-lining occurs when a respondent marks the same response in almost all the items.

Structural equation modeling is a multivariate data analysis technique used to measure relationships between constructs, as well as between constructs and their associated indicators.

Survey non-response occurs when entire responses are missing. Survey non-response rates are usually 75%–95%.

Surveys are often used for gathering primary data. Designing surveys involves a six-step process: (1) Determine the survey goal, (2) determine the type of questionnaire required and the administration method, (3) decide on the questions and (4) the scale, (5) design the questionnaire, and (6) pretest and administer the questionnaire.

Suspicious response patterns are issues in response styles in respect of straight-lining and inconsistent answers that a researcher needs to address in the analysis.

Syndicated data are data sold to multiple clients, allowing them to compare key measures with those of the rest of the market.

Telephone interviews allow researchers to collect data quickly and facilitate open-ended responses, although not as well as personal interviews.

Test markets are a type of field experiment that evaluates a new product or promotional campaign under real market conditions.

Test statistic is calculated from the sample data to assess the strength of the evidence in support of the null hypothesis.

Test-retest reliability is a type of reliability assessment in which the researcher obtains repeated measurement of the same respondent or group of respondents, using the same instrument and under similar conditions. Also referred to as *stability of the measurement.*

Ties are identical values in a distance matrix used in cluster analysis.

Total sum of squares quantifies the difference between the observations and the line indicating the average.

Transforming data is an optional step in workflow of data, involving variable respecification and scale transformation.

Treatments are elements in an experiment that are used to manipulate the participants by subjecting them to different situations. A simple form of treatment could be an advertisement with and without humor.

***t*-test** is the most popular type of parametric test for comparing a mean with a given standard and for comparing the means of independent samples (independent samples *t*-test) or the means of paired samples (paired samples *t*-test).

Tukey's honestly significant difference test is a popular post hoc test used in an ANOVA that controls for type I errors, but is limited in terms of statistical power. Often simply referred to as *Tukey's method.*

Tukey's method See *Tukey's honestly significant difference test.*

Two-sample *t*-test is the most popular type of parametric test for comparing the means of independent or paired samples.

Two-tailed tests are a class of statistical tests frequently used when the hypothesis is not expressed directionally (i.e., $\neq$). The region of rejection is on two sides of the sampling distribution.

Two-way ANOVA is a type of ANOVA that involves a single metric dependent variable and two factor variables with three (or more) levels.

Type I error occurs when erroneously rejecting a true null hypothesis. Also referred to as α error.

Type II error occurs when erroneously accepting a false null hypothesis. Also referred to as *β error.*

Unbalanced scale describes a scale with an unequal number of positive and negative scale categories.

Uniqueness is a statistic used in principal component analysis and factor analysis that indicates the proportion of a variable's variance that the factors do not capture. The uniqueness equals 1 – communality.

Unit of analysis is the level at which a variable is measured. Typical measurement levels include that of the respondents, customers, stores, companies, or countries.

Univariate statistics are statistics that describe the centrality and dispersion of a single variable.

Unstandardized effects express the absolute effects that one-unit increases in the independent variables have on the dependent variable in a regression analysis.

Validation sample is a random subsample of the original dataset used for validation testing.

Validity is the degree to which a researcher measures what (s)he wants to measure. It is the degree to which a measure is free from systematic error.

Variable represents a measurable characteristic whose value can change.

Variable names should be clear and short so that they can be read in the dialog boxes (e.g., *loyalty1*, *loyalty2*, etc.).

Variable respecification involves transforming data to create new variables or to modify existing ones.

Variance a measure of dispersion computed by the sum of the squared differences of each value and a variable's mean, divided by the sample size minus 1.

Variance inflation factor (VIF) quantifies the degree of collinearity between the independent variables in a regression analysis.

Variance ratio criterion is a statistic used in cluster analysis to determine the number of clusters. The criterion compares the within- and between-cluster variation of different numbers of clusters.

Varimax rotation is the most popular orthogonal rotation method used in principal component analysis and factor analysis.

Verbatim items See *Open-ended questions*.

Visual aids include overhead transparencies, flip charts, or slides (e.g., PowerPoint or Prezi) that help emphasize important points and facilitate the communication of difficult ideas in a presentation of market research results.

Visual analogue scale is a type of answering scale in which respondents use levers that allow scaling on a continuum. This scale does not provide response categories.

Ward's linkage is a linkage algorithm in hierarchical clustering methods that combines those objects whose merger increases the overall within-cluster variance by the smallest possible degree.

Web surveys are less expensive to administer and can be fast in terms of data collection, because they can be set up very quickly. Also referred to as *computer-assisted web interviews (CAWI)*.

Weighted average linkage is a variant of the average linkage algorithm used in cluster analysis that weights the distances according to the number of objects in the cluster.

Welch correction is a statistical test used in an ANOVA to assess the significance of the overall model when the group variances differ significantly and the groups differ in size.

White's test is a statistical test that detects the presence of heteroskedasticity in regression analysis.

Wilcoxon matched-pairs signed-rank test is the nonparametric equivalent of the paired samples *t*-test.

Wilcoxon signed-rank test is the nonparametric equivalent of the independent samples *t*-test.

Wilcoxon test is the nonparametric equivalent of the one sample *t*-test.

Workflow is a strategy to keep track of the entering, cleaning, describing, and transforming of data.

***z*-standardization** is a type of scale transformation in which the values of a scale are standardized to a zero mean and unit standard deviation.

***z*-test** is any statistical test for which the distribution of the test statistic under the null hypothesis can be approximated by a normal distribution.

Index

A
Acquiescence, 72, 101
Adaptive questioning, 68
Adjusted R^2, 189, 234
Agglomerative clustering, 322, 327
Aggregating data, 33
Aggregation, 122
Agresti, A., 118
Akaike information criterion (AIC), 235, 280
α error, 158
α-inflation, 180
Alternative hypothesis, 156
American Customer Satisfaction Index, 60
Analysis of variance (ANOVA), 166
Anti-image, 272
Arithmetic mean, 113
Armstrong, J.S., 385
Autocorrelation, 230
Average, 113
Average linkage, 323, 324

B
Back-translation, 72
Balanced scale, 77
Bar chart, 111
Bartlett method, 284
Bayes information criterion (BIC), 235, 280
Before-after design, 88
Before-after experiment with a control group, 89
Before measurement effect, 89
β error, 158
Between-group mean squares, 186
Between-group variation, 182–183
Big data, 57
Binary logistic regression, 220
Bivariate graphs, 115
Bivariate statistics, 117
Bonferroni correction, 187
Box-and-whisker plot, 112
Box plot, 112
Breusch-Pagan, 228
Bubble plot, 116

C
Calinski/Harabasz pseudo F, 341, 356
Canberra distance, 335
Case, 28
Causality, 18
Causal research, 18, 86
Census, 43
Centroid linkage, 323, 324
Chaining effect, 324
Chebychev distance, 334
χ^2*-test*, 108, 116
CIA World Fact Book, 54
City-block distance, 334
Closed-ended questions, 71
Cluster analysis, 314
Clustering variables, 314
Clusters, 314
Cluster sampling, 45
Codebook, 123
Coefficient of determination, 232
Collinearity, 221, 276
Communality, 276, 277
Company records, 53
Complete linkage, 323, 324
Composite measure, 29, 121
Computer-assisted personal interviews (CAPI), 66
Computer-assisted self-interviews (CASI), 66
Computer-assisted telephone interviews (CATI), 67
Computer-assisted web interviews (CAWI), 67
Confidence interval, 189

E. Mooi et al., *Market Research*, Springer Texts in Business and Economics,
DOI 10.1007/978-981-10-5218-7

Confirmatory factor analysis, 266
Constant, 28, 217
Constant sum scale, 74
Constructs, 28, 121
Construct score, 121
Construct validity, 39
Consulting firms, 55
Content validity, 40
Contingency coefficient, 119
Contingency tables, 116
Convenience sampling, 46
Conversion rate, 56
Correlation, 117
Correlation residuals, 283
Covariance, 117
Covariance-based structural equation modeling, 289
Cox, N. J., 378
Cramer's V, 119, 360
Criterion validity, 40, 319
Cronbach's alpha, 289
Crosstabs, 116
Cross-validation, 238
Customer relationship management, 53

D
Data entry errors, 102
Degrees of freedom, 171
Dendrogram, 341, 354
Dependence of observations, 34, 35
Dependent variables, 35, 216
Depth interviews, 15
Descriptive research, 17
Directional hypothesis, 157
Directly observed qualitative data, 82
Discriminant validity, 40
Disturbance term, 218
Divisive clustering, 322
"Don't know" option, 77
Double-barreled questions, 71
Duda-Hart index, 342, 343, 357
Dummies, 120
Dummy variables, 120, 256
Dunnett's method, 188
Durbin-Watson (D-W) test, 230

E
Effect size, 188
Eigenvalues, 276
Eigenvectors, 275
Enterprise Resource Planning, 53
Equidistance, 37
Equidistant scale, 78
Error, 218
Error sum of squares, 232
Estimation sample, 237
Eta squared, 188
Ethics, 388
Ethnographies, 16
Ethnography, 63
Euclidean distance, 333
Existing research studies, 54
Experimental design, 87
Experimental research, 86–89
Experiments, 86
Expert validity, 40
Exploratory factor analysis, 266
Exploratory research, 14
External secondary data, 54
External validity, 87
Extraneous variables, 86
Extreme response styles, 101

F
Facebook, 56
Face-to-face interviews, 66
Face validity, 40
Factor analysis, 266
Factor-cluster segmentation, 320
Factor extraction, 275–276
Factor levels, 180
Factor loading, 276, 295
Factor rotation, 280
Factors, 266
Factor scores, 283
Factor variable, 163, 180
Factor weights, 275
Familywise error rate, 180, 187
Field experiments, 21, 87
Field service, 7
Finlay, B., 118
Focus groups, 15, 84
Forced-choice scale, 76
Formative constructs, 30
Free-choice scale, 76
Frequency table, 113
F-test, 232
F-test of sample variance, 164
Full factorial design, 86
Full service providers, 6
Funnel approach, 78
Furthest neighbor, 323
Fused market research, 34

G
Global representativeness, 42
Governments, 54
Gower's dissimilarity coefficient, 339, 340

H
Hershberger, S.L., 284
Heywood cases, 299
Hierarchical clustering, 322
Histogram, 111
Hybrid market research, 34
Hypothesis, 16, 154

I
Inconsistent answers, 101
Independent observations, 35
Independent samples, 163
Independent samples t-test, 165, 175
Independent variables, 35, 216
In-depth interviews, 82
Index, 29, 121
Index construction, 29
Indicators, 28
Indirectly observed qualitative data, 82
Instagram, 56
Interaction effect, 190
Intercept, 217
Internal consistency reliability, 41, 288
Internal secondary data, 53
Internal validity, 87
Internet data, 56
Interquartile range, 114
Inter-rater reliability, 41
Interval scale, 37
Interviewer bias, 66, 68
Interviewer fraud, 100
Item content, 70
Item non-response, 105
Items, 28
Item wording, 71

J
Jaccard coefficient (JC), 338, 339
Judgmental sampling, 45

K
Kaiser criterion, 278, 295
Kaiser–Meyer–Olkin (KMO), 272
Kendall's tau, 119
KISS principle, 370
k-means, 329, 330
k-means++, 333
k-medians, 333
k-medoids, 333
Kruskal-Wallis H test, 167
Kruskal-Wallis rank test, 167

L
L1, 334
Lab experiments, 21, 87
Laddering, 83
Latent concepts, 28
Latent root criterion, 278
Latent variable, 28
Lead users, 103
Left-skewed, 122
Left-tailed hypothesis, 158
Levels of measurement, 35
Levene's test, 164
Likert, 98
Likert scale, 73
Limited service providers, 7
Line chart, 116
Linfinity, 334
Linkage algorithm, 323
LinkedIn, 56
Listwise deletion, 105, 108, 109
Literature databases, 55
Little's MCAR test, 107
Local optimum, 332
Log transformation, 122
Lower adjacent value, 112
L2 squared, 333

M
Mahalanobis distance, 335
Mail surveys, 68
Main effect, 190
Manhattan metric, 334
Manipulation checks, 86
Mann-Whitney U test, 167
Marginal effect, 206
Marginal means, 155
Marketing opportunities, 12
Marketing symptoms, 12
Market research firms, 54
Market segmentation, 17, 314
Matching, 337
Matching coefficients, 337
Mean, 113

Means-end approach, 83
Measurement error, 38
Measurement scaling, 35
Measure of sampling adequacy (MSA), 272
Measures of centrality, 113
Measures of central tendency, 113
Measures of dispersion, 114
Median, 112, 113
Middle response styles, 101
Minto principle, 386
Misresponse rates, 72
Missing at random (MAR), 106
Missing completely at random (MCAR), 105
Missing data, 104
Mixed methodology, 34
Mixed mode, 69
Mobile phone surveys, 67
Moderation analysis, 237
Multicollinearity, 221
Multi-item constructs, 30
Multi-item scale, 28
Multinomial logistic regression, 220
Multiple imputation, 105, 106, 108, 109
Multiple regression, 218
Mystery shopping, 63

N
Negative skew, 122
Nested models, 234
Net Promoter Score (NPS), 30
Noise, 184
Nominal scale, 36
Non-directional hypothesis, 158
Non-hierarchical clustering methods, 330
Nonparametric tests, 154
Non-probability sampling, 45
Non-random missing (NRM), 107
Normal probability plot, 165
Null hypothesis, 156
Number of answer categories, 74

O
Oblimin rotation, 281
Oblique rotation, 281
Observation, 28
Observational studies, 15, 62
Omega squared, 189
One-sample t-test, 165, 169
One-shot case study, 87
One-tailed test, 168
One-way ANOVA, 180
Open-ended questions, 71
Operationalization, 29
Oral presentation, 368
Ordinal scale, 36
Ordinary least squares (OLS), 223
Orthogonal rotation, 281
Outliers, 102, 231
Outside values, 112

P
Page requests, 56
Paired samples, 163
Paired samples t-test, 165, 178
Parallel analysis, 279
Parametric tests, 154
Partial least squares structural equation modeling, 289
Partial sums of squares, 183
Partitioning clustering methods, 329
Path diagram, 287
Pearson's chi-squared, 360
Pearson's correlation coefficient, 117
Personal interviews, 66
Phi, 119
Pie chart, 113
5-Point scales, 75
7-Point scales, 75
Population, 41
Positive skew, 122
Post hoc tests, 187
Potential Ratings Index by Zip Markets (PRIZM), 346
Power analysis, 160
Power of a statistical test, 159
Practical significance, 159
Practice effects, 41
Predictive validity, 40
Presentations, 386
Pretest, 80–82
Primary data, 31, 62
Principal axis factoring, 266
Principal component analysis (PCA), 266
Principal-component factor, 294
Principal components, 275
Principal factor analysis, 266
Probability sampling, 43
Profiling, 344
Projective techniques, 15, 84
Promax rotation, 281
Pseudo T-squared, 343
p-value, 172
Pyramid structure, 386

Q
Qualitative data, 32
Qualitative research, 34, 82
Qualtrics, 67
Quantile plot, 165
Quantitative data, 32
Quantitative research, 34
Quartile, 114
Questionnaire design, 78
Question order, 78
Quota sampling, 46

R
R^2, 189, 232
Ramsey's RESET test, 227
Random error, 38
Random noise, 184
Range, 114
Rank order scales, 73
Ratio scale, 37
Recoding, 120
Reflective constructs, 30
Regression method, 284
Regression sum of squares, 232
Related factors, 190
Reliability, 38
Reliability analysis, 267
Representative sample, 42
Research design, 13
Research problem, 12
Residual, 218
Respondent bias, 66
Response categories, 76
Response category labeling, 77
Response rates, 81
Reverse-scaled items, 72, 100
Right-skewed, 122
Right-tailed hypothesis, 158
Robust regression, 226
Russell and Rao coefficient, 338, 339

S
Sales reports, 53
Sample size, 47
Sampling, 41
Sampling error, 44, 158
Sampling frame, 43
Scale development, 29
Scale properties, 74
Scale transformation, 121
Scale types, 73
Scanner data, 17
Scatter plot, 104, 115
Scheffé's method, 188
Scree plot, 278
Screening questions and Screeners, 78
Search engines, 58
Secondary data, 31
Segment specialists, 7
Self-selection, 88
Semantic differential scales, 73, 98
Sensory variables, 274
Sentence completion, 84
Sequential sums of squares, 183
Sessions, 56
Set the Scale, 73–78
Shapiro-Wilk test, 164
Shelf tests, 22
Significance level, 154
Simple matching (SM) coefficient, 337
Simple random sampling, 44
Simple regression, 218
Single-item constructs, 30
Single linkage, 323, 324
Skewed data, 122
Snowball sampling, 45
Social desirability bias, 70
Social media analytics, 56
Social network analysis, 56
Social networking data, 56
Solomon four-group design, 89
Spearman's correlation coefficient, 119
Specialized service, 7
Specific representativeness, 42
Split-half reliability, 288
Split-sample validation, 237
Squared Euclidean distance, 333
Stability of the measurement, 41
Standard deviation, 115
Standard error, 169, 218
Standardized effects, 236
Standardizing, 336–337
Standardizing variables, 121
Stata, 124
Statistical significance, 154
Straight line distance, 333
Straight-lining, 100
Strata, 44
Stratified sampling, 44
Structural equation modeling, 267
Survey non-response, 104
Surveys, 64
Suspicious response patterns, 100
Syndicated data, 6, 54

Systematic error, 38
Systematic sampling, 44

T
Tables, 115
Telephone interviews, 66
Ten-point scales, 75
Testing effect, 89
Test markets, 22, 63
Test-retest reliability, 41, 288
Test statistic, 155, 168
Text mining, 56
Ties, 345
Total sum of the squares, 232
Tracking cookie, 56
Trade associations, 54
Transforming data, 120
Treatments, 86
Trend analysis, 56
Tufte, E.R., 378
Tukey's honestly significant difference test, 188
Tukey's method, 188
Tukey's statistic, 188
Two-sample t-test, 165
Two-tailed test, 168
Two-way ANOVA, 180
Type I, 158
Type II, 158
Types of research problems, 13

U
Unbalanced scale, 78
Unexplained variation, 183
Uniqueness, 276, 277, 295
Unit of analysis, 33
Unit non-response, 104
Univariate graphs, 110
Univariate statistics, 113
Univariate tables, 110
Unstandardized effects, 236
Upper adjacent value, 112

V
Vague quantifiers, 72
Validation sample, 237
Validity, 38
Variable, 28
Variable coding, 98
Variable names, 98
Variable respecification, 120
Variance, 114
Variance inflation factor (VIF), 221
Variance ratio criterion, 341
Varimax rotation, 281
Verbatim items, 71
Visual analogue scale, 75

W
Ward's linkage, 323, 324
Web surveys, 67
Weighted average linkage, 323
Whisker, 112
White's test, 228
Wilcoxon–Mann–Whitney test, 167
Wilcoxon matched-pairs signed-rank test, 167
Wilcoxon rank-sum test, 167
Wilcoxon signed-rank test, 167
Within-group mean squares, 186
Within-group variation, 183–184
Workflow, 96
Written report, 368

Z
z-scores, 121
z-standardization, 121
z-test, 170

Printed by Printforce, the Netherlands